The Macmillan book of the
MARINE
AQUARIUM

The Macmillan book of the
MARINE AQUARIUM

A definitive reference to more than 300 marine fish and invertebrate species
and how to establish and maintain a reef aquarium

Nick Dakin

Foreword by Julian Sprung

Macmillan Publishing Company
New York

Maxwell Macmillan Canada
Toronto

Maxwell Macmillan International
New York Oxford Singapore Sydney

©1992 Salamander Books Limited

Macmillan Publishing Company
866 Third Avenue
New York, NY 10022

Maxwell Macmillan Canada, Inc.
1200 Eglinton Avenue, Suite 200
Don Mills, Ontario M3C 3N1

Library of Congress Cataloging-in-
Publication Data

Dakin, Nick.
    The Macmillan book of the marine
aquarium : a definitive reference to
more than 300 marine fish and
invertebrate species and how to
establish and maintain a reef
aquarium/Nick Dakin ; foreword by
Julian Sprung.
        p.   cm.
    Includes indexes.
    ISBN 0-02-897108-6
    1. Marine aquariums.   2. Marine
aquarium fishes.   3. Marine
Invertebrates as pets.   I. Title.   II.
Title: Marine aquarium.
SF457.1.D35   1993
639.3'42--dc20   92-45126
                                CIP

CREDITS

Managing Editor: Anne McDowall
Editor/Consultant: John Dawes
Design: John Heritage
Picture Editor: Tony Moore
Index: Stuart Craik
Color reproductions: Regent
Publishing Services
Filmset: SX Composing Ltd.
Printed in Hong Kong

Half-title page: The Mandarinfish
(Synchiropus splendidus).
Title page: A Pistol Shrimp *(Synalpheus*
sp.) at rest on a *Tridacna* clam.
Left: The Blue-girdled or Majestic
Angelfish *(Euxiphipops navarchus).*
Page 9: A spectacular and colorful
tropical marine invertebrate aquarium.

THE AUTHOR

Nick Dakin set up his own independent aquatic consultancy after leaving the teaching profession. Years of research and observation have brought him a great deal of success with his own marine fish and invertebrate systems and a high reputation in the fishkeeping world. A regular contributor to the aquatic press, Nick is well known for his monthly 'troubleshooting' column in the UK magazine *Practical Fishkeeping*.

Julian Sprung is a marine biologist, diver and consultant, with many years of experience of keeping marine fish and invertebrates. He regularly lectures on aquatic subjects and is contributing editor of the US magazine *Freshwater and Marine Aquarium*.

THE CONSULTANTS

John Dawes is the Editor of *Aquarist and Pondkeeper*, a leading UK aquatic magazine. He travels extensively on consulting, judging, lecturing and writing assignments and has written over 500 publications for educational, scientific and popular journals, as well as several books on aquatic subjects.

Dr Keith Banister is an independent consultant, author and broadcaster on fish and fisheries, environmental concerns and aquatic animal welfare. As well as many reports on aquatic subjects, he has written six books on fishkeeping and underwater life.

THE CONTRIBUTORS

Adrian Exell (What is a Marine Fish?; Health Care and Disease Treatment)

Martyn Haywood (Part Five: Tropical Marine Invertebrates)

Les Holliday (The Coral World of Fishes)

Stan Kemp (Collecting and Conservation)

Dick Mills (Part Four: Tropical Marine Fishes; Part Six: The Coldwater Aquarium)

Sue Wells (What is an Invertebrate?)

ACKNOWLEDGMENTS

In addition to all the people mentioned above, the author and publishers would like to thank the following for their help in preparing this book: Terry Evans of Wetpets, Max Gibbs of The Goldfish Bowl, Richard Sankey of The Tropical Marine Centre.

CONTENTS

PART FIVE: TROPICAL MARINE INVERTEBRATES 302

PART SIX: THE COLDWATER AQUARIUM 362

APPENDIX ONE: MARINE ALGAE 374

APPENDIX TWO: SPECIES TO AVOID 380

GLOSSARY 384

GENERAL INDEX 386

SPECIES INDEX 392

PICTURE CREDITS 400

FOREWORD

A marine aquarium is a living work of art, an attraction for the eyes and the imagination. Seeing the fabulously coloured marine creatures within an aquarium, we are astonished that these little gems are actually alive. The brilliance of their colours is matched equally by improbable shape, texture and pattern, while the way these creatures move is fascinating. The delicate undulations of the Mandarinfish's pectoral fins, the fireworks display of pulsing *Xenia*, a soft coral, or the effect of the motion of water over the flowing fins of say, a *Pinatus* Batfish, or a field of coral polyps, is hypnotic.

The marine aquarium affords a window to the sea far away from its pounding shores, and a window to the unknown. It is precisely this element of mystery that holds the most enduring attraction for hobbyists, who learn about animal behaviour, biology, chemistry, and ecology from exposure to an aquarium. When it is a marine aquarium, there is a particular satisfaction in the knowledge that you are taming a little piece of the sea.

Sometimes the aquarium will not be tamed, however, and every hobbyist suffers moments of frustration because aquarium keeping is not, and never will be, an exact science. Living systems do not respond like machinery. Only experience and patience will allow you to achieve reproducible success.

Our hobby has undergone periods of enthusiasm and periods of waning interest. Every time a new product or filter is introduced, one hears claims that *the* solution has been found to the successful, maintenance-free marine aquarium, and this attracts new hobbyists and re-attracts the old salts who gave up. Patience, knowledge, and experience will always succeed where the exaggerated claims fall short. Still, there is a purpose to such claims. They dispel a myth and widely accepted view that has long been an obstacle to the growth of the hobby: the perception that marine aquariums are impossible to keep. I can't recall how many times after I mentioned that I was a marine aquarist, a new acquaintance would blankly utter, "Oh, I've heard that marine aquariums are impossible . . ." My reply is that a marine aquarium is as difficult or easy as you make it.

With that in mind, you can use this book as a guide to help you decide just what kind of marine environment to create, and how to do it. You can start with a small, simple system, a painless dip of a toe to test the water, but, eventually, you might just dive right in. Don't drown! You have many options, but you must be patient and avoid taking on too much responsibility too quickly.

The 'fish only' tank, or one that primarily emphasizes fish, has long been the mainstay of the marine aquarium hobby. Showy angelfishes, butterflyfishes, tangs and wrasses cruising among the skeletons of dead or fake coral is the stuff of most public displays. This type of display also typifies the first marine aquarium for most hobbyists. Whole displays may also be made with the fascinating partnership of anemones and clownfishes. And few hobbyists escape the attraction of the seahorse. Those fond of danger choose lionfishes and moray eels.

A more recent trend in marine aquarium keeping is the fascinating creation of whole ecosystems. This aspect of the marine hobby focuses primarily on the duplication of tropical coral reefs and lagoon settings, but also includes temperate or 'coldwater' reef environments. Both are covered in this book. 'Reef tanks' are especially fascinating because the decor is dynamic – ever changing as it grows. Each time you view the aquarium there may be a new discovery, a new pet, or growth. While older systems achieve success through a sort of clinical maintenance, the modern trend toward duplicating a natural environment achieves stability through the cultivation of rich populations of animals, plants, and micro-organisms, so that little external filtration may be required at all.

Now, as there are more technical gadgets and sophisticated forms of filtration available to hobbyists than ever before in the history of aquarium keeping, an appreciation of the simplest forms of creating a successful aquarium is returning. While the idea of using live rock to create a naturally balanced aquarium is not new, only now is it popular and widely accepted. One of the most fascinating things about the hobby of marine aquarium keeping is that, as we learn more about the environments we create and the creatures we keep, the techniques and technology continue to evolve.

The future of our hobby is bright. Advances in the care, propagation and aquaculture of marine species have brought an exciting new aspect of marine aquarium keeping within our reach. The trend towards environmental awareness and concern about depleting the natural resources makes captive propagation of marine life a positive alternative. Much progress is being made in this endeavour through the work of expert aquarists and active aquarium societies.

These organizations of aquarists now have a multiple duty: conservation, regulation, education and captive propagation are presently on club agendas. If we are to continue to enjoy and learn from our hobby, we must actively ensure that the methods of capture and care of the creatures we keep are responsible, for while our impact on the marine environment is minuscule compared with the destruction from industrial development and pollution, to proceed wrecklessly in our endeavour is counter-productive. Most people who enjoy the rewards of marine aquarium keeping are also sensitive to environmental issues. As marine aquarists, we know the value of increasing public awareness of the marine environment by exposing people to the beauty and wonder of a healthy marine aquarium.

PART ONE

THE MARINE ENVIRONMENT

The stunning beauty and the staggering diversity of animals to be found in and around the world's oceans often defies the imagination. Rich tapestries of life are delicately interwoven, providing a balanced ecosystem virtually unrivalled in the natural kingdom. In this opening section we shall be trying to gain an overview of life on the coral reef, looking at such widely ranging topics as the use of camouflage, cleaning symbiosis and reef activity by night and day. This vital background information can be of invaluable assistance when you are trying to recreate a successful mini-environment. We also examine the biological make-up of marine fishes and invertebrates; understanding how each creature has developed and adapted to survive a life in seawater can teach us important lessons in the care of marine livestock. We consider fish senses and patterns of reproduction, as well as finding out how and why they differ from their freshwater cousins, and take a look at the amazing diversity of invertebrates – the most adaptable of creatures, filling every conceivable niche in the marine environment.

The final chapter in this section – Collection and Conservation – seeks to demonstrate that the aquarist can play an important part in the conservation of the world's seas by demanding that this important resource is managed respectfully. It is encouraging to note that the marine hobby has made positive and practical contributions to various Third World economies by providing much-needed jobs in the collection of fish and invertebrates and in the production of artificial corals. With proper organization, the reefs of the world can be regarded as a renewable resource – it is up to the hobbyist to use it wisely.

Left: *Who could fail to be impressed by this inviting underwater scene captured photographically off the island of St Lucia. Squirrelfishes shelter beneath coral ledges as sea fans sway gently in the warm sea current; all is harmony.*

The Coral World of Fishes

The first and overwhelming impression of any coral reef is the sheer brilliance, abundance and diversity of the reef fishes. Brightly coloured, gaudily patterned fish of every conceivable body shape dart among the corals or hang in huge shoals in midwater. It is not unusual to find 20 or more species living closely together on just one small coral outcrop, and large reef areas, such as the Great Barrier Reef of Australia, can be home to a remarkable 2000 species.

In such a crowded environment, the competition for food and space is ever present. In order to survive and to get the most from their surroundings, reef fishes have adapted in shape and behaviour to live in various parts of the reef and become dependent on different food sources. Feeders on coral polyps and the tiny worms and crustaceans that live in cracks and holes in the reef have needlelike snouts and highly compressed body shapes to allow them easy access to deep crevices that other fishes cannot reach. Teeth may be fused together to form a parrotlike beak for scraping off algae covering the coral or combined with powerful jaws to feed on the coral itself. The main food source for the majority of reef fishes, however, is other fish, and 'eat or be eaten' is the general rule for survival.

Offence and defence on the reef

To overcome the immense difficulties of surviving in such a competitive situation, reef fishes have developed an armoury of weapons and defence strategies and these are used by predator and prey alike. Here, we look briefly at a number of these strategies for survival on the reef.

Cryptic camouflage

Camouflage is a universal strategy on the coral reef. Predators use it to conceal themselves as they lie in wait to surprise any unsuspecting prey coming within their reach and many less aggressive species hide behind the cloak of camouflage as a means of evading predation. A good

Below: *Animals that use camouflage as a defence often have the amazing ability to rapidly change colour and markings to match their surroundings. The Yellow-spotted Stingray is one of these creatures and is almost undetectable at times.*

example of the former is the Indo-Pacific Giant Moray Eel (*Gymnothorax javanicus*), which reaches over 2m (6.6ft) in length and hides within crevices and holes in the reef, perfectly camouflaged to mimic the surrounding coral. This formidable fish feeds on a wide variety of reef fishes and its gargantuan proportions indicate just how successful its hunting technique can be. A good example of defensive camouflage is seen in the Yellow-spotted Stingray (*Urolophus jamaicensis*). This small, placid ray from shallow sandy Caribbean waters hunts by excavating depressions in the sand to expose the small shellfish and various other invertebrates on which it feeds. To elude its predators, the body of the ray is covered in yellow spots that blend with the seabed and it has a chameleonlike ability to change colour, shade and pattern to match its surroundings. If all else fails, a further weapon is provided by a sharp venomous spine at the base of the tail, which is known to deter the ray's most ardent predators, including large Lemon and Hammerhead Sharks.

Behavioural camouflage

Behavioural camouflage is seen in the tropical Atlantic Trumpetfish (*Aulostomus maculatus*), a common reef fish that often hovers vertically, nose down, among gorgonians and sea whips, cleverly hiding in wait for passing small fishes and shrimps. Closely related to seahorses and pipefishes, this bizarre creature has an elongated body with a head extended into a long, trumpet-shaped snout. It can often be observed adopting a further subterfuge, using other, non-aggressive fishes, such as herbivorous surgeonfishes, as cover

Right: *The Atlantic Trumpetfish is a clever hunter. Blending in with the surrounding sea whips, it adopts a nose-down position and hovers realistically, moving gently to and fro with the current, ready to engulf any small fish or shrimp that ventures too close.*

to sneak up on its prey. Almost invisible within a shoal of slow-moving surgeonfishes, the Trumpetfish will suddenly dart out and pounce on a small fish or shrimp, sucking the unsuspecting prey into its tube-shaped mouth.

The power of advertising

Camouflage coloration is one method to deceive a predator or gain a meal, but by far the largest number of coral reef fishes are gaudy and colourful. Bright colours and patterns can serve purposes other than simply to adorn. It often pays to advertise, and if you are a fish that evades predation by having poisonous or distasteful flesh it is important that predators know this before deciding to sample you for themselves. Bright yellow, especially combined with black spots or bars, is a universally recognized indication of a poisonous species, and is used by terrestrial as well as aquatic animals. The juvenile phase of the Indo-Pacific Cube Boxfish (*Ostracion cubicus*) is an excellent example, using this type of livery to advertise the poisonous mucus covering its body.

Above: *When combined with black spots, the bright yellow body of the Cube Boxfish provides an effective way of warning would-be predators of the poisonous mucus covering its flesh.*

The art of deception

Butterflyfishes (family Chaetodontidae) are among the most attractive reef fishes and have a 'state of the art' ability to use colour and pattern to aid their survival. They have a disclike, highly compressed body and often display bold patterns that disguise the fish by breaking up its body outline or masking conspicuous features such as the eyes. Such disruptive coloration serves as an effective means of evading capture, and many eye-masked species take the illusion one stage further by employing conspicuous false eyes on the base of the tail or on the dorsal fin. This confuses predators into attacking the 'wrong end', while the fish dashes off in the opposite direction.

Drastic measures

Not all reef fishes employ disguise or deception to survive; many have evolved other strategies to protect themselves. This is shown to perfection by members of the porcupinefish family (Diodontidae). Porcupinefishes are easily recognized by the prominent spines that cover the head and body. These erect spines are a major deterrent and are made even more emphatic by the fishes' extraordinary ability to inflate themselves with water into spiky balls at least double their original size, effectively preventing attacking predators from swallowing them whole. In addition to this impressive defence mechanism, the horny skin and poisonous flesh act as a further discouragement, and the parrotlike beak is capable of delivering a nasty bite.

Unfortunately, porcupinefishes have not evolved a defence mechanism effective against their worst enemy – man! The very ability that protects porcupinefishes in the wild has resulted in their downfall. Inflated specimens are popular as souvenirs and many are collected for the curio trade, dried, varnished and offered for sale as mantleshelf ornaments or lampshades. The Bridal Burrfish (*Chilomycterus antennatus*), one of four

Above: *The Bridal Burrfish has an impressive armoury of weapons. Its sharp spines, horny skin and beaklike mouth deter most predators.*

Above right: *More ardent predators would soon discover the Burrfish's second line of defence, an ability to inflate to double its size or more.*

Below: *Fully inflated, the once tiny Burrfish becomes a spiky ball; its size, and its now erect spines preventing predators from swallowing it whole.*

porcupinefish species represented in the Caribbean, is rapidly becoming threatened due to the large numbers taken for sale, and there is an urgent need to protect this particular species.

Living in harmony

Not all living interactions on the reef are based on aggressive predator/prey relationships. Within the teeming diversity of plants and animals that live together on a coral reef, there are many examples of widely differing life forms involved in intricate and interdependent liaisons. After all, the very existence of a coral reef depends upon a symbiotic relationship between the stony coral polyps and the microscopic zooxanthellae algae that live within their tissues.

One of the most interesting of the mutually beneficial associations is that between clownfishes and their host anemones. These brightly coloured fish are found singly or, more often, as a pair or small group hovering above their host anemone and seek a safe haven within its venomous tentacles at the first hint of danger. The immunity enjoyed by clownfishes has only recently become properly understood. The clownfish acquires its immunity from the otherwise deadly stinging cells of the anemone by the dual strategy of manufacturing a sugar-based mucus to disguise its natural protein body composition and by slowly covering itself with a layer of mucus from the anemone. (In fact, young 'unprotected' clownfishes have been observed 'dashing' through the tentacles of an anemone to pick up some of the mucus.) The main trigger mechanism that activates the anemone's nematocysts, or stinging cells, is the protein-based mucus covering most fishes. As the anemone is naturally equipped to avoid stinging itself by recognizing its own mucus, it is deceived by the clownfish's 'cloak of disguise' into assuming that the fish is part of itself.

The clownfish obviously benefits from the safe protection offered by the anemone's tentacles but any

Above: *The association between the clownfish and its host anemone is well known. The dependence the clown attaches to its host is such that the numbers of anemones available on a reef can directly effect the size of the clown population.*

advantage to the anemone is less clear. One theory is that the bright coloration of the clownfishes acts as a warning to would-be predators of the deadly consequences of approaching too closely to the tentacles. Of course, this also keeps potential prey fish at bay, but it seems that anemones only feed on fish 'by chance', with juvenile 'inexperienced' fish being the most common victims. The bulk of the anemone's food source is composed of other invertebrates and zooplankton.

A life of service

Cleaning symbiosis is a further type of mutually beneficial arrangement, which is common on land as well as underwater. Coral reefs have many examples of this fascinating practice; over 50 coral reef species of fish and quite a number of invertebrates have given up the conventional means of hunting for food and live by cleaning parasites, small pieces of dead tissue and fungus from the fish living on or visiting the reef.

Cleaner species often set up business at particular locations on the reef, known as 'cleaning

stations'. The large numbers of fish they attract often form into orderly queues as they patiently wait their turn to be groomed. The cleaners provide an essential service in relieving their 'clients' of parasites and minor infections, while earning themselves a meal at the same time. The tactile experience resulting from the cleaning process often seems to be enjoyed, some clients living in close association with, and becoming very protective towards, their cleaners. The relationship between client and cleaner usually works to a defined set of rules. The cleaner is allowed to perform its duties all over the body, even inside the mouth and in the gill cavities, and in turn can trust its host not to make a meal of its defenceless companion in the process.

Cleaning symbiosis has evolved to become a very specialized

arrangement. There are full-time cleaners, those that mix normal feeding patterns with part-time cleaning, and the juveniles of certain species which act as cleaners only when young. The list of juvenile cleaners includes many species, of which young angelfishes and butterflyfishes are perhaps the most familiar examples. It is not uncommon to find the juveniles of a number of different species working in combination to service a cleaning station or joining forces with a full-time cleaner.

Cleaning gobies (*Gobiosoma* spp.) and cleaner wrasses (*Labroides* spp.) are the two main groups of full-time cleaner fishes, the gobies performing the service in the tropical West Atlantic and the labrids filling the same niche in the Indo-Pacific. There are six species of the cleaner goby, of which the Neon Goby (*Gobiosoma oceanops*), which employs a dazzling, electric blue lateral stripe to advertise its cleaning service, is perhaps the best known. The Cleaner Wrasse (*Labroides dimidiatus*) sports a similar blue lateral stripe, a trademark of many full-time cleaners, and is one of a

Above: *Several Cleaner Wrasse* (Labroides dimidiatus) *perform cleaning duties on a large Queen Angelfish, which has visited their station.*

Below: *This Cube Boxfish does not object in the least to being cleaned by a specialist shrimp, the brightly coloured* Lysmata amboinensis.

number of cleaner wrasse species distributed over the Indo-Pacific.

Cleaner shrimps head the list of invertebrate cleaners and are represented by the Caribbean Red-backed, or Painted Lady, Shrimp (*Lysmata grabhami*), the Indo-Pacific *L. amboinensis* and the circumtropical Coral-banded

Shrimp (*Stenopus hispidus*). The cleaning methods used by cleaner shrimps are very similar to those adopted by cleaner fish. They also establish cleaning stations and attract clients by sporting conspicuous patterns and coloration and by waving their white, threadlike antennae.

Daytime on the reef

During the daytime, the coral reef is often a crowded place. From the early grey of morning to the bright, sunlit hours of midday, there is a progressive increase in the numbers of fish and other animals venturing into activity until, finally, the water and the surface of the reef are filled with movement. Hundreds of colourful fishes hang in shoals or busily forage for food. Small damselfishes hover just above outcrops of staghorn corals, never straying far away and always ready to dash to safety at the first sign of danger. Suddenly, a large predator in the form of a grouper or Barracuda moves in for the kill, the distressed victim relaying a warning message across the reef and causing momentary panic as fish dive for cover in all directions.

These daytime predators arrive at dawn to feed, and are succeeded by huge populations of fish that graze and browse. These grazers fully exploit the many food sources provided by the reef. The herbivores feed on the algal turf growing in the niches between the living coral or use their chisel-like teeth to scrape off the thin film of algae adhering to coral surfaces. Parrotfishes are the most common of these algal grazers and can form large shoals, systematically grazing large areas of submerged reef flat. Surgeonfishes also form aggregations and are the true farmers of the reef, reserving large areas of reef flat for the shoal and protecting their fields of algae from other herbivores. The tiny herbivorous damselfish *Stegastes nigricans*, found in the Red Sea and Indo-Pacific, follows a similar practice, tending and cleaning its square metre or so of algal turf and, despite its diminutive size, pugnaciously defending its adopted territory from other much larger herbivores that try to take over.

The coral polyp grazers are represented by members of the butterflyfish family, with their long, forcep-shaped snouts perfectly designed for reaching into crevices in the coral and their laterally compressed bodies allowing access through the narrowest of gaps. Well named, these colourful marine 'butterflies' flit among the coral heads like their airborne counterparts fluttering from flower to flower. The angelfishes, closely related to the butterflyfishes, form a further group of grazers that devote the daylight hours to browsing on sponges and algae. In the Caribbean and tropical West Atlantic, where sponges feature as one of the largest constituents of the reefs, angelfishes are often prolific, flourishing on the rich pastures available to them.

The major remaining group of animals of the daytime can also be distinguished by their feeding patterns. These are the midwater feeders searching for food in the currents laden with zooplankton that often sweep along the reef edge. Indo-Pacific communities of these fishes would include various damselfishes and the Golden Jewelfish (*Anthias squamipinnis*), which hang in clouds close to the reef. The Pennant Butterflyfish (*Heniochus diphreutes*), a species quite unusual for members of the butterflyfish family, also patrols along the reef in aggregations of many hundreds to feed in this manner. The Caribbean counterparts of this group would include the Creole Wrasse (*Clepticus pharrai*) and the Sergeant Major Damselfish (*Abudefduf saxatilis*).

Carnivorous corals, with tentacles withdrawn, rest and await the richer pickings of the night, when the zooplankton will rise from the depths to feed and in turn provide a meal for the millions of hungry coral polyps. Bathed in sunlight, the corals by day are transformed into tiny greenhouses, their symbiotic microscopic algae harnessing the energy of the sun to benefit their coral hosts. Many other reef animals have entered into a similar partnership with these tiny zooxanthellae, including anemones, sponges and clams. The huge Giant Clam (*Tridacna gigas*) of the Indo-Pacific is an excellent example, which can reach more than 1m (3.3ft) across and weigh 254kg (560lb). Living in shallow, brightly lit water, the clam exposes its colourful fleshy mantle that houses large numbers of zooxanthellae. By

Below: *Coral reefs are busy places during the day. A variety of different fish species may share a good feeding position in the current.*

maintaining just the right levels of algae in its tissues, the clam is often able to rely upon them for nutrition. The clam will control the algae population by consuming only sufficient of their numbers to maintain optimum levels.

By mid-afternoon, the search for food has subsided and the reef becomes a more relaxed place. The major predators have returned to deeper water and, as the light wanes, the animals of the daytime prepare for darkness, scurrying to safety in the recesses of the reef.

The reef at night
The pressures on food sources and space are such that activity continues throughout the 24-hour period, and the reef at night is a lively place. Before the night's proceedings can take place, however, the members of the daytime community must withdraw to safe retreats.

The midwater feeders are the first to leave, realizing how vulnerable they are to attack from dusk predators, such as large mackerel or

Above: *A dainty damselfish* (Amblyglyphidodon *sp.) is well concealed among the branches of a soft coral during a long Red Sea night.*

jacks, in the fading light. They are followed quickly by the herbivores. Disc-shaped surgeonfishes disperse from their shoals and wriggle into incredibly small clefts in the reef flat. Large solitary parrotfish become very defensive at this time and may

either nervously take flight if approached or hold their ground, making mock aggressive displays. Presently, some species of parrotfish will select a hole in the reef and settle inside it for the night, enveloping themselves in a cocoon

Below: *This parrotfish is enveloped in a mucous shroud. This is thought to discourage predators by masking the fish's shape and scent.*

of mucus. This mucus shroud deceives predators, both by disguising the fish's outline and also by masking its taste to confuse those that hunt by using this sense. (A highly developed sense of taste can assist underwater predators in much the same way as a keen sense of smell serves terrestrial predators.)

Many reef fish change colour at night, predators as well as their potential victims. They may take on more muted shades or, surprisingly, use bright red coloration as effective camouflage in the nightly battle for survival. Soldierfishes and squirrelfishes have large eyes and a pink or red coloration – a good indication that these fishes are nocturnal predators. Their squirrel-like eyes are an adaptation to their nocturnal mode of life and the gaudy pink or red livery really does act as low-light-level camouflage. This is because water quickly filters out red wavelengths of light, making pink and red appear grey or black in the gloomy depths.

One nocturnal reef fish, the aptly named Flashlight Fish (*Photoblepharon palpebratus*), confounds all recognized theories of night-time coloration. This tiny, sombre grey fish, common in the Red Sea and Indo-Pacific, sports a pair of elliptical 'flashlights', one beneath each eye. These luminescent organs are filled with bacteria that can generate light by a series of biochemical reactions. The light produced not only helps the fish to hunt but also actively attracts the zooplankton on which it feeds. It is easy to assume that these points of light act as beacons to attract would-be predators, but in reality the reverse is true. Flashlight Fishes hunt in shoals, using their glowing lights as an important means of communication, and this tends to confuse predators rather than entice them. These extraordinary fishes can turn their 'flashlights' on and off at will by manipulating muscles beneath the eye.

As darkness falls, fish are not the only predators in action. The reef itself springs into life, as corals spread their delicate tentacles to trap passing zooplankton with batteries of lethal stinging cells. Crinoids, feathery armed relatives of the starfishes, make for the highest points on the reef and spread their arms to take the plankton soup.

Herbivorous feeders are generally associated with the daytime reef community, but *Diadema* sea urchins, slow-moving, open-reef animals that defend themselves with long, needle-sharp spines, are an exception. Each night, they migrate from the reef slope up onto the reef flat pastures, now vacated by their daytime occupants, and feed undisturbed on the algae. Brightly coloured sea slugs appear from beneath the coral rubble at the base of the reef edge and join the sea

Below: *A shoal of fish, mainly squirrelfish, occupy a cave at night, where it matters little whether they swim the right way up or upside down!*

urchins, some species grazing on algae while the carnivorous ones – the nudibranchs – feed on sponges and coral polyps, seemingly immune to the stinging nematocysts.

The nightly saga of the reef unfolds and continues until dawn, while daytime feeders rest, to live and feed another day.

The coral web of life

The coral reef and adjacent areas of mangrove swamp and lagoon seem able to support an endless diversity of life. Living organisms claim two essential requirements from their environment: a habitat and a food supply, both available in abundance on the reef and in nearby waters. The fundamental interaction between the members of the huge complex community of animals, plants and bacteria that make up the reef ecosystem is based on the food chain. In fact, this chain is so complex that it is more appropriate to think of it as a food web. The two main ingredients that fuel this system are the energy of the sun and the supply of nutrients in the form of decomposing organic matter. Since these nutrients are a product of the organisms living within the ecosystem, they form the bond that makes the food chain a never-ending process.

The first level of feeders are the plants, the so-called primary producers that use the energy of the sun to convert the nutrients into living matter by the process of photosynthesis. (Just to complicate matters, there are some fishes and invertebrates, called benthic feeders, that feed directly on detritus before it is broken down into nutrients and processed by the plants.) The many herbivorous fishes and other reef animals occupy the next level since they take advantage of the food source represented by the plants. Such herbivores include browsers and grazers that feed on algae and on plant material growing on the reef itself.

The connection between plant life and the next group of feeders is by way of the zooplankton, simple, tiny animal life forms that float in the

water and feed on microscopic plants known as the phytoplankton. The zooplankton thus form the basis of a wide range of different feeding strategies. For example, animals such as corals feed directly on the zooplankton (but also rely heavily on their symbiotic relationship with

Above: *Microscopic zooplankton – simple animal life forms – are the staple diet of a multitude of fish and invertebrate species.*

Above: *Moray eels are voracious predators, and will quickly capture any fish that is foolish enough to pass too close to their lair.*

the zooxanthellae in their tissues). Other carnivorous creatures feed on small fishes and invertebrates that have in turn fed upon the zooplankton. At the top of the 'feeding tree' are the large carnivores that feed on a wide range of animals in the other levels of the coral web of life.

This very simple classification does not reflect the true complexity of the reef feeding pattern. There are omnivorous fishes and other animals, for example, that feed on both animal and plant material. Also, feeding patterns and food preferences often change during the life cycle of some members of the reef community; zooplankton-feeding juveniles may become herbivores or carnivores as adults and some species may move from shallow waters to feed in deeper waters at a later stage in their lives.

Reef fishes can be classified into five groups by their adult feeding pattern:

Feeders on algae, such as parrotfishes, surgeonfishes, some damselfishes, and blennies.

Feeders on zooplankton, such as cardinalfishes, fusiliers, some damselfishes, jewelfishes, manta rays and whale sharks.

Feeders on sessile invertebrates, which include the coral and sponge feeders and are represented by butterflyfishes and angelfishes.

Feeders on large invertebrates (e.g. molluscs, crustaceans and echinoderms), such as snappers, stingrays, puffers, boxfishes, triggers, some wrasses and emperors.

Feeders primarily on fish, such as many sharks, barracudas, groupers, moray eels, cornetfishes, needlefishes and scorpionfishes.

A number of fish families, such as the triggerfishes, damselfishes, wrasses, puffers and boxfishes, are omnivorous and where these are included above, the category shows their main feeding preference.

What is a Marine Fish?

Seawater covers 71 percent of the earth's surface. However, huge tracts of the ocean are fathomless deeps, largely devoid of life. These are the aquatic desert lands, lacking in essential nutrients and thus unable to support a food chain without an input from the overlying 'layers'.

The thousands of species of marine fish mostly inhabit the narrow bands of the shallow, nutrient-rich waters – the continental shelves, which teem with life. These species of fish have evolved over millions of years to fill every conceivable niche, each species supremely suited in form and function to its own particular aquatic existence.

There are two main groups of fish: elasmobranchs, with their cartilaginous skeletons include sharks, dogfish and rays, but these do not, on the whole make suitable home aquarium subjects. Of more interest to the fishkeeper are the larger, more highly developed group of teleosts, or bony fish.

What is a marine teleost?

It is actually very difficult to give a completely watertight definition of a teleost. However, the following features are shared by most species: they spend all their life in water and would die out of it; they have a bony skeleton, including a skull; they have fins with spines and rays (see *Locomotion*); they possess a swimbladder used in buoyancy control; they have outwardly orientated gills, sited in a cavity (buccal cavity), which is covered by a bone flap (operculum, see *Respiration*); their bodies are typically covered with protective scales; they have a unique sense organ called a lateral line (see *Senses*); they are cold-blooded, i.e. poikilothermic, (which means their body temperature matches that of the water surrounding them).

Building on that basic definition, this section of the book will look briefly at how fish are adapted to live in the demanding aquatic environment. A basic knowledge of marine fish anatomy and physiology will improve your understanding of the creatures in your care.

Body shape

Marine fish come in all shapes and sizes. Their external body form is a function of the environment in which they have evolved and of the way they live. Some of the factors dictating a fish's evolved shape are: how active it is; how it feeds; whether it is a predator or potential prey (this is a relative term in the sense that a predator may, in turn, be preyed upon); and its defence and attack systems necessary for survival. This is best illustrated with a couple of examples. The Leopard Moray Eel (*Gymnothorax tesselatus*) has a long, serpent-like body, the only finnage being a dorsal fin extending the length of the body.

Below: *The Moray Eel has a snake-like body, which is well adapted to slipping into rock crevices, where the creature lurks ready to ambush passing prey.*

They swim laboriously with a sinuous undulation of the body. They have traded locomotory efficiency for a powerful form, which hides easily in crevices and cracks in the coral. Here a moray will lie in wait, ready to ambush unsuspecting fish swimming past, which it grabs in its massive tooth-filled jaws. Compare this with the bizarre appearance of the Lionfish (*Pterois volitans*). This has a body form much more typical of our concept of a fish, but has developed elaborate finnage with long flowing fins tipped with venomous spines. This gives the illusion of a much larger fish to predators, which are further dissuaded by the venomous spines. The fish also uses its outspread fins to manoeuvre small prey fish into a corner with no hope of escape from its large, well developed jaws.

Locomotion

Seawater is over 800 times denser than air. Moving through it is fraught with problems, such as drag, negative buoyancy and the sheer effort required to force a body through such a dense medium. Locomotory adaptation, like body form, is closely linked to a fish's lifestyle and its need for locomotion. We have seen above two examples of the moray eel and lionfish, where predatory feeding tactics dictate the species' body form at the expense of locomotory efficiency. The moray eel's body has hardly any locomotory adaptations, although the powerful muscles are ideal for the final uncoiling lunge that completes the ambush. The lionfish's flowing fins create drag which limits swimming speed. However, lionfish have a well-developed swimbladder, which is an adjustable air sac that allows them to hang motionless at any level in the water. The careful manoeuvring of a lionfish is accomplished by subtle fin movements, which prove an effective steering mechanism. Most fish dwelling on the reef, like the lionfish, have two types of muscle; a small portion is brown muscle, which is used for continuous activity because it is very well supplied with oxygen-rich blood, while the larger proportion of body mass is made up of white muscle, which can be used only for short bursts of emergency, high-speed swimming because of the poorer oxygen supply.

Continuously swimming midwater fish, which cruise over or on the edge of the reef, such as Barracuda (*Sphyraena barracuda*) have a body form supremely suited to a high-speed locomotion in water. Their bodies are optimally streamlined to minimize energy for reducing drag and aiding displacement. Swimbladders are either absent or much reduced, cutting down cross sectional area, and, therefore, drag. Muscle is primarily brown for constant speed swimming. Fins are minimal size and only used for turning. They are usually retracted during swimming.

Above: *Although slow swimmers, Lionfish possess well-developed swimbladders, enabling them to hang almost motionless in the water.*

Below: *Barracuda* (Sphyraena barracuda) *are adapted for speed. The body is streamlined and muscle developed for prolonged speed swimming.*

Senses

Marine fish need to be aware of what is happening around them. Sensory equipment is needed for communication, navigation, attack, defence and food location.

Sight In the crystal-clear waters of a tropical reef, sight is a key sense. Fishes' eyes are somewhat similar in structure to those of most other vertebrate animals, but usually possess a spherical lens. Fish can focus selectively on both near and far-away objects, their field of view

Above left: *The butterflyfish's eyes are set well to the side of the head, giving good all-round vision and early warning of possible dangers.*

Above right: *Coral Trout are predators and need forward-set eyes to concentrate all their attention on their potential prey.*

being determined by the position of the eyes on the head. Butterflyfish (*Chaetodon* spp.) with eyes on the side of the head have a wide field of view on both flanks, which is good

for defence. Predators such as Coral Trout (*Cephalopholis miniatus*), on the other hand, have forward-facing eyes, which enable better detailed focussing on the prey ahead. Examination of eye structure and behaviour suggests fish have better colour vision than we do. The glorious colours of most reef fish are used in territorial and reproductive behaviour to communicate with each other.

Sound Water is a denser medium than air and sound travels further and faster as pressure waves through the aquatic world. Fish use two systems to sense these pressure waves. Firstly, they have a sensitive lateral line system. This consists of a series of canals and pits set just below the skin surface, the main lateral line running along the mid

Eye position and field of view

Predatory fishes

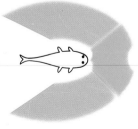

Non-predatory fishes

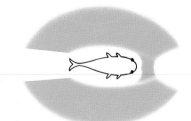

■ High-definition hunting sight

■ Low-definition warning/alert sight

Left: *The eyes of a predator are set well to the front of the head, thus concentrating most vision in a forward direction and on to the prey ahead. A grazing fish has a wide all-round field of vision due to eyes set to the side of the head – an essential development as a defence against predators.*

line of each flank in most species. Background noise is tuned out and only low-frequency sounds 1/10 to 200 Hertz are picked up by the lateral line. This mechanism can be used for pressure wave echo sounding, which allows the fish (like a bat with its echo sonar) to navigate in the dark around obstacles. Fish also have an inner ear, which picks up higher frequency sound up to 8000 Hz. Sound is used in communication by fish and ultra-sonic equipment used on a reef will reveal a cacophony of grunts, squeaks and clicks as fish attract, threaten and 'talk' to each other.

Orientation Sense receptors in the inner ears and closely related structures enable the fish to orientate themselves in a three-dimensional aquatic environment.

Smell and taste Most chemicals dissolve into water and readily diffuse through it. The difference between taste and smell is blurred in the aquatic environment so they are both encompassed in the term 'chemoreception'. Fish have special chemoreceptor sites concentrated in their nasal openings, scattered in the mouth around the head and, in some species, in other parts of the body. Fishes' sense of smell is over one million times more acute than our own. The olfactory sense is vital to fish both for discovering food and for communicating. To communicate with smell, fish release chemical messages, 'smells' called pheromones.

Co-ordination and control The co-ordination and control of body processes in response to external and internal stimuli is achieved by the brain in co-operation with highly developed nerve and endocrine (gland/hormone) systems. The brain receives and assimilates information from the sensory organs and then co-ordinates and stimulates the correct response from the appropriate organs in the body. The brain also integrates reflex actions, such as breathing, and is the site of learning and memory.

Nutrition
Fish have evolved the ability to exploit many different food sources. There are species that are solely carnivorous, such as fish-eating groupers (*Cephalopholis* spp.), and vegetarian species, such as many tangs (*Acanthurus* spp.) which graze macro- and micro-algae. Many species are omnivores, eating a combined diet of vegetable and meat matter. Individual species have

Below: *Tangs spend much of the day grazing on macro- and micro-algae, especially in highly oxygenated surge areas, where growth is most prolific.*

evolved specialized mouth structures that allow them to cope with specific food sources, and digestive systems tuned to deal effectively with their diet. For instance, fish eaters have a short gut with a stomach, which prolongs contact with special enzymes in an acid environment to encourage protein breakdown.

Osmoregulation

Fish are literally parcels of fluids within a fluid environment. In both marine and freshwater fish there is a difference between the salt concentration of the environment and their body fluids. Since the two are separated in places by very thin membranes, notably the gills, it is not surprising that there is a constant tendency for salt and water to flow in or out of the fish's body. The processes at work here are called diffusion and osmosis. If two solutions of different concentration are separated by a permeable membrane, such as those that form the biological boundaries of a fish, the ions of the salt will move by diffusion through the membrane from the more concentrated to the weaker solution, while the water molecules will move in the opposite direction by osmosis to dilute the stronger solution. As in many natural processes, there is a tendency towards equilibrium.

For a fish's body to work efficiently, it is essential that it maintains its internal salt/water balance at a constant level, in spite of the salt concentration of the water in which it lives. Fish counteract the natural forces of diffusion and osmosis by means of a process called osmoregulation. Here we see how the 'priorities' differ in marine and freshwater fish.

In freshwater fish, the body fluid has a higher concentration of salt than the surrounding environment. The tendency is therefore for water to flow in and for salts to be lost from the tissues. To counteract the first process, freshwater fish have very efficient kidneys, which excrete water very rapidly. Salt loss is minimized by efficient reabsorption of salt from urine before it is excreted, and active uptake of salt through special cells in the gills and from the food ingested by the fish.

Since seawater has a higher concentration of salt than the body fluids in marine fish, there is a constant tendency for water to be lost from the tissues and salt to flow in. Marine fish solve the dehydration problem by drinking large amounts of water and excreting little urine. The influx of salt is counteracted by selectively absorbing only a few salts from the sea water they drink and using energy to eliminate salts through special cells in the gills.

Respiration

Fish require oxygen for life. The vital processes by which they remove it from their aquatic environment and transfer it to the cells is called respiration. Freshwater contains only five percent of the oxygen present in air and saltwater contains 20 percent less oxygen than freshwater. Therefore, a marine fish's respiration system needs to be very efficient. It is essential to move large volumes of relatively oxygen-deficient water over the absorption surfaces to allow sufficient oxygen to be taken up. This transport mechanism also has to be energy-efficient because, as we have seen, water is 800 times denser than air.

To generate the necessary water flow, fish use the structure of their mouth and buccal cavity, plus the gill covers and their openings, the opercula, to produce a very effective low-power pump. This produces a constant flow of water over the gas-absorption surfaces of specialized respiration structures called gills. To absorb oxygen efficiently, the gill structures need to present a large surface area and a thin wall between the oxygen-carrying water and the blood. These parameters are limited by the fact that the more ideal gills become for gaseous exchange, the more likely they are to promote osmoregulatory problems, since

Osmoregulation in marine fishes

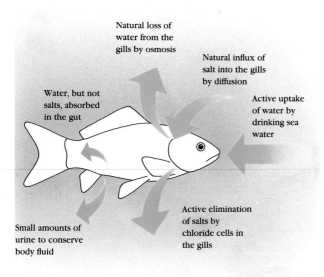

Natural loss of water from the gills by osmosis

Natural influx of salt into the gills by diffusion

Water, but not salts, absorbed in the gut

Active uptake of water by drinking sea water

Small amounts of urine to conserve body fluid

Active elimination of salts by chloride cells in the gills

Osmoregulation in freshwater fishes

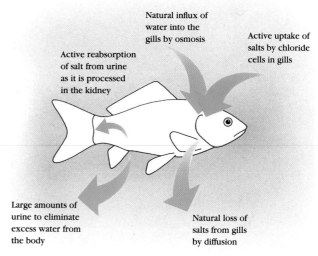

Natural influx of water into the gills by osmosis

Active uptake of salts by chloride cells in gills

Active reabsorption of salt from urine as it is processed in the kidney

Large amounts of urine to eliminate excess water from the body

Natural loss of salts from gills by diffusion

Pumping cycle in fish respiration

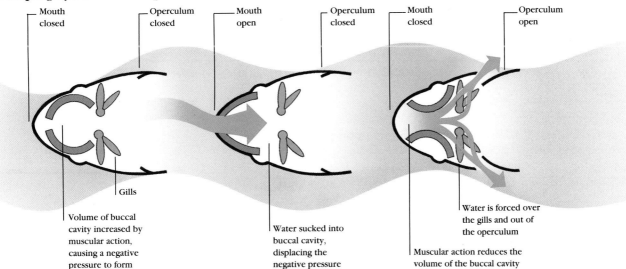

Mouth closed — Operculum closed

Mouth open — Operculum closed

Mouth closed — Operculum open

Gills

Volume of buccal cavity increased by muscular action, causing a negative pressure to form

Water sucked into buccal cavity, displacing the negative pressure

Muscular action reduces the volume of the buccal cavity

Water is forced over the gills and out of the operculum

they provide an ideal site for water influx or loss. Gill structures, therefore, represent a compromise between the requirements for respiration and osmoregulation.

Oxygen is absorbed into the blood by simple diffusion. The blood flowing into the gills has a lower oxygen concentration than the surrounding water, so oxygen moves into the blood to redress the balance. This process is further improved by the fact that blood is pumped in the opposite direction to the water moving over the gills. This counter-current system ensures that the blood oxygen level remains below that of the water right across the gills and allows many fish to remove up to 80 percent of the water's oxygen content. The oxygen is actively picked up by haemoglobin in the red blood cells and transported to the body tissues, where a relatively high carbon dioxide level, allied to relatively lower oxygen concentrations in the fish's tissues, causes the oxygen to be given up for use in the cells' essential functions.

Carbon dioxide is a waste product of metabolism but, since this is soluble in the blood, it poses no problems in removal, diffusing out easily through the gill walls.

Reproduction

Survival of fish species clearly depends on their ability to reproduce themselves. Therefore,

Above: *A pair of Fire Clownfish* (Amphiprion frenatus) *tend their eggs.*

with around 27,000 living species in our seas and fresh waters, it is hardly surprising to find that they have developed a wide variety of successful strategies. All reproductive strategies are different means of applying the energy allocated to producing offspring in the most efficient way; that is, balancing the number and size of eggs or young produced with the amount of effort applied in parental care. For instance, damselfish and clownfish pair up and lay relatively small clutches of fairly large demersal eggs, which are guarded and tended by the parents.

Contrast this with Yellow Tangs (*Zebrasoma flavescens*), which come together in large shoals and

breed above the reef at the full moon when currents run strong. They broadcast millions of small eggs – and even larger numbers of sperm. Later, the juveniles hatch out and become part of the vulnerable plankton where they are left to their own devices.

Sex on the reef is a fascinating subject. Many species, such as clownfish, even have the ability to change sex. All members of a group are male, except for the dominant specimen which becomes a female. When she dies the next most dominant male changes sex and takes on her role.

What is an Invertebrate?

The term 'invertebrate' is a convenient catch-all title, loosely covering all those animals that do not have a backbone (or vertebral column) and an internal skeleton. Invertebrates range in size from microscopic planktonic animals to giant squid, fierce predators that roam the deep, colder waters of the world. There are so many species of invertebrate animals that it is difficult to estimate the total number involved. Of the two million species of animals in the world, about 97 percent are invertebrates. Land-living insects, spiders and worms make up a large proportion of these, but there are probably just as many invertebrate species in the seas as on the land.

The range of marine invertebrates is so vast that there is no sea habitat in which they do not occur. The species found in aquarium shops come primarily from the shallower, warm waters of the tropics, but the cold waters of the Poles contain massive populations of shrimps, anemones, sponges and the like.

So adaptable are invertebrates that they have even overturned one of the major, and previously unassailable, scientific precepts, namely that in the first instance all life on earth is primarily dependent upon the sun's energy. (For example, a cabbage captures light energy through photosynthesis and is then eaten by a rabbit that in turn becomes food for a fox.) Now, however, scientists have discovered a small but flourishing ecosystem of bacteria, sponges, molluscs, crabs and filter-feeding worms that exist at depths impenetrable by the sun's rays. The primary energy source for these creatures is submarine volcanic activity in the form of heat and emitted chemicals.

Naturally, these creatures are beyond the scope of the home aquarist, but even so, the invertebrate keeper is faced with selecting and housing creatures with widely different lifestyles and requirements. In this chapter we take a closer look at each phylum (major group) of animals of interest to the aquarist. As well as illustrating and describing the structure and behaviour of species suitable for the aquarium, this overview features some of the other fascinating animals that make up the diverse world of invertebrates. Clearly, keeping marine invertebrates is not a challenge to be taken up lightly, but the reward is a window onto an absorbing and often unseen world.

Phylum PORIFERA

In evolutionary terms, sponges are the most primitive animals likely to be of interest to the marine hobbyist. When alive, they look very different from the familiar dried sponges used in the bath. Unfortunately, many are difficult to transport – exposed to the air they soon die – and only a few are regularly available. However, many species can be found in well-established 'living reef' tanks, where they usually arrive as accidental introductions with living rock (see page 106), and some of the encrusting tropical species are very attractive. Deep-water sponges are generally white, pale yellow or green, but there are a number of brightly coloured species – green, yellow, orange, red and purple – particularly from shallow tropical waters. The purpose of these colours, caused by pigments, is unknown but it has been suggested that they could have a warning function or could protect the sponge from the sun's rays.

At a conservative estimate, there are at least 5,000 sponge species, but 10,000 may be nearer the number. Most are marine, with one freshwater family of about 150 species.

Habitat

Sponges are found in all seas wherever there is a suitable substrate for their attachment, such as rocks, shells, corals, plants, boats, pilings, oil rigs and all the other objects that man provides. They are particularly abundant in shallow waters of the continental shelf. In some areas, sponges are so plentiful that they make up 80 percent of all living matter. Commercial bath sponges may occur in large beds, which make them easy to collect. Sponges have been described as living hotels, their chambers providing temporary or permanent accommodation for a vast number of other organisms, ranging from

Left: Sponges take on many shapes, colours and forms, enabling them to colonize most situations.

algae to fish. For example when the 'residents' of some Caribbean loggerhead sponges were counted, there were several thousand shrimps in each sponge. And the sponge threadworm can occur in tens of thousands in a single sponge, making up a significant proportion of its weight.

Although some sponges are eaten by sea slugs, turtles and some fish, many are toxic, particularly if their preferred habitat is in an exposed site. The toxins deter predators, help to keep the sponge free of larvae, etc. and protect the sponge from colonization by corals and other sponges.

Shapes and sizes

Most sponges are irregular in shape; the shapes often depend on the water current, the space available and the substrate to which the sponges are attached. In strong water currents they often grow as rounded or flattened clumps, but in calm water they may take on a branching appearance. A number of species form encrustations over any solid object, taking its shape. Confusingly, sponges of the same species may look very different under different conditions, making identification difficult.

Sponges vary greatly in size, ranging from less than 1cm (0.4in) to approximately 2m (6.6ft) in height and diameter. Deepwater species tend to be particularly enormous, with the largest species being found in the Caribbean and Antarctic. Some may not grow once they are mature, and some Antarctic sponges are known not to have grown for ten years. Given a good food supply in the tank, many small sponges grow quite rapidly.

Structure

Sponges are unlike any other group of marine invertebrates in that they are simply aggregations of cells and have no true tissues or organs. They are also incapable of movement. Not surprisingly, early naturalists assumed they were plants and it was

Above: *These dramatic sponge 'chimneys' (*Verongia *sp.) expel water and waste material through the exhalent opening at the top of each stack.*

not until the 1800s, when a sponge pumping water was observed through a microscope, that it was finally realized that they are animals.

The name Porifera means 'pore-bearer'. This reflects the fact that cells making up a sponge enclose a system of canals and chambers that open to the surface through many small openings, or pores. The simplest sponges are vase-shaped, with a central cavity surrounded by a wall containing the pores. Special cells, known as collar cells, line the inner wall and draw a current of water in through the pores by means of their whiplike hairs, or flagella. The water current supplies the sponge with oxygen and food particles before passing out through the large opening at the top – the osculum – taking waste material with it. Sponges are very efficient filter feeders, the collar cells straining and ingesting bacteria and minute organic particles from the water. Because these particles are too small for many other species to use, sponges are well adapted to living in the nutrient-impoverished waters of the tropical seas.

Larger sponges need a more efficient filtering system to supply all their needs, so the body wall becomes increasingly convoluted. In the more complex species, the central body cavity disappears completely and is replaced with

A simple sponge

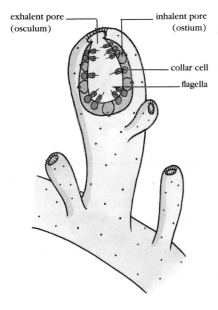

small chambers housing the collar cells. A number of large volcano-shaped processes may develop, each bearing an osculum out of which the current passes. Sometimes, the water current can be detected as much as 1m (3.3ft) above large sponges. The volume of water pumped through a sponge can be remarkable; for example, in one species it has been calculated that a sponge 10cm (4in) high and 1cm (0.4in) in diameter can pump 22.5 litres (5 gallons) of water a day.

Sponge cells often lie in a gelatinous medium, supported on a 'skeleton' of spicules or fibrous material called spongin, which is why sponges feel firm to the touch. Sponge spicules, usually invisible to the eye, look like slivers of glass and their different shapes are an important aid in identifying species. if you rub a sponge between finger and thumb, you can feel the spicules, but take care; in some sponges the spicules penetrate the surface and can irritate the skin.

Sponge classification

Sponges are divided into four main groups according to their type of skeleton. The Calcarea, the simplest sponges, have calcareous spicules. The Hexactinellida have six-rayed siliceous spicules and include the beautiful deep-water glass sponges, such as the Venus Flower Basket, *Euplectella aspergillum*. The Sclerospongiae have massive limy skeletons composed of calcium carbonate, siliceous spicules and organic fibres and can easily be mistaken for corals.

Above: *The skeleton of the Venus Flower Basket* (Euplectella aspergillum) *is often prized as a decorative ornament.*

The Demospongiae is the largest group, encompassing species with variously shaped siliceous spicules, a spongin skeleton or a mixture of both. This group includes the sponges of interest to the marine hobbyist, as well as the bath sponges and the boring sponges. The bath sponges have been harvested in the Mediterranean for centuries (the Romans used them for padding their helmets, as well as for bathing) and, more recently, in Florida waters. They are allowed to dry in the sun until the soft tissues rot, leaving the spongin skeleton. Boring sponges are in fact quite interesting; their name arises from the fact that they bore through rocks, stony corals and bivalve shells, probably by secreting an acid to dissolve the calcium carbonate. They eventually cause the death of their host and can be quite a pest in commercial oyster beds.

Reproduction

Like the majority of marine invertebrates, sponges reproduce by releasing eggs and sperm which, after fertilization, form larvae that float in the plankton before settling in a suitable place and developing into a new sponge. Sponges also have remarkable powers of regeneration; complete new animals will grow from small detached or broken pieces.

A complex sponge

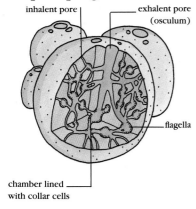

Sponge spicules

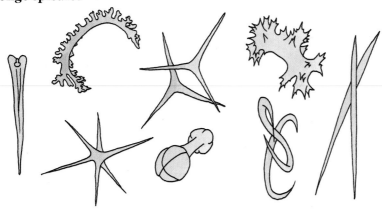

Phylum CNIDARIA

The phylum Cnidaria consists of a vast group of animals, the majority of which are marine. They include jellyfish, sea anemones, sea fans and corals, and are found throughout the world, from the coldest depths to the sun-baked shallows of tropical lagoons. Cnidarians play an immensely important role in marine communities and, like sponges, often provide a habitat as well as food for other invertebrates and fish. Many cnidarians are of interest to the aquarium hobbyist especially in 'living-reef' aquariums, i.e. those aiming to recreate a section of coral reef. However, it is only fairly recently that many aquarists have succeeded with any but a very limited number of species. In the last ten years, a higher quality of salt mixes, a recognition of the importance of trace elements and stable specific gravities, and improved lighting, filtration and nitrate reduction techniques have all contributed to simplifying the maintenance of corals and anemones in the aquarium. But given that these animals populate such widely different habitats, it is impossible to accommodate them all within one set of environmental conditions.

Structure

Cnidarians seem very variable in appearance, ranging in size from tiny *Hydra* to coral heads (colonies of polyps) several metres across. However, they are all characterized by a radially symmetrical body plan. This means that if you slice horizontally through a cnidarian you will find the body organs arranged in an even circle around a central axis. The body is basically a simple sack, or stomach, with a single opening used both as a mouth and as the exit through which waste is ejected. This is usually surrounded by tentacles armed with tiny stinging cells called nematocysts, used for catching food. Each nematocyst ejects a hollow thread, like a harpoon, into the body of the prey and injects a paralyzing poison.

Sessile (non-moving) cnidarians also use nematocysts to protect the living space around them from other animals, including encroaching individuals of the same species.

Polyps and medusae

All cnidarians exist in one of two alternative forms: the polyp, usually attached to rocks or other objects, or the free-swimming medusa. Some groups occur only as polyps and some only as medusae, but

Above: *Cnidarians such as this anemone play an important role in marine communities, providing a protective habitat for other animals.*

some pass through both phases, starting as polyps before budding off medusae as part of their life cycle. A sea anemone is a typical polyp, cylindrical in shape, attached to the substrate at the base end, and with mouth and tentacles at the free end. Polyps often have an external or

Stinging cells (nematocysts)

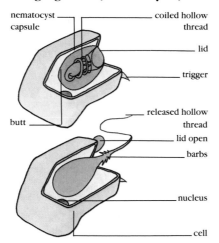

nematocyst capsule — coiled hollow thread — lid — trigger — released hollow thread — lid open — barbs — nucleus — butt — cell

A jellyfish

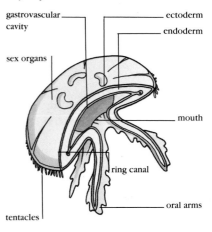

gastrovascular cavity — ectoderm — endoderm — sex organs — mouth — ring canal — oral arms — tentacles

internal skeleton, as in the corals, and many form colonies by budding off new polyps from the parent polyp. Jellyfish are typical free-swimming medusae, the bell-shaped body having a convex upper surface, below which hang the mouth and tentacles. Medusae are rarely colonial. They swim by alternate contractions of two sets of muscle, which cause them to pulsate. The arms and tentacles often develop into complex shapes to deal with a variety of prey.

Classification
There are approximately 10,000 species of cnidarian, divided into three main groups: the Hydrozoa, the Scyphozoa and the Anthozoa.

The Hydrozoa are mainly marine, but a few species occur in fresh water. The group contains the tiny sea firs (hydroids) and also the complex floating colonies that make up the Portuguese men-of-war. The fire corals that are found on coral reefs and that produce painful rashes if touched, also feature in this group. Hydrozoans characteristically have both polyp and medusa phases.

The Scyphozoa include all the jellyfish, and the medusa phase if predominant. Jellyfish are well named – even the firmest ones contain 94 percent water. They are highly mobile, and very graceful in the water, swimming by contracting their umbrella-shaped bodies, but they are largely at the mercy of the sea currents. Most species are carnivorous and catch animals with their tentacles, which are armed with nematocysts. They are still virtually unknown in aquariums, perhaps because so many of them are venomous. Tropical species can be extremely dangerous and one jellyfish has tentacles up to 20m (66ft) long. A few species feed by wafting currents through their mouths and trapping food in strands of mucus. Jellyfish reproduce by releasing fertilized eggs into the sea. These develop into small larvae that float among the plankton in the ocean currents.

The Anthozoa are the largest group of cnidarians. They have no medusa phase and include some 6,000 species. These are of greatest interest to the aquarist, as they include the sea anemones and corals, once known as 'flower animals'. Many corals and anemones are filter feeders or rely on zooxanthellae, but a large number benefit by being fed directly with small pieces of fish or shrimp, sprinkled over the animal or lightly pushed among the tentacles once or twice a week.

The body structure of sea anemones is based on a simple polyp, with multiples of six tentacles around the mouth. In some species, the base is modified for burrowing, but more usually anemones are attached to hard objects by means of a suckerlike disc. Although most seem to remain rooted to the spot, some can move by creeping over rocks. Species of *Stomphia* even leave their rock and swim away when touched by a starfish or

Above: *The graceful beauty of the jellyfish belies the fact that it is often highly carnivorous in nature, catching animals with its tentacles.*

predatory sea slug. Tropical anemones are generally larger than their temperate relations and can reach up to 1m (39in) in diameter. Sea anemones are often brightly coloured and a single species may have several different colour forms. Some anemones have zooxanthellae but most catch living prey, including fish, using the nematocysts (see page 31) on their tentacles.

Others trap organic particles in the water in mucus streams propelled towards the mouth by the tentacles. Several of the large anemones are host to clownfishes, *Amphiprion* sp., that live in their tentacles. The fish are protected by the anemone, and themselves provide protection for their host by deterring predatory fish. They also provide a cleaner service for the anemone. Anemones can reproduce

What are zooxanthellae?

Like several other groups of marine invertebrates, many cnidarians have small single-celled plants or algae, called zooxanthellae that live in the body tissues. Both animal and plant appear to benefit from this symbiotic association, which is particularly common in corals. The cnidarian probably uses the carbohydrates and oxygen produced by the zooxanthellae and the latter use the animal's waste products and assist in the assimilation of vital trace elements from the surrounding water. In the United States and West Germany there is much intensive study into the functions of these algae and the more that is discovered, the more vital they appear to be. Although it has been said that up to 90 percent of their food energy is obtained from these zooxanthellae, hard corals, for example, still need to capture planktonic organisms to survive. The *Tridacna* clams also play host to algae within their tissues but, unlike the corals, these clams will ingest the algae if they are hungry. With the advent of suitable lighting, it is now possible to satisfy the needs of the zooxanthellae in the domestic aquarium. Generally speaking, cnidarians with beige, brown, green or blue colouring have zooxanthellae and thus require strong lighting of the correct spectrum if these algae are to function correctly; like all plants they produce food through the process of photosynthesis, for which light is essential. In contrast, cnidarians within the colour range of purple, through red to orange and yellow usually lack zooxanthellae. They are often deep-living and generally do not require, or appreciate, intense lighting.

by budding off from the polyp, but they also reproduce sexually by releasing sperm and eggs.

Also of interest to aquarists is a small group of anthozoans that are effectively halfway between anemones and corals. These are the zoanthids, such as the green polyps (see page 316). They resemble small anemones but are colonial. There is no skeleton or basal disc, but the polyps have one or two rings of smooth slender tentacles. Zoanthids encrust rocks and even other animals, such as sponges and corals. The false corals, which include the mushroom polyps (see page 316) are also in a halfway position. Their polyps resemble those of true corals, but have no hard external skeleton. The tentacles often have clubbed tips and are arranged in rings around the mouth.

Anthozoans with eight tentacles

Corals that do not build reefs are classified in another group of the Anthozoa and have an eight-tentacle body plan, unlike the six-tentacle plan of sea anemones and hard corals. This group includes soft corals, sea pens, sea feathers, sea whips, sea fans and also the precious red and pink corals. They are generally colonial, with an internal calcareous or horny skeleton and tentacles that are often branched or featherlike. The precious corals are deep-water or cave-dwelling species and are not suitable for the aquarium, but their beautiful coloured calcareous skeletons are used for jewellery.

Below: *The delicate and intricate tracery of the sea whip's tentacles are appreciated to best effect when seen in close-up, as here.*

A sea anemone

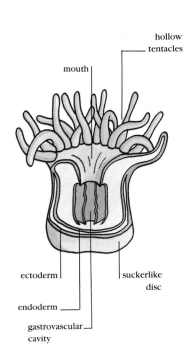

mouth

hollow tentacles

ectoderm

suckerlike disc

endoderm

gastrovascular cavity

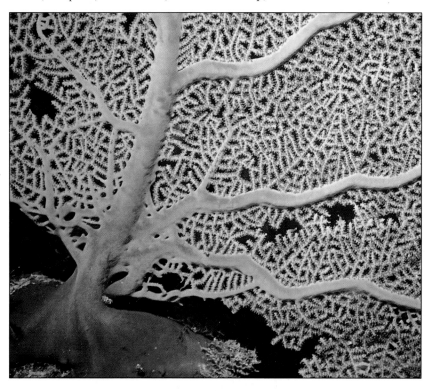

Soft corals include the leather corals, pulse corals and cauliflower corals (see pages 324-325). They are not as demanding to keep in the aquarium as the hard corals, but still need good water quality and husbandry. The polyps protrude from a fleshy mass, which is sometimes lobed, and is strengthened by calcareous spicules. Deep-water species tend to have more spicules and so are more rigid than shallow-water forms, which are subjected to a greater wave force. The polyps can be completely withdrawn into the body.

Sea pens and sea feathers (see page 323) are fleshy colonies, with short polyps arising from the sides of a central polyp. The lower end of this main polyp forms a stalk that is buried in soft sediments and several species can retract into the mud if disturbed. The skeleton is made of calcareous spicules and may reach 1m (39in) in height. Some sea pens emit waves of glowing phosphorescence when disturbed. They are usually found on the sandy and muddy bottoms of sheltered bays and harbours.

Horny corals (see page 314), a group of species that look rather like plants, includes sea whips and sea fans. The main stem is firmly attached to a hard surface by a plate or tuft of creeping branches. The stem has a central strengthening rod that is generally made out of a horny material called gorgonin (these species are often referred to as 'gorgonians'), although some species have a calcareous skeleton. The short polyps are found all over the branches of the colony but are absent from the main stem. Colonies are often brightly coloured and may reach a height of 3m (10ft). They often provide a home for sponges and hydroids, bryozoans and brittle stars, which stick to their branches.

Hard or stony corals Many corals have calcareous external skeletons and the biggest group of these – the stony corals – are responsible for building coral reefs. They are

Above: *The feathery tips of these polyps 'pulse' in a rhythmical action, capturing tiny food particles and absorbing oxygen.*

most often shades of beige and green, although some are blue or pink. The stony corals are in the same group as sea anemones and have a six-tentacle body plan. The coral animal is basically a tiny sea anemone sitting in a chalky cup, but colonies of these animals can build structures as enormous as the Australian Great Barrier Reef, some 2000km (1260 miles) long and consisting of over 2500 separate reefs. In colonial stony corals, individual polyps may be as small as 5mm (0.2in) in diameter, but in some solitary forms, such as *Heliofungia*, the polyp may be as much as 50cm (20in) in diameter. Brain corals (see page 306) are so-named because the polyps are arranged in continuous rows, so that the skeleton has longitudinal fissures in a brainlike mass. The polyps in colonial forms are connected laterally and lie over the limestone skeleton that they secrete. Coral reefs are built up over thousands of years; as old ones die, new colonies form on top.

A coral reef provides a habitat for sea anemones and other corals, as well as for a wide variety of other marine invertebrates, fish and plants. Reef-building corals need warm, clear water – the temperature should rarely drop below 21°C (70°F) – and are easily suffocated by sediment. Coral reefs are often

described as oases in an oceanic desert, because the tropical waters in which they occur are very poor in nutrients compared with temperate waters. As a result, if too much food is introduced into a tank, many stony corals will retract their tentacles. Clear water is also needed by the zooxanthellae, on which stony corals heavily depend. To build their skeletons, stony corals need a high pH level and a good reserve of calcium in the water. Without this, they do not flourish and appear to come 'unstuck' from their bases. They can, however, survive in the sea in colder, darker waters, but in such situations their capacity to secrete limestone is reduced and reefs are not formed.

Reef corals grow in many different shapes, depending on their preferred position and water depth. Deeper corals and those in sheltered, still waters tend to form branches, while corals in exposed positions are usually compact. New coral colonies can grow from broken fragments of larger colonies if the conditions are right; this is one way in which coral reefs recover from damage inflicted by storms and hurricanes. Stony corals also reproduce sexually by releasing sperm and eggs. Recently, it has been discovered that corals on the Great Barrier Reef all spawn on the same night, once a year. The reef becomes covered with a mass of swirling eggs and sperm; the reason for this extraordinary event is not yet clearly established, but it could be that it confuses predators.

Phylum PLATYHELMINTHES

The large group of wormlike creatures known as flatworms includes the various parasitic flukes and tapeworms well known to aquarists and studiously avoided by them! However, most fishkeepers would be delighted to come across some of the large and colourful flatworm species found on coral reefs. Unfortunately, these are rarely imported, even though they are fairly common in the wild.

Above: *This species,* Pseudoceros bajae, *is a typically colourful and interesting flatworm, which is much sought after by marine aquarists.*

Below: *In common with many species in the animal kingdom, the black and yellow warning colours of this flatworm indicate that it may be poisonous.*

Structure

Flatworms are the most primitive worms and their ancestors occupied a key position on the evolutionary tree leading to the higher animals. The flatworm gut is still a simple, blind-ended tube, and respiration occurs throughout the body surface, but there is an excretory system, muscles along the lines of those found in higher animals and a centralized nervous system, with a tiny brain.

Flatworms are generally small and sombre in colour, and a few are green due to the presence of zooxanthellae. However, the most attractive species – and those of interest to the aquarist – are extremely brightly patterned, often with some form of banding or striping. These species reach about 5cm (2in) in length and are in the family Pseudocerotidae from the Indo-Pacific. This family is in a group characterized by their branching guts. Like most flatworms, the body is flattened, but in this group it is wafer thin and has a leaflike shape, being almost as broad as it is long. They have a recognizable head, usually with numerous pairs of eyes and often a pair of sensory tentacles.

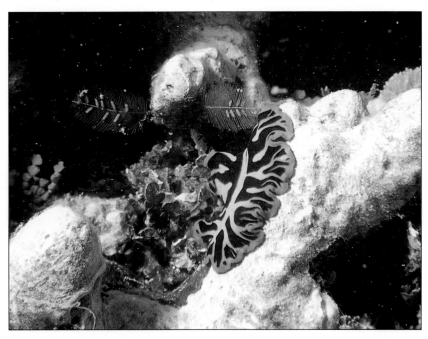

Function and behaviour

The unpleasant flukes and tapeworms belong to the class Trematoda and Cestoda, whereas the free-living flatworms are in the class Turbellaria, of which there are about 4,000 species. Free-living flatworms differ from the parasitic forms in that their body is covered with cilia, which, with the muscles, produce the characteristic gliding movement and create the 'turbulence' that has given rise to the group's name. Most of the free-living flatworms are marine and live on the bottom of shallow waters in the intertidal zone, although a few swim freely in the water. They are generally more active at night. The majority are carnivorous. The mouth opens on the underside of the body and sometimes has a muscular tube or funnel-like pharynx that can be extruded through the mouth to grasp or pierce food. The animal digests its prey by releasing enzymes over it and sucking the softened food into the mouth.

Reproduction

Free-living flatworms are hermaphrodite, i.e. they have both male and female organs, but they do not normally fertilize themselves. After copulation between two individuals, a large egg mass is laid from which hatch small larvae. Many flatworms also reproduce asexually by dividing, and freshwater flatworms can regenerate complete animals from small pieces. It is not known whether this also applies to the tropical marine flatworms.

Phylum ANNELIDA

This large group of invertebrates contains the segmented worms, many of which are marine. They have long, soft bodies and are oval in cross-section. The rings on the body, from which the name Annelida derives (in Latin, 'anulus' means a ring), are not merely external, but involve many of the internal organs. These animals have no solid skeleton, but gain rigidity from hydraulic pressure in the fluid-filled body cavity. Alternate contraction of two sets of muscle – one circular and the other longitudinal – allows the animal to move. Most species, apart from the leeches, have bristles or chaetae, protruding from each segment. These also help them to move and may be adapted for other purposes as well. Annelids have a more or less straight gut, with a mouth at one end and an anus at the other. Other characteristics indicate their much higher level of evolution in comparison to the invertebrates

discussed so far, namely a good circulatory system and the presence in each segment of excretory organs and a compact mass of nerve cells called a ganglion. There are three main groups of annelids.

Oligochaetes

The class Oligochaeta consists mainly of the familiar terrestrial earthworms, but also includes some important marine species, although none of these is of interest to the aquarist. The diversity and abundance of marine oligochaetes in estuaries and sheltered coasts is only just being realized. Many are able to tolerate high levels of dissolved organic matter and occur in polluted habitats.

Leeches

Hirudinea – or leeches – are mostly bloodsucking external parasites. There are only a few marine species; most of them are in a single family and feed on fish body fluids.

Polychaetes

The segmented worms of interest to the aquarist are all in the class Polychaeta, the bristleworms. This is the largest and most primitive group of annelids, and the majority are marine. They are often strikingly beautiful and very colourful and, unlike the other two groups of annelids, they show enormous variation in form and lifestyle. Apart from the head and terminal segments, all the segments are identical, each with a pair of flattened, fleshy lobelike paddles called parapodia, which are used for swimming, burrowing and creating a feeding current. The bristles, or chaetae, on the parapodia are immensely variable between species. In the sea mice, for example, they form a protective mat

Below: *The Sea Mouse* (Aphrodita *sp.) does not much resemble a worm, owing to its extended bristles, which form a protective covering over the animal.*

over the back of the worm and give the animal a furry appearance. The bristles of fireworms, on the other hand, are long and poisonous for defence, and are shed readily if a worm is attacked. Fireworms are voracious predators that usually feed on corals, but are known to attack animals, such as anemones, ten times their size. The species *Hermodice carunculata* is sometimes accidentally introduced into the home aquarium; take great care when handling it.

Polychaetes can be split into two groups on the basis of their behaviour. The mobile polychaetes are free-living forms with well-developed parapodia used for swimming and, often, for burrowing in the sand and mud. Many of them live under boulders and coral heads, and are common on coral reefs. This group also includes tube dwellers that leave their tubes to hunt for food. By contrast, the sedentary polychaetes are tube dwellers that rarely, if ever, leave their tubes, obtaining food with their tentacles.

In tube dwellers, the tentacles usually form a crown that catches food particles in the water. These are carried to the mouth by cilia on the tentacles. Mobile polychaetes often have an eversible pharynx, ending in fierce jaws that seize other animals or suck their body fluids, or grasp large pieces of plant material. The body surface provides sufficient area for respiration in small polychaetes, but larger ones need gills. In tube dwellers, these are usually situated near the tentacles and water is drawn past them by special movements of the body.

Ragworms, which live in the intertidal zone in muddy habitats, are mobile polychaetes, and well known to fishermen who often use them as bait. They live in U-shaped burrows, emerging only to feed on plant and animal debris on the surface. Lugworms, also familiar as bait, spend their entire lives in burrows, filtering sediment through the gut and leaving worm casts on the surface. Both worms respire by directing a current of water through

their burrows by regular body contractions.

The tubeworms are the only polychaetes of real interest to the aquarist. Both the fan or featherduster worms in the family Sabellidae and the 'Christmas tree' worms in the family Serpulidae make excellent aquarium inhabitants and are recommended for beginners. They are sedentary,

A fanworm

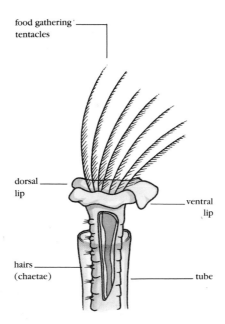

spending all their lives in their tubes, and are the most attractive of all the polychaetes.

In tubeworms, the parapodia are degenerate, there are no jaws and the heads are reduced. Instead, feathered, rather stiff tentacles radiate from the head to form an almost complete crown, which is used both as a gill and for feeding. Particles of food are trapped on the branches of each tentacle and channelled to the central rib, from where they flow in a stream of mucus to the mouth. Fanworms have colourful orange, green or purple tentacles forming the crown. Serpulids often have more brightly coloured blue and red tentacles, which may act as a warning device. Fanworms can contract their crown with startling rapidity, thanks to giant nerve fibres, which run from one end of the body to the other within the main nerve cord. The tentacles are extremely sensitive and will respond even to the shadow of a hand passing over the tank.

The cylindrical lower part of the

Above: *The main body of the Christmas Tree Worm remains deep in the coral, but the two branches of stiff tentacles extend above the surface.*

body is protected by a tight tube, secreted by the animal and made of a parchmentlike mucus. Serpulid worms are smaller and produce a tube of a stony calcareous material. As extra protection, a calcareous 'plug' may be present to block the entrance of the tube after the tentacles have been withdrawn. The serpulid group of worms also includes the tiny *Spirorbis* worms, which are often found on rocks and the bottom of boats as calcareous tubes arranged in a flat spiral about 3.3mm (0.12in) in diameter.

Annelid reproduction

Most annelids have sex cells in each segment; some species are hermaphrodite, but in others there are animals of both sexes. At certain times of year, the sex cells are shed into the sea, where fertilization takes place and a ciliated larva is formed. In some cases, the worms die after shedding the sex cells.

Reproduction in polychaetes often involves swarming, which helps to ensure that sperm and eggs are released at the same time, rather like the corals on the Great Barrier Reef. There is often a special sexual phase that looks quite unlike the normal burrowing individual, with

Above: *Red-banded Fanworms* (Potamilla fonticula) *may be found both in small colonies, as seen here, or as isolated individuals.*

Below: *The parts of a fan worm are clearly seen here – feathery tentacles, outer tube and (at the bottom of the picture) the normally unseen worm.*

enlarged eyes and parapodia for swimming.

The palolo worm from the South Pacific, though not an aquarium species, is worth mentioning because it is so strange. These worms have sex cells in the rear

segments only. These segments change shape and colour and, when ready, this rear section breaks off and rises to the surface, where the eggs and sperm are shed. All the palolo worms do this on one night, near dawn, at full moon in November. When all the segments have risen to the surface, the sea takes on the appearance of vermicelli, turning milky white when the eggs and sperm are released. Meanwhile, the front part of the worm remains in the coral and rocks and regenerates the missing parts. But perhaps strangest of all is the behaviour of the people who live on the islands in this part of the world; the palolo worm is considered a great delicacy and on the appointed day people venture out in boats at dawn and collect buckets of the worm segments as they rise to the surface!

Another interesting polychaete is the fireworm from Bermuda. When the worms come to the surface, the females start to emit a greenish phosphorescent glow. This attracts the males, which dart towards the females, emitting flashing lights at the same time. As the different sexes approach each other, the sex cells are shed.

Phylum CHELICERATA

King, or horseshoe, crabs, sea spiders and a few mites are the only marine species in this large group that is dominated by the spiders, mites and ticks. The chelicerata are related to crustaceans and insects, all animals with an external skeleton, or cuticle, made of a material called chitin. The body of a chelicerate has just two sections: the cephalothorax at the front, which includes the head and legs, and the abdomen at the back, which may have some appendages. The name chelicerata arises from the fact that instead of the antennae found in insects and crustaceans, these animals have a pair of pincerlike mouthparts called chelicerae.

The horseshoe crabs are the only species in this group of interest to aquarists. They are extremely primitive animals; over 300 million years, ago, horseshoe crab ancestors looked much like their modern descendants. They were once widespread and included many species. Today, however, there are only four species, which are considered as 'living fossils'.

Structure

Horseshoe crabs are light greenish grey to dark brown in colour and can reach a length of 60cm (24in). Males are generally smaller than females. They have a heavily armoured body, and look as if they should be taking part in a science fiction film! The front section (cephalothorax) is covered by a horseshoe-shaped carapace, hinged to the abdomen; the domed shape of the carapace helps the animal to burrow through the mud. Horseshoe crabs have been described as 'walking museums', as the carapace is often covered with large numbers of other organisms, including algae, coelenterates, flatworms, bryozoans and molluscs. The crab has a long mobile tail spine, a telson, which helps it to move and to right itself if it is accidentally overturned. Immature crabs have spines along the top of the telson, but these do not grow and so appear proportionately smaller as the crab gets older; they may help the young crab to burrow, and act as a deterrent to predators.

There are five pairs of walking legs on the front section. The first four pairs have pincer tips and heavy bases that are used for moving food into the mouth. The fifth pair has an even heavier base that is used to crush thin-shelled molluscs, and a whorl of spines to sweep away silt as the animal burrows. The back section of the crab has several pairs of appendages, five of which form gill books – an adaptation unique to horseshoe crabs. These look like the leaves of a book and act as gills for respiration. The gill books are also used for movement by young horseshoe crabs, which swim along upside down.

Behaviour

Horseshoe crabs live on the bottom of sandy and muddy bays and estuaries, and only return to the beach to breed. They generally move along the bottom with a stiff-legged gait, but are capable of swimming in quieter water. They feed on a wide variety of other invertebrates, including worms and molluscs, digging in the mud and passing the food to their mouths with their pincer-tipped front legs. The American horseshoe crab, *Limulus polyphemus*, can consume at least 100 young soft-shelled surf clams a day and seems able to detect this prey up to 90cm (36in) away.

Reproduction

The breeding period is characterized by the migration of huge numbers of crabs into shallow waters along the shores of bays and estuaries. This is best known in North America, where the massive emergence of horseshoe crabs on the beaches of the Atlantic and Gulf of Mexico is one of the most spectacular phenomena of the coast. The crab makes a useful food for poultry and pigs, a good fertilizer, a bait for other fisheries, and has recently become very important in biomedical research. As a result, large harvests are taken during the breeding period. Thousands of crabs may be taken in each session and, at the peak of the fishery in the 1920s and 1930s, four to five million crabs were being collected annually.

The crabs migrate to the beaches, usually either at full or new moon and within two hours of high tide. The males move sideways to the shore and intercept females heading directly for the beach. The couples then proceed to the beach. The males fertilize the eggs as they are released, and the female lays them in an excavation in the sand, from several hundred to several thousand at a time. The adult crabs leave the beach as the tide ebbs and the eggs hatch in about five weeks. The young crab larvae emerge only at an appropriate high tide. Juveniles moult several times a year, burying themselves in the sand to protect themselves from predators; adults may moult once a year or even less.

A horseshoe crab

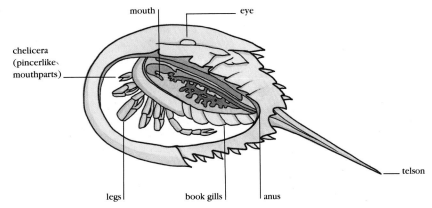

mouth

eye

chelicera (pincerlike mouthparts)

legs

book gills

anus

telson

Phylum CRUSTACEA

Crustaceans have aptly been called 'the insects of the sea'. Although there are some freshwater and terrestrial species, the majority are marine. They have invaded every possible habitat in the sea, from deep cold abysses to warm shallows, and have exploited every way of life in the same way as insects have on land. There are nearly 40,000 species, ranging in size from microscopic parasitic and planktonic animals to giant spider crabs from Japan with a leg span approaching 3m (10ft) and lobsters weighing up to 20kg (44lb). Many species are commercially important, such as the edible prawns, lobsters and crabs. Others are a major food source for higher animals. Crustaceans themselves are often active and efficient predators on invertebrates and fishes.

Structure

Crustaceans have a distinct head, thorax and abdomen, although in some species the two front sections fuse together. Most have a telson, or tail piece, and some have a rostrum, or spine, which projects between the eyes. The number of body segments, limbs and other appendages varies between species, but there are always two pairs of antennae on the head, often a pair of stalked, compound eyes, and at least three pairs of mouthparts. Whereas smaller crustaceans respire through the body surface, the larger crustaceans have gills, which are more complex in the more active species and often associated with the leg appendages. The appendages of crustaceans show a marked division of labour, different pairs being adapted for walking, feeding, respiration or reproduction. The tailfan found in many species – and used for swimming backwards – is also developed from appendages.

'Suits of armour'

Like insects and the horseshoe crabs, crustaceans have a cuticle of chitin that serves both as a suit of armour and a skeleton, and it is this 'crust' that has given rise to the name of the group. In most large crustaceans, it is strengthened with carbonate and other calcium salts to keep it rigid. In some species, this carapace, as it is known, can be thick enough to make the animal almost invulnerable to all but the most determined predators. The disadvantage of an exoskeleton is that it must be shed at intervals as the animal grows. At moulting time, the cuticle splits at the thorax and the animal squeezes itself backwards, leaving behind a perfect hollow replica of itself, which is sometimes eaten by its owner to reduce energy losses. The animal swells rapidly in size with fluid, so that the new cuticle is larger when it hardens than the old one. The body is then deflated, leaving space for growth. After moulting, the animal is soft-skinned and vulnerable; in a tank, be sure to provide plenty of hiding places for it to retreat into during this phase. Check that the pH level of the water is not too low,

Below: *A small colony of Gooseneck Barnacles* (Lepas ansifera) *is a common sight in the tropics. Note the stalks by which the barnacles are attached.*

otherwise the new shell may not harden properly and the animal will become deformed. Any leg or claw lost before moulting will be regenerated at the next moult, but make sure that the animal is not at a disadvantage in the tank and isolate it if necessary.

Minor crustacean groups

There are eight groups of crustaceans, but not all are considered here. One of the more primitive groups, the Branchiopoda, is largely freshwater, but may be familiar to aquarists because it includes the water flea, *Daphnia*, often used as food for small fish. The brineshrimp, *Artemia*, found in salt pans and ponds, also belongs to this group. The Copepoda include the tiny, usually transparent, herbivores with no carapace that make up most of the sea's plankton and are an important food source for many fish.

Barnacles

It is often not appreciated that barnacles – both the common conical ones and the stalked or gooseneck barnacles, such as *Lepas* (see page 330) – are crustaceans. All barnacles are marine and pass their adult lives attached to rocks or other suitable substrates. These include manmade objects – barnacles are major fouling organisms – and also living animals, such as crabs, turtles and whales. Although they could not

look more different from shrimps, crabs and lobsters, dissection shows them to be closely related.

A barnacle has been described as 'an animal standing on its head within a limestone house and kicking food into its mouth with its feet'. It is attached to the substrate by a secretion from its antennae and is enclosed in a shell, developed from the cuticle, consisting of a number of plates. It feeds by circulating water through this and filtering out minute particles with the long, finely branched feet that now act as a filter-feeding mechanism. Goose barnacles are attached by stalks, and are so named because in medieval times it was believed that geese hatched from the egg-shaped shells. This confusion also gave rise to the common name of the barnacle goose! Barnacles have the strange habit of accumulating heavy metals, particularly zinc. Studies in the River Thames in the UK show that up to 15 percent of the dry weight of barnacles may be zinc. Since such high concentrations are easy to measure, barnacles are very useful for environmental monitoring.

Malacostracans

The largest group of Crustacea is the Malacostraca. It contains almost three quarters of all known crustaceans and many of the most interesting, popular and colourful of the marine hobbyist's invertebrates.

It also includes many groups that are not kept in the aquarium, but which are important for other reasons. The isopods are usually flattened and include the familiar terrestrial woodlice, as well as many marine worms, which are usually bottom dwelling and rather dull in colour for camouflage purposes. The amphipods are laterally compressed and curved, with no carapace. The male is often larger than the female and rides on her back before mating. This group includes the sandhoppers that are found almost anywhere in the world on beaches when the tide is out. The tropical mantis shrimps (*Odontodactylus* sp.) have front legs adapted for seizing prey, similar to the terrestrial praying mantises; they lie in wait in front of their burrows for unsuspecting prey – usually fish – before striking. Another important group includes the shrimplike crustaceans of the open sea known as krill, which form the food of the great whales, seals, birds and squid in the southern oceans.

Decapods

The largest and most familiar crustaceans in the Malacostraca – the crabs, lobsters, crayfish, shrimps and prawns – are all in the Decapoda – literally 'ten feet'. They have five pairs of leglike appendages on the thorax and several rows of gills at the base of the legs, covered by the carapace. Most species are easy to

A typical decapod crustacean

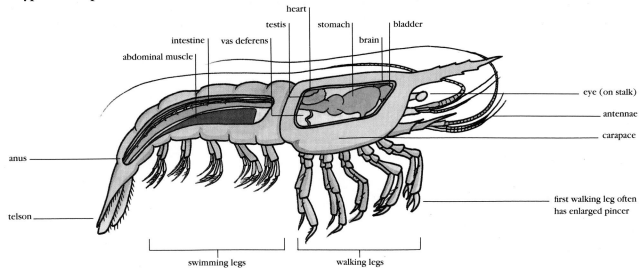

heart

testis | stomach | bladder

intestine | vas deferens | brain

abdominal muscle

eye (on stalk)

antennae

carapace

anus

first walking leg often has enlarged pincer

telson

swimming legs | walking legs

maintain in a domestic aquarium, where they tolerate less than perfect water conditions and accept almost anything remotely edible. They are a particularly interesting group of invertebrates as many are very active. Furthermore, they often display commensal behaviour, which literally means 'sharing food at the same table'. In the crustacean world, it means that they live in close association with other animals for mutual benefit. For example, shrimps and crabs, such as *Neopetrolisthes obshimai* (see page 332) and *Periclimenes brevicarpalis* (see pages 338-9), are often found in the tentacles of sea anemones, between the spines of sea urchins or within the shells of molluscs. They benefit by 'stealing' scraps of their neighbour's food but many also help in keeping the neighbour and the surroundings clean. The Boxing Crab, *Lybia tessellata*, goes one step further and holds anemones in its claws for defence.

Shrimps and prawns

The shrimps and prawns are often laterally compressed, usually with light external skeletons, and include swimming and bottom-dwelling animals. By flexing the abdomen they can move fast enough to escape danger. People often think that shrimps and prawns are different species, but the names have no scientific meaning, although larger species are often called prawns. Some 350 species of this group are used by man for food. They also include the cleaner shrimps, some of which make good aquarium species, such as *Lysmata amboinensis*. Their characteristic coloration may help their 'clients' to locate them. Many other shrimps are suitable for the aquarium.

Lobsters and crabs

The lobsters and crabs are all bottom-dwelling walking animals. They move on only eight legs, the front pair being modified as huge claws. These are used for seizing and shredding prey, for protecting themselves and to proclaim their territories, as in fiddler crabs.

The spiny lobsters, such as *Panulirus versicolor* (see page 335), have rather less well-developed but spiny claws. Spiny lobsters often migrate in enormous numbers. Observations have shown, for example, that *Panulirus argus* in the Caribbean may travel as much as 50km (31 miles) in the autumn, covering a distance of 15km (9

Below: Panulirus versicolor, *the Purple Spiny Lobster, is a nocturnal feeder, rarely observed during daylight hours.*

Above: *The Cleaner Shrimp* (Lysmata aboinensis) *makes a colourful and useful addition to the reef aquarium.*

miles) each day. Up to 100,000 animals make the journey travelling in groups, one behind the other, and keeping together by touching and perhaps by using the row of spots clearly visible on the abdomen of the animal in front.

Lobsters with heavy claws are extremely long-lived and some may survive 100 years. They live in holes

on rocky bottoms and are mainly scavengers, but will eat live food. Some have an unpleasant tendency to cannibalize their weaker relatives. The larger lobsters have asymmetrical claws; a large one with rounded teeth that is used for crushing, and a smaller one with sharper teeth used for seizing and tearing their prey. The edible parts of the lobster are the well-developed muscles, particularly the ones in the abdomen that are used for swimming. Lobsters are most active at night and their eyes are comparatively poorly developed. Instead, they have sensory bristles all over the body and legs; some of these are sensitive to touch and others are sensitive to chemicals.

Hermit and porcelain crabs

The hermit crabs, porcelain crabs and squat lobsters are all intermediate in shape between lobsters and true crabs and are scavengers. Squat lobsters have large symmetrical abdomens usually flexed below their bodies rather like crabs, and porcelain crabs, such as *Neopetrolisthes obshimai*, look very like their crab relatives. Hermit crabs, such as *Dardanus megistos* and *Pagurus prideauxi* (see pages 331 and 333), live in empty mollusc shells in order to protect their soft abdomens. They carefully choose a shell of exactly the right size, and change the shell as they grow. They usually choose right-handed shells, although sometimes they will use the rare left-handed shells. The shell is gripped by their specially adapted legs and its opening blocked by one or more claws.

'True' crabs

The 'true' crabs are usually carnivorous walkers on the seabed, although in some species the limbs are adapted for swimming. They have a very reduced abdomen, held permanently flexed below the front segments, which are fused. The carapace is large for the animal's size and extended at the sides. Crabs often have long eyestalks and can move sideways. All these adaptations make the crab a very efficient mover. They are found at all levels, from the deep ocean trenches – where a blind crab preys on the strange animals recently discovered 2.5km (1.6 miles) below sea level around hot vents – to the intertidal zone, beaches and even far inland on large islands. Crabs have a rapid escape reaction and can burrow backwards into mud or sand. Several species of crab, such as

Above: *The tiny Porcelain Crab* (Neopetrolisthes ohshimai) *lives among anemones' tentacles for protection. It has a complicated filter-feeding apparatus at the front of its head.*

Calappa flammea (see page 330), cover their carapaces with living organisms, such as algae, sponges and other encrusting animals to provide camouflage or even a source of food.

Reproduction

Most crustaceans have separate sexes, although some are hermaphrodites. Terrestrial crabs and hermit crabs return to the sea to breed; their mating is usually seasonal, sometimes involving an elaborate courtship ritual. Many crabs mate while the female is still soft from moulting, the fertilized eggs being retained by the female until they hatch, either in a brood pouch or attached to the appendages. The eggs usually hatch into free-swimming larvae that metamorphose through various stages into the adult form. Several stages of larvae look very different from the adult; in fact, one stage – the zoea – looks so different that early naturalists classified it as a completely different animal!

Phylum MOLLUSCA

The Mollusca is one of the largest phyla in the animal kingdom, with more than 100,000 species. This group has long been important to man, both as a source of food and for its beautiful shells, used for a variety of decorative purposes. Molluscs are found in almost every habitat, and about half are marine. From the aquarist's point of view, many of the most attractive and interesting invertebrates are found within this group. All require good water conditions and, although some should be left strictly to experienced hobbyists, many species are well within the scope of the novice.

Structure

The body of a mollusc consists of a head (although this has been virtually lost in the bivalve group), a muscular foot and a 'visceral mass', which contains the digestive and other organs. The foot is used for gliding over the sea bottom, over rocks or vegetation. In some sea slugs, lobes of the foot are developed for swimming and in a great many species it is used for burrowing. Marine molluscs have gills, which in several groups – particularly the bivalves – are adapted for feeding. Many molluscs have a 'radula', a kind of tongue covered in hundreds of teeth, used for rasping at food. This is made of chitin and is secreted continuously, old rows of teeth at the front dropping off as they become worn.

The name Mollusca means soft-bodied but, like crustaceans, most molluscs (but not squids or octopods) have an external skeleton to protect and support the body. This is the shell, which is made of a material called conchiolin impregnated with calcium carbonate. Shells come in a wide variety of colours, patterns, shapes and textures that usually reflect the lifestyle of the animal. They may have regular lines or marks indicating interruptions of growth, which can occur in cold weather. Thus, some species can be aged in much the same way that trees can be aged from their growth rings. The inner layer of shell is often made of tiny blocks of crystalline calcium carbonate and is called nacre or, when the layer is thick, mother-of-pearl. A few species, such as the chambered nautilus and the pearl oysters, consist almost entirely of nacre. Some molluscs have very reduced shells, often internal, or have lost them altogether.

Shells have a variety of functions in addition to acting as a skeleton. Some species, such as the abalones and limpets, have shells that can be clamped tightly to windswept rocks to avoid dessication when the tide is out, or to avoid being removed by predators. Others have streamlined shells for burrowing. The razor shells are a good example; when alarmed, these can plunge into the sand by as much as 1m (39in). Perhaps the strangest shells of all are the carrier shells, *Xenophora*. Not content with the natural form of their own shells, they attach stones, bits of coral and other empty shells to the surface, perhaps to camouflage it.

All molluscs have a mantle – a fold of skin enclosing the gills, anus and various other glands and organs. Sometimes it is brightly coloured, perhaps as a warning; at other times it may be coloured to provide a camouflage. The mantle of a cowrie is extended back over the shell when the animal is active and its mottled colour provides good camouflage; it also means that the shell remains shiny and lustrous, unlike many other species whose shells become worn and covered with encrustations over time.

The mantle also secretes the shell and forms what is probably the most famous product of the mollusc – the pearl. Pearls are formed naturally when layers of mother-of-pearl are laid down around particles of grit lodged in the mantle cavity. Some molluscs can be persuaded to form 'cultured' pearls by inserting a tiny hard object, such as a minute piece of shell, into the animal. Commercial pearls come mainly from the pearl oysters, but many people are surprised to discover that other molluscs can form them. The queen conch can produce a delicate pink pearl, although this is

Below: *The Sea Hare* (Aplysia *sp.), a large Opisthobranch, releases a purplish poisonous dye from a gland in its mantle when it is attacked.*

not very valuable, and the largest pearl in the world – the pearl of Allah – came from a giant clam and weighs 6.4kg (14lb). The mantle may also secrete poisons for defence, in the form of acids, as in some cowries, or inks as in some sea slugs and the cephalopods.

Reproduction

Marine molluscs either have separate sexes or are hermaphrodites. Apart from the cephalopods and a few other species, such as the Caribbean Queen Conch, which copulate, most shed their eggs and sperm into the sea at the same time and fertilization takes place in the water. In some cases, the eggs are laid on the bottom or on vegetation in clutches surrounded by a jellylike material. This can be spectacularly colourful, as in some sea slugs. As in many other marine invertebrate larvae, planktonic larvae called veligers develop and float in the ocean currents, often over huge distances. Oysters are known to have been dispersed in this way over 1300km (780 miles). Some species, such as

Below: *The spotted shell of the Tiger Cowrie (seen here with egg capsules) is kept lustrous by the mantle, which expands when the cowrie is active.*

A predatory univalve

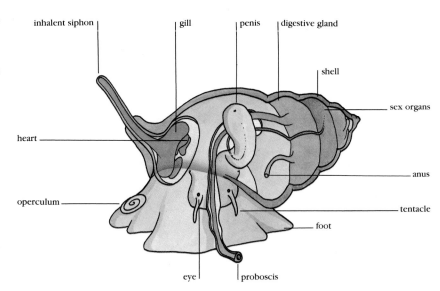

the giant clams, have shorter planktonic lives and their ranges are therefore smaller. A few species have no planktonic stage and produce young that look identical to their parents; these species may have very narrow distributions.

Classification

Molluscs are an extremely varied group, ranging from small parasitic clams that burrow into the arms of starfishes, to the giant squid that

roam the deep waters of the oceans. Of the various groups of mollusc, the three largest are of interest to the aquarist.

Gastropods

The largest group, with 60,000-75,000 species, is the Gastropoda, or univalves. It includes the terrestrial slugs and snails, as well as vast numbers of marine and freshwater species. Gastropods have single shells, often strongly coiled, which usually open on the right-hand side. In species where the adults appear to have differently shaped shells, the juveniles practically always have coiled shells. The shape of the cowrie, for example, is obtained by the final large whorl of the shell enclosing the earlier smaller parts of the spiral within it, as can be clearly seen if an old shell is cut open. Some species have an operculum that operates as a trapdoor to close the shell. This is usually made of a horny material, but in some molluscs it is calcareous. The beautiful green operculum of the turban shell is used in jewellery and is known as a cat's eye.

Marine gastropods include herbivores, detritus feeders, carnivores and a few ciliary feeders, in which the radula is reduced or absent. The radula is usually closely adapted to the food that a species eats. The simplest gastropods are the limpets and abalones, both herbivores that use their hard radulas to rasp at seaweeds on rocks. Other herbivores are the cowries, spider shells and conchs. The Caribbean Queen Conch lives in sea-grass beds and propels itself over the bed using a modified operculum, resembling an outsize fingernail, that it jams into the sand, while the muscular foot heaves the animal forward.

Carnivorous gastropods have fewer, larger, more pointed teeth on a narrower radula than herbivores, and a proboscis that carries the mouth and radula. Whelks feed on dead organic matter, but will also prise open live bivalves by wedging the valves open with the edge of their shells to obtain the flesh inside. Dog whelks bore holes through the shells of other molluscs and barnacles using special teeth and then suck the tissues out. Cone shells have a long proboscis and harpoonlike teeth on the radula. They impale their prey, which includes small fishes, worms and other molluscs, on the radula, paralyze them with a nerve poison and swallow them whole. The poison can be fatal to man.

Many marine gastropods are burrowers and have siphons or tubes that extend from the mantle and sometimes the shell. These act as snorkels, enabling the animal to continue to draw a water current containing oxygen and food into their bodies. The siphons are also used to detect prey from a distance. Many marine gastropods have tentacles on the head, with eyes at the base.

Opisthobranchs are one group of gastropods of particular interest to the aquarist. They include the bubble snails, which have a very thin, almost translucent shell; the sea hares with a very reduced shell; and the sea slugs, with no shell at all, which can be considered the marine equivalents of land slugs.

Some opisthobranchs creep slowly along the sea bottom or over seaweed and corals, but many are agile and beautiful swimmers, such as the Spanish Dancer (see page 347), which swims in the surface of the oceans. The opisthobranchs are all hermaphrodites.

Sea hares (see page 346), with their prominent tentacles, are among the largest opisthobranchs and have internal gills and a simple internal shell plate.

Below: A pair of colourful Chromodoris lubocki *are responsible for laying a string of eggs, a fairly common occurrence in the marine aquarium.*

Above: The bright colours of sea slugs, such as this Nembrotha *sp., are used to camouflage them among corals, or to warn predators that they are poisonous.*

Sea slugs are often flamboyantly coloured, either as a warning if they are poisonous, or to camouflage them on the corals and seaweeds on which many species are found. The gills are often in the form of feathery plumes on their backs, and give rise to their other name – nudibranchs, or naked gills. The dorid group of nudibranchs, such as *Chromodoris* (see page 346), have gills in a small cluster at the rear, while the gills of the aeolid group are irregularly placed along the back. Aeolids can withdraw their gills into the body for defensive purposes, while those of dorids are permanently exposed. To counter this weakness, many of the dorids, which feed on anemones and stinging hydroids, 'pirate' the stinging cells (nematocysts) of their cnidarian food and re-use them in their gill tufts as protection against predators. Nudibranchs with smooth or warty backs have no visible gill mechanism and, in some cases, respiration may take place directly through the skin. Although these species tend to be less attractive than many of the dorids, they often prove hardier than the latter and more adaptable to aquarium life.

A bivalve

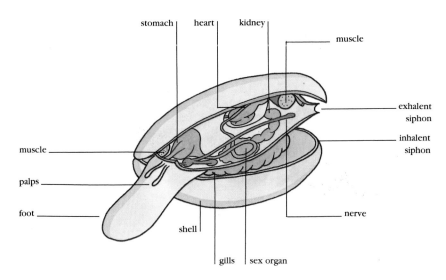

stomach | heart | kidney

muscle

exhalent siphon

inhalent siphon

muscle

palps

foot

shell

nerve

gills | sex organ

Many also have fleshy extensions of the digestive system on their backs.

A few sea slugs are herbivores, but many are 'grazing carnivores', which may seem a contradiction in terms until one realizes that they graze on sedentary animals such as corals, sponges and other invertebrates. Many have distinct dietary preferences and regularly occur in association with certain species. For example, *Chromodoris quadricolor* is found with the sponge *Latrunculia*. Unless the exact food preference is known, keeping these species in an aquarium is very difficult. As a general rule, it is a good idea to house all sea slugs in a well-stocked 'living reef' aquarium in the hope that they will find a suitable food source. Unfortunately, although sea slugs tend to be very attractive, they cannot, with a few exceptions, be recommended to beginners.

Bivalves

The bivalves are the second largest group of molluscs after the gastropods, with 15,000-20,000 species. They include many commercially important species, such as mussels, clams, oysters, scallops and cockles. Their shells, or valves, are in two, usually symmetrical, hinged parts, held together tightly by a pair of powerful muscles. As in the gastropods, bivalve shells can be very variable in shape, colour and texture, the largest being those of the giant clams (see page 343). In bivalves, the head has been lost and there is a pair of large gills, generally used for feeding as well as respiration. These are shaped like leaves or curtains and are covered with cilia that beat continuously to draw in a current. Plankton is trapped by mucus on the gills and carried by the cilia to the mouth. Many bivalves require a high concentration of organic matter in the water and tend to be found in coastal areas. Few can survive long out of water; intertidal species, such as mussels must close their shells tightly at low tide to retain water.

Most bivalves live a sedentary life on or in the seabed. Some, such as mussels and some of the oysters are attached to rocks and other hard substrates by strong elastic fibres, known as byssus threads. Bivalves feed on filtered phytoplankton or take in detritus with siphons that reach to the surface in burrowing species. Burrowers have a large flattened foot that digs through the sand or mud by a combination of muscle action and blood pressure.

Some bivalves are surprisingly mobile, particularly the cockles, which are capable of leaping, and the scallops, such as the Flame Scallop (see page 342), which swim by opening and closing their shells and expelling water from the mantle cavity so forcefully that they move under a form of jet propulsion. They have a row of tiny eyes around the mantle edge to detect danger and can therefore make a rapid escape from a predator. The boring molluscs, like the boring sponges, are aptly named for their lifestyle; as soon as the larvae settle, they start excavating into rock or coral, either by using their shell valves, which often have serrated edges, or drills, or by secreting an acid. The shipworms have long cylindrical bodies and bore into timber, using the excavated sawdust as food.

Below: *The giant clams (this is* Tridacna gigas) *are the largest bivalves.*

Cephalopods

With 650 species, the Cephalopoda is the third largest group, and includes squid, cuttlefish and octopi. They are the most highly developed molluscs and include some of the most intelligent and fascinating, if also rapacious, invertebrates in the world. The majority are quite large, and many species are totally unsuited to the home aquarium. These include the famous giant squid, which reaches a length of 20m (66ft)! Furthermore, comparatively few species are regularly available.

The shell is very reduced or even absent, and a complete shell is found only in the chambered nautilus. However, unlike gastropods, the nautilus only lives in part of its shell, the rest being divided into chambers which are filled with gas and used as a buoyancy organ. Cuttlefish and squid are unusual molluscs in having an internal shell. The flattened cuttlefish 'bone', often found on the beach, is comparatively soft; the squid 'pen' is very thin and reduced.

The mouth of a cephalopod is surrounded by tentacles, derived from the foot found in other molluscs. Octopi have eight tentacles, squid and cuttlefish have ten, and the chambered nautilus has 38. The tentacles have well-developed senses of touch and taste: those of the octopus (see pages 344-345) are extremely sensitive and can discriminate texture and pattern. All cephalopods are active predators on fish and crustaceans and use their tentacles to locate and capture prey. The mouth has a strong beak for tearing pieces from the prey, which is then pushed into the mouth by the radula. Their voracious appetites are an important consideration when you come to choose suitable tankmates. They are safe with most sessile (non-moving) invertebrates, but will catch and kill any moving animal available. They put a heavy demand on filtration systems, producing a lot of waste products and, at the same time, demanding ideal water conditions. Despite this,

A cuttlefish

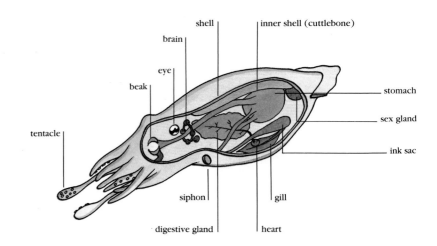

they are justifiably popular and can become quite tame.

Unlike most molluscs, cephalopods are able to move very rapidly (they travel backwards, trailing their tentacles). The octopus usually crawls using the suckers on its arms, but it is capable of swimming. In the squid and cuttlefish, the mantle cavity has become a pump that squirts water through a funnel in a form of jet propulsion. The torpedo-shaped squid move around in shoals and reach the greatest swimming speeds of any marine invertebrates. There are even flying squid that can shoot out of the water and glide for some distance, sometimes travelling through the air at 25.5kph (16mph).

The cephalopods have well-developed eyes and the largest brains of all invertebrates. They are responsive to external stimuli – made possible by their giant nerve fibres, similar to those found in some worms – and can respond extremely rapidly to events signalling danger. Octopi even have a memory and can be trained. Many cephalopods are capable of rapid colour change and use this for camouflage, defence and copulation. Pigment cells, called chromatophores, in the skin expand or contract rapidly, sometimes producing stripes and patterns. By contracting completely they can make the squid almost invisible.

All cephalopods, except nautilus, have an ink sac containing the dark brown ink known as sepia. The cephalopod discharges the sac to produce a 'smoke screen' of ink that hangs like a cloud in the water, fooling the predator while the cephalopod escapes. However smoke screens are of little help to the squid, which form the main food of sperm whales; it has been estimated that the whales consume over 100 million tonnes in a year.

Reproduction

Unlike many other molluscs, cephalopods have separate sexes. Octopi go through an elaborate courtship ritual in which the male changes colour and arouses the female by stroking her. Then he transfers a sperm 'packet' on the tip of one of his arms into her mantle cavity, where it fertilizes the eggs. The eggs, as in other cephalopods, are large and yolky and are usually laid in the shelter of a crevice or shell. They are guarded by the female until they hatch as miniature versions of their parents. In some species, the female does not feed during this period of guardianship and dies after the eggs have been hatched. Squid generally lay their eggs in sticky clusters on rocks in open water. Cuttlefish eggs resemble bunches of grapes and may be washed up on the shore after a gale.

Phylum ECHINODERMATA

This group of entirely marine animals consists of about 6000 species and includes many that are of great interest to hobbyists, such as starfishes, sea urchins, sea lilies, feather stars and sea cucumbers, all of which show a huge variation in structure. However, they share certain constant features. One of the most striking common characteristics is the five-rayed radial symmetry, shown most clearly in the starfish. Like the lower invertebrates, they lack a distinct head, brain and complex sense organs. The nervous system consists mainly of nerve cords along the arms and simple receptor cells over the animal's surface, which respond to touch and chemicals in solution. However, many other aspects of their structure indicate that they are highly evolved invertebrates.

Structure

The skeleton is internal and consists of calcareous ossicles, or plates, that usually bear spines and ridges, from which the name Echinodermata – meaning spiny skinned – is derived. The skeleton is perforated by numerous tiny spaces, which makes it very light while remaining strong.

Echinoderms are also unique in possessing a water vascular system. This consists of five radiating canals containing sea water that connect by side branches to many hundreds of pairs of tube feet. Each tube foot, which in many species has a sucker at the end, can be moved by means of valves and muscles. There is a bladderlike reservoir at the base of the tube foot and when this contracts, water is forced into the foot and it extends. The tube feet are used in locomotion, respiration, feeding and sensory perception. On its own, a tube foot is a very weak structure, but by working with its neighbours in relays, sufficient pressure can be exerted to enable starfishes to pull apart the two shells of a mussel or cockle.

Starfish and sea urchins have extraordinary tiny protuberances over the body that look like minute

tongs or forceps. These were once thought to be parasites on the animals, but are now known to be part of the body and are called pedicellariae. They may be on stalks or attached directly to the skeleton. They have two or more pincers, and their detailed structure is very variable. Their main function seems to be to remove sand and debris from the surface of the animal, but in some species they are used for defence, and may be capable of injecting poison.

A starfish

Above: Pentagonaster duebeni, *the Biscuit Starfish, detects food by smell and can move towards it at a surprisingly fast pace.*

Behaviour

The echinoderms are mainly bottom-living marine animals. Some, such as starfishes, which feed on molluscs, are active predators. Others are filter-feeders, comparable with featherduster worms, that trap small particles of food on feathery appendages.

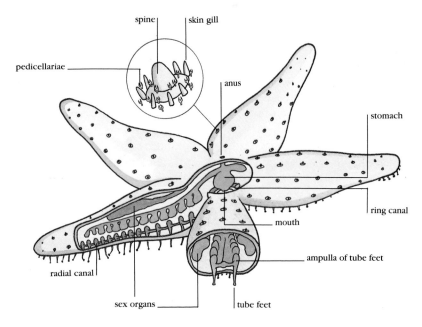

spine | skin gill
pedicellariae
anus
stomach
ring canal
mouth
ampulla of tube feet
radial canal
sex organs | tube feet

Others survive by sifting through the accumulated detritus on the seabed. A number of species are diurnal (active during the day), but many more confine their activities to the hours of darkness, when their, soft, unprotected bodies are less at risk from predators.

The very bright colouring of the tropical species, and the easy maintenance and often cosmopolitan diet of echinoderms, justifies their popularity among invertebrate keepers. Given good water conditions and a suitable diet, many species can be expected to live in the aquarium for several years. Like many invertebrates, echinoderms often live in close association with other animals. Small starfishes, shrimps and gobies live a well-camouflaged existence among the arms of crinoids, for example, or even inside the bodies of sea cucumbers, other starfishes and sea urchins, emerging only to feed. They are infrequent but welcome bonuses to the aquarium.

Echinoderms reproduce by shedding their eggs and sperm into the sea, where fertilization takes place. Surprisingly enough, the larvae are bilaterally, rather than radially, symmetrical and swim by means of ciliated bands on the body surface. They float in the currents before settling and metamorphosing into an adult.

Feather stars and sea lilies
The free-living feather stars that inhabit shallow seas are most abundant in the tropics. Both they and the stalked sea lilies of deeper waters are in the group known as crinoids. They are the most primitive echinoderms and their history is well documented from the large numbers of fossils, dating back about 500 million years, when crinoids were among the commonest animals in warm shallow seas. Thick beds of limestone in Derbyshire in the UK owe their existence to the accumulation of stalk segments from the ancestors of crinoid species found today. Only the few rare sea lilies now retain this stalk as adults,

but the larvae of all crinoids are initially anchored to the substrate by a small stem before they break free and drift onto the reefs.

Crinoid arms are usually forked and branched and their tube feet lack suckers. The tube feet lie in a double row along the upper side of each arm and are used for respiration and feeding, small particles of matter sticking to the mucus on the feet. The sea lilies are fixed to the bottom and look rather like palm trees with upturned fronds. They feed on fine particles sieved from the water.

Feather stars, such as the Red Crinoid *Himerometra robustipinna*, have a central disc, or cup, from the underside of which grow the short, spiky 'cirri', or hooked appendages, that they use to grip rocks and corals. They can walk rather clumsily on their cirri or swim in a rather spectacular fashion, each arm beating up and down independently with undulations. The cup extends upwards and outwards into five arms that are usually repeatedly branched. These are edged with small extensions called pinnules, and the end result is an animal that looks not unlike a feathery shuttlecock. Their arms have a great degree of vertical flexibility but very little lateral movement, which makes them very brittle.

Above: *The delicate tracery of the arms of the basket stars is well illustrated in this Red Sea species. At rest, the arms are curled into a tight ball.*

Brittle stars and basket stars
Brittle stars, such as *Ophiomastix venosa* and basket stars, such as *Astrophyton muricatum* (see page 350), are in the group known as ophiuroids. Like the feather stars, these have a central disc and their tube feet lack suckers and are used for respiration and feeding only. The skeleton is made up of many ossicles, which fit tightly together. In brittle stars, the disc is flat and the long, thin and very mobile arms are clearly set off from it, unlike those of the starfish. In most brittle stars the arms are smooth, but a number of commonly imported species have short spiky extensions along the arms that may offer some protection from predators. The arms are capable of lateral movements and limited vertical movements, but cannot be coiled around objects. Brittle stars can move remarkably quickly, the arms propelling the animal with snakelike movements.

They feed either by collecting tiny edible particles on their arms as they wave them about, or by tearing off pieces of seaweed or dead fish. In the sea, they often live in huge beds and can form a seething carpet

Above: *A Red Feather Star clings to a rock, having emerged at night to filter feed on small particles carried in the water current.*

up to five animals deep. Like feather stars, brittle stars are very fragile, as the name implies, but they are able to regenerate lost arms very rapidly.

In basket stars, the five arms are divided and subdivided to form a greatly branched 'net' and they can be coiled around objects. During the day, they rest on sea fans or rock pinnacles, looking like loose balls of string, but at night, they spread their arms to produce a roughly circular trap, or 'basket', to catch whatever small food items the water currents bring their way.

Starfishes

The starfishes are the most familiar echinoderms and many are brightly coloured. They usually have five or more well-developed, stout arms radiating from the centre of the body. Like brittle and basket stars, these can regenerate relatively easily if damaged. The tube feet are on the underside of the arms and the mouth is in the centre of the underside. Many starfishes feed by everting the stomach through the mouth, encircling and digesting large food items, then retracting the stomach and moving on.

Although not as fast as brittle stars, they can detect food by smell, and crawl towards it at a surprisingly fast rate. Some feed on minute particles but most prey on live animals, forcing open shellfishes or chewing sponges. They can be serious pests of commercial oyster and mussel beds and the infamous Crown-of-thorns Starfish, *Acanthaster planci*, (see also page 58) can wreak havoc on a coral reef if large numbers congregate to feed on the coral polyps.

A sea urchin

Above: *Although the similarities are not obvious, the Long-spined Sea Urchin* (Diadema antillarum) *is a close relation of the starfishes.*

Sea urchins

The sea urchins and sand dollars in the group known as echinoids have globular or flat bodies with an internal shell, or 'test', of closely fitting plates. At first sight, they appear to bear little relation to the starfishes, but there are clear similarities. They can be thought of

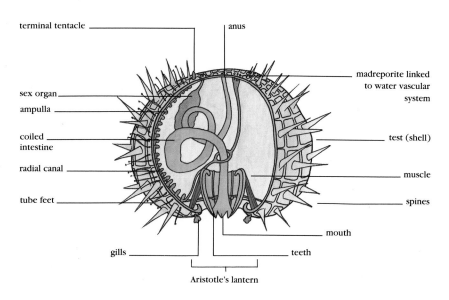

terminal tentacle — anus

sex organ

ampulla

coiled intestine

radial canal

tube feet

madreporite linked to water vascular system

test (shell)

muscle

spines

mouth

gills — teeth

Aristotle's lantern

as starfish with their arms bent over their backs and their skeletal plates fused together. The tube feet vary in shape and function, and may be used for movement or creating currents in tunnels. Some urchins burrow in sand and mud, keeping a vertical shaft open to supply water for respiration. Others burrow into rock using the spines attached to the test to dig with. The shape of the spines is adapted to the species habitat; urchins that live on surf-beaten shores have short, stout spines, whereas those from calmer waters, such as *Diadema savignyi*, have longer spines. In addition to providing protection from predators, the spines may be used with the tube feet for climbing rocks.

Many urchins have a unique organ known as Aristotle's lantern, named after its discoverer. This consists of five hard calcareous teeth, suspended from a complex chewing apparatus, and ringing the mouth on the underside of the body. The teeth are used to scrape algal material from rocky surfaces. Sand dollars and heart urchins have no lantern. They burrow in the sand, collecting particles of detritus on their modified spines and tube feet.

Sea cucumbers

The sea cucumbers, or holothurians, are elongated, sausage-shaped echinoderms. The tube feet around the mouth have become sticky tentacles, which vary slightly from one species to another according to the size of the detritus particles on which they feed. Some sea cucumbers sweep the surface of the sand or mud with the tentacles, but others rely on water currents to bring food to them. Other tube feet over the body are used for locomotion. The skeleton is reduced to small crystals embedded in the skin and these produce a rough leathery feel.

When attacked, and possibly as a result of chemical changes in the

Right: Pseudocolochirus *sp. This vividly coloured sea cucumber is much in demand with marine aquarists, despite its invariably high price.*

habitat, sea cucumbers can discharge the stomach and its contents through the anus to help them escape. In the wild, the stomach and intestine are regenerated quickly, but avoid very deflated specimens when buying stock. An alternative defence

adopted by the sea cucumber is to squirt out sticky threads. Amazingly enough, sea cucumbers are a gastronomic delicacy in the Far East, where they are known as *trepang*, or *bêche-de-mer*, the latter name coming from the Portuguese *'bicho-do-mar'* or 'worm-of-the-sea'.

A sea cucumber

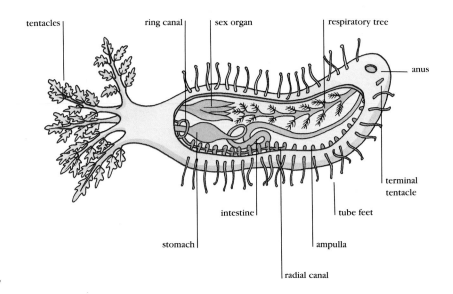

Phylum CHORDATA

The majority of – and the most familiar – chordates are the vertebrates, or animals with backbones. However, there are some species within this phylum that represent the link between the vertebrates and the invertebrates and these are often known as the protochordates. They lack a true backbone, but have a stiff rod, or notochord, in their bodies in at least one stage in their life cycle, and a single hollow dorsal nerve cord. The sea squirts, or ascidians, are the only protochordates of interest to the hobbyist and even these usually arrive in the aquarium by accident.

There are over 1,000 species of sea squirts, many of which live on reefs. Some are solitary, and large individuals may be up to 50cm (20in) high. Others are colonial and form mats composed of many small individuals. They have a stiff, jellylike or leathery bag-shaped body called a tunic, with large inlet and outlet siphons, although colonial forms have a single communal outlet. Water is drawn in and passes through a strainer, where small particles of food are filtered out before the water is discharged. Their common name comes from their habit of squirting water if they are squeezed.

Sea squirt larvae are like tiny tadpoles and show the typical chordate characteristics. They have a notochord, sense organs and

A sea squirt

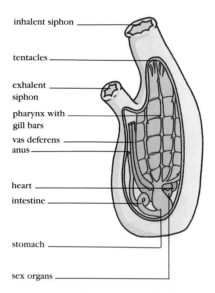

- inhalent siphon
- tentacles
- exhalent siphon
- pharynx with gill bars
- vas deferens
- anus
- heart
- intestine
- stomach
- sex organs

nervous system, all of which are lost when the larva settles and turns into the adult form. Salps are an interesting, free-swimming group of sea squirts that float in the open waters of the oceans. They are jellylike and reproduce by budding, the new individuals often remaining attached to the old ones, thus forming long strings. Many sea squirts are very colourful, such as *Distomus* spp., but they are inactive and have never become popular.

The other group of simple chordates contains less than 20 species. These have a notochord and are called lancelets, or amphioxus. They look like little transparent fish, and they burrow in coarse sand in shallow water, lying with their mouths exposed to sieve food from the water current. They sometimes aggregate in large numbers, and are fished and eaten in some parts of the world, particularly China.

Below: *The inlet and outlet siphons of this small colony of sea squirts can be seen here. Note how some fuse together, while others remain as individuals.*

Collection and Conservation

In 1980, the International Union for the Conservation of Nature and Natural Resources published a seminal document on the relationship between just one species – humans – and all the others on this planet. It was written over several years, as a result of the slow appreciation of the increasing dominance of the one species over all the rest. This imbalance had to be redressed if possible or, to take the worst scenario, humankind had to be informed of the disastrous consequences if our current behaviour persisted. *The World Conservation Strategy* was not meant to be a document that advocated reversing time, eliminating sprawling cities and about half of the human population. Rather, it tried to establish the terms for as harmonious a balance as possible between dominant humans and the rest of the world's species, to the mutual benefit of both camps.

The document argued that the frequently misused and misunderstood concept 'conservation' simply means the management of human use of the world in such a way that it may yield the greatest sustainable benefit to the present generation and maintain this into future generations of both humans and other species (the majority). Humans are a part of the environment. It is not there for us. Therefore, we must live with it, not let it live and die for us, thereby effectively signing our own death warrants. (It is perhaps worth remembering here that a parasite can only be successful and survive if it does not kill its host.)

It may at first seem difficult to reconcile the keeping and selling of marine fishes and invertebrates with the conservation of the world's coral reefs for future generations, but these two aims are not mutually exclusive. It is possible for the marine conservationist and the aquarist to work hand in hand in examining different facets of conservation and collection. There can even be a two-way feedback in that the observations of one may be of benefit to the other. An observation of a detail of a fish's life by an aquarist may help the biologist understand one part of the jigsaw of reef life. Equally, the conservationist's or biologist's report that two particular species are, say, always found in close association at night can help the hobbyist improve the quality of life for the animals in his or her care.

The trade chains

In any type of retailing there must be cooperation at all levels of the chain, from before collecting on the coral reef to the final sale of a healthy fish to the hobbyist. There are, however, many limiting factors to consider, which start with the local fishermen and culminate in establishing concordance in all the worldwide trade chains.

There are two basic trade chains. Both start with the fisherman who collects the desired species. These are then sold on to a person variously described as a transporter, gatherer or middleman, who will visit a number of collectors and take the entire consignment to a holding facility. The holding facilities, which are often those of exporters, are usually close to international airports. Here, the fishes should be given time to recuperate from the stress of collection and transportation. Although the decision to provide a rest period for the fishes ought to be a deliberate one, it can also occur inadvertently while the exporter waits for enough consignments of the required species to build up to fulfil orders. The orders are then made up, the fishes packed appropriately, and flown out to a receiving wholesaler in Europe, America or Japan. A good wholesaler then quarantines the fishes and sells them on to a retailer, who sells directly to the public.

The other basic trade chain is one in which the retailers commission

Left: *The retail shop is the last link in a chain that starts with local fishermen, who collect the desired species and sell them to a transporter.*

or employ a consolidator. The consolidator is an agent who will work on behalf of several retail outlets to obtain the most competitive carriage rates from the airlines. This cuts out the need for the wholesaler so that the retail shop can offer the livestock for sale at a most competitive price. The disadvantage of the consolidator chain is that the range of species is generally more limited than that obtainable from wholesalers, and quarantine facilities have to be provided by the retailer.

As there are thousands of miles of collectable coral reefs, transport conditions and logistics are of fundamental importance. Coral fish are, by nature, delicate and do not respond well to poor treatment in catching, handling or transporting. The nature of transport from the point of collecting to the holding facility is a limiting factor in this stage of the chain. Most third world countries – where most fish are caught – do not have motorways or rail terminals near the coral reefs affording easy access to the holding facility. This may seem obvious, but it is a fact often overlooked by those examining the trade in corals and reef fishes. It has been calculated that over 99 percent of the world's coral reefs are logistically

impossible, or at least financially unworthwhile, to collect from, as the transport overheads are so high.

Freight space, or more importantly the lack of it, will adversely affect the transport of coral reef fishes. As over 80 percent of the invertebrates and coral reef fishes sold in Europe originate in the Far East, space is at a premium. The 'mega-top' 747s that some airlines are using on a non-stop Far East or Europe service have a smaller volume of freight space available and, as the ageing fleet of 747 Jumbos is phased out, the situation may well get worse, the advantage of a non-stop flight being nullified by the smaller volume of freight capable of being transported. This man-made limitation could well be with us sooner than wished, and the law of supply and demand means that if goods cannot be shipped, and thus cannot reach the market in the volume demanded, there is no point in collecting them.

Twenty years ago, the holding facilities at many of the exporters were little more than concrete vats with primitive aeration systems

Below: *Large tanks, good filtration and effective monitoring are essential in a well-run holding station. Butterflyfish are sensitive and require extra care.*

(mostly, individual pumps with airstones to provide the oxygen and a weak water current). Biological filtration systems were unknown. Happily, this situation is now changing. There are several reasons for this. Firstly, the initial cost of the fish is higher; collectors are now encouraged to take more care in their collecting so that the fishes are in perfect condition. Secondly, some species of fish are less abundant, or live further from the collecting sites, than before. Thirdly, greater time is taken in obtaining and ensuring the fishes reach the exporter in good condition. More time is therefore expended by the collectors, who then demand a greater return from the exporter. The exporters' capital investment is therefore higher and if they do not take advantage of the latest technology to ensure the health of their stock they will lose more money than before.

Not all exporters in the Third World have become rich through their efforts. Many have to be helped financially to install sophisticated and beneficial holding systems. Often, this financial help comes from the importers in the northern hemisphere, who use the greater capital, to which they have more access, to provide better systems. The financial burden is spread, but

a self-contained underwater breathing apparatus (SCUBA). A hooker is a low-pressure compressor at the surface in a boat or canoe moored above the diver. The airhose is fixed to the seabed and the divers take a lungful of air when necessary. The divers using the hooker system may work with ordinary goggles to aid their vision underwater. There are also those who use a full-face mask. In this case, the mask is filled with air from the hooker pipe which gives them a prolonged period of freedom. It is not uncommon for up to half a dozen divers to work off the same hooker pipe. Even when a face mask is used, the diver cannot go much deeper than the hooker hose can reach. SCUBA divers use a high-

the importers gain a benefit from the security of receiving high-quality, healthy fish. To this end, fish are now given temporary housing in all-glass aquariums, often individually, and provided with water-circulation plants with water being filtered through cartridge filters. There are also ultra-violet sterilizers and trickle filters, and ozone can be added via a protein skimmer, while a level of veterinary help is available if required.

Collecting methods

The three main methods of collecting fish are by net, trap or cyanide. Most fish are caught at depths of less than 10 metres by divers, using either a hooker pipe or

Below: *A collector carefully disentangles a fish caught in the soft mesh of the barrier. This is done quickly and skilfully.*

pressure compressor to recharge their cylinders. This gives the divers greater freedom but is more expensive, both initially and in running costs.

On the seabed a fine-meshed nylon mono-filament net about 10 metres (33ft) long and 1 metre (40in) high is placed along a part of the sea floor. Weights keep the bottom of the net in place, while floats at the top keep it upright so that it resembles a curtain rising up from the bottom of the sea. The fish are chased into this barrier, become enmeshed, and are then carefully removed and placed in a 'goodie' bag. A variant of this technique is to bait the net with pieces of fish. The bait attracts some species, and others, discerning a feeding activity, come to investigate and are caught. The collectors then work along the net, picking out the fishes and putting them into the 'goodie' bag. The 'goodie' bag is usually of plastic about 60-75cm (24-30in) long and 45cm (18in) wide. The collected fish, of whatever species, are placed in the bag and the neck held closed by the diver. It is not normal for there to be interspecific aggression within the bag. The reason for this is unclear, but it may be that taking a fish away from its territory disorientates it and, having no familiar landmarks to defend, it just stays there with formerly antagonistic bedfellows.

Fish rarely try to bite their way out of the bag. Triggerfish will occasionally perforate the bag, but the holes are not big enough to allow the escape of the inmates and the triggerfish's dentition is not geared to producing anything more than nip-sized holes. What is interesting is that the surgeonfish do not seem to erect their caudal peduncle spines when they are in the 'goodie' bag. Netting bags are rarely used in the Philippines.

Drop nets are also used to collect fish. A drop net is a circular net with weights around the edge and a float at the centre. This is dropped over an area of reef and the fish chased into the raised central portion from which they are collected.

Fish traps vary in style in different parts of the world but, in essence, follow the same basic pattern. A trap is made of a net on a wooden frame measuring approximately one metre by two metres (40 × 80in). A cone, usually made of bamboo, leads into the net. Bait, of fish or meat, is fastened in the holding part of the net. Fish are attracted into the net by the bait and, once in, cannot get out. Quite why they cannot navigate out the way they came in is not understood, but there is deep fishy psychological reluctance to go downstream headfirst. It is possible that as water carrying the vital oxygen goes in through the mouth and out through the gill covers, facing the 'wrong way' may induce some feeling of breathlessness. However, whereas that logic seems to work for the common minnow trap, in which the netted neck is placed facing into the current and the fish enter by the knocked out punt, it does not seem to be valid for fish traps set around the coral reefs from Thailand to the Philippines. There appears to be no particular orientation. They are usually placed about 10-12 metres (approximately 30-40ft) down, well below low water mark and the surf zone (according to locality), on a

Below: *The final stage of transferring the catch of fish into a large plastic 'goodie' bag. Spare bags are tucked into the diver's waist belt.*

firm bottom and close to coral bosses. The position of the trap is rarely changed; it remains anchored to a rock or coral head. The trap is then visited once or twice a day and its contents collected.

Originally, these traps were used to catch food fish, but, of late, their efficacy in trapping aquarium fish in perfect condition has been realized. The bait and mesh size can be adjusted to obtain the desired species of the local fauna. All fishes caught at the deeper part of the diver's depth range, or in deep-set traps, are brought slowly to the surface to allow the fish time to adjust to pressure differences. Rapidly hauling a fish up from as little as 10m (33ft) can cause it damage and distress.

The techniques listed above have been used at an artisanal level for centuries, and the balance between the needs of the human population and the productivity of the accessible parts of the reef has been maintained. The most damaging technique is the third in the list – sodium cyanide poisoning.

Sodium cyanide is a poison of considerable efficacy, but some animals are much more sensitive to it than others, and those most affected may not be the species sought. Sodium cyanide is commonly used for collecting in reefs around the Philippines, and the technique has allegedly spread to Malaysia and Indonesia. In this method of fishing, the collector dissolves the cyanide in water. The fish are then chased into the coral and the cyanide solution squirted into the region. The fish become stupified and can be collected by hand. They are subsequently taken in containers to an unpoisoned area (upcurrent) and left to recover.

There are two drawbacks with this method of fish capture. Firstly, the fish that are caught fairly quickly, or on the edge of the cyanide cloud, and rapidly recuperate are most

Right: Cyanide used to catch fish has killed these corals. This is one reason why this method of catching should be strongly discouraged.

likely already destined for ill-health. Their liver and kidneys will have already suffered irreversible damage, the degree of which will depend on the duration of contact with the cyanide solution. Almost every fish stupified enough to be caught in this way will be damaged. At first, they may appear healthy, but after about a month, they will start to show signs of distress, swimming in circles, stopping feeding and becoming so emaciated that their ribs can be seen through the skin. Then they will die. An oddity of the cyanide poisoning symptoms is that, even at their end, the fish have not lost colour.

The second disadvantage in using cyanide is that the fish not collected, as well as the corals, remain in contact with the cyanide long enough to be killed. This is especially serious for the corals, as they are very sensitive to the cyanide. Coral does not regenerate as rapidly as the coral reef fish can repopulate the collected area. Dewey (1979) estimated that the mortality rate of fish at sodium cyanide collecting sites was 75 percent, and that there was a further mortality rate of 20-25 percent of the recovered fishes before reaching the exporting centre at Manilla. Rubec (1986) noted that even seemingly unaffected coral heads in regions that had been cyanided had died within two months. It has further

been suggested (Schiotz 1989) that the Crown-of-Thorns Starfish (*Acanthaster planci*) has become more abundant because of collecting by this method.

Although damage has been done in the past, the future looks brighter as a result of close ties that wholesalers have with their collectors. Customers, too, are no longer prepared to buy fish collected by cyanide and this message has gone through the retailers to the wholesalers and thence to the suppliers.

The only reasons for using cyanide in the first place are that it is cheap and less labour-intensive than using nets. Despite the fact that net-caught fish are going to be slightly more expensive, their greater viability allied to the lower level of damage that net-fishing causes to the environment, are compensating factors. Indeed, the lack of cyanide-caught specimens has become an advertising policy for a number of UK-based wholesalers.

It was direct pressure from aquarists, retailers and wholesalers that started the movement in the Philippines away from cyanide and towards nets. It is encouraging that in October 1990, over 10 percent of the cyanide fishermen in the Philippines had changed to using nets. But there is still a long way to go, and the hobbyist must continue to be a driving force.

Farming the reef

The coral reefs can provide mankind with a production yield far above some terrestrial ecosystems. Taken from various sources and expressed as grams per square metre per year (dry weight), the open ocean gives 125, the continental shelf 250 and agricultural land 600. In dramatic contrast, the figure for a coral reef is 3,500. There ought, therefore, to be no difficulty in regarding a coral reef ecosystem as a sustainable resource of a high order. However, this does not mean that it is infinite; no natural resource is infinite.

Foward planning and enlightened management can provide initiatives whereby the reef can be cropped of fish and corals to the financial benefit of the local population and the ultimate satisfaction of the aquarist. The best people to farm a reef in this way are the local people themselves. They have the advantage of being there all the time and can rapidly detect changes in the same way that an agricultural farmer can

know when to let fields go fallow or when to change the crops grown on a particular field to maintain the fertility of the soil. The local people become both guardians and beneficiaries of their local resource, and it is in their own interest to maintain the ecological health of the reef system. The guardian function they fulfil is especially important, as they would prevent the dynamiting or poisoning of their asset by outsiders, and they are most unlikely to destroy 'the goose that lays the golden egg'.

There is also scope for reef expansion that can be most effectively practised on a local basis. For example, off the Bahamas, a research team based in Miami found a most useful purpose for old car tyres. They strapped them up in large bundles and put them close to existing reefs. Surprisingly quickly they became covered with algae and, within a few years, a rich reef invertebrate fauna was present, which, in turn, attracted reef fish.

Even the best-managed and best-

Above: *Farming reefs on a sustainable basis is in the interests not only of these fishermen, but also of all those concerned with the aquatic industry.*

intentioned projects can suffer and local depletions can occur, but if the area affected is small and surrounded by a vigorous ecosystem, these deficiencies will soon be made good. It should be remembered, though, that large-scale natural disasters can overwhelm any population. For example, until March 1882 there was a thriving fishing industry off the east coast of North America for the Tilefish (*Lopholatilus chamaelionticeps*). After that March there were no Tilefish to be caught and the industry collapsed and fishing communities were badly affected financially. During March and April millions and millions of these fishes died. One steamer reported sailing for two days through Tilefish corpses. It seemed as if that species had become extinct. About 20 years later, a few were caught and, by 1915, the species had become sufficiently abundant for a small fishery programme to be started in the area. The cause of this dramatic decline was not man-made; but quite natural; the Gulf Stream had moved. The recovery was also natural, but doubtless speedier than it might have been had the disaster occurred in a centre of human interference.

In more recent years, parts of, particularly, the Australian Great Barrier Reef were plagued by the Crown-of-Thorns Starfish (*Acanthaster planci*). The reason for the population explosion is unknown. It has been argued that, as species often undergo natural cycles in population numbers, it was a natural population explosion at the beginning. But then it is possible that it became exaggerated by the scarcity of the corals on which it fed for preference. It is also highly likely that, if the stories are not apocryphal, one 'control' measure

greatly increased the population numbers. In some areas, when the starfish were collected, they were cut into pieces and thrown back. Happily for starfish, but less fortunately for the custodians of the reef, each arm of the starfish can regrow into a whole animal.

This seemingly foolish act has a precedent. Towards the end of the last century, the famous oyster beds at Whitstable, on the north coast of Kent in south-east England suffered a plague of Common Starfish (*Asterias rubens*). Common Starfish like oysters as much as humans do. They open the shells by exerting a strong, gentle pull on both valves until the oyster's adductor muscle tires and the shell opens a little bit. The starfish then effectively inserts its stomach and the digestion process starts. The oystermen were so enraged by their loss of livelihood, that they dredged and raked up starfish by the thousand, cut them into pieces and threw

them back into the sea as a warning to other starfish. Needless to say, this was not an effective method of control. (The starfish plague did eventually decline, however, and the oyster population slowly built up, though not quite to its former numbers, as pollution later became a limiting factor.) The point of this seeming digression is to illustrate the fact that natural regeneration is possible, but may be slow. The chance of it happening is, however, greatly diminished when natural conditions are altered.

No matter whether the source of depopulations is caused by over-collecting locally, dynamiting reefs for food fish, or explosions of the magnitude of Krakatoa in 1883, if the environment remains unchanged, recolonization will occur. If it can

Below: *The curio trade collects shells in vast quantities. Such indiscriminate collecting is both uncontrolled and wasteful of a natural resource.*

occur after Krakatoa, when a large part of an island disappeared, it can occur anywhere. But conditions must remain unchanged, and here is where we have to extend our thinking beyond the reef as an isolated ecosystem.

Conservation and tourism

Although coral reefs encourage tourism, tourism is not good for coral reefs; or at least, it is a double-edged sword. Yes, a reef may attract tourists to the benefit of the local community, but too many tourists walking over the reef will destroy it. Each footstep can break a piece of coral or tread a polyp into oblivion. Too many footsteps, and there is no reef. In some of the more popular resorts, local patches of reef have been destroyed in this way.

The indiscriminate collection of tourists' curios and souvenirs can also have an uncontrolled effect upon the ecosystem that is the reason for the tourists' presence. 'Uncontrolled' is the vital word. Tourists will hardly ever have the longterm awareness of the sustainable bounty of the reef

Above: *As the boom in SCUBA diving continues, it is important that divers are properly trained and observe a strict 'look but don't touch' policy.*

possessed by the local people. Ironically, there are instances of tourists having their souvenir confiscated by Customs and Excise on their return from holiday. In their ignorance they had brought back a species protected under the regulations of CITES (The Convention of International Trade in Endangered Species).

Tourists also bring waste, which must be disposed of in a sensible fashion. Just jettisoning empty beer cans and plastic bags in the sea is not the way to conserve the ecosystem. It can be argued that beer cans and plastic bags can be removed from the reef, but the other form of tourist waste – sewage – must be disposed of with even greater care.

Tourists' feet treading over exposed coral at low tide allowed for, what about the sea-borne threat of SCUBA divers? Here, luckily, the more responsible training schemes inculcate into the divers a 'look but don't touch' policy. Sadly, not all SCUBA divers are responsible but, slowly, this element is being shamed and persuaded into behaving responsibly and becoming a guardian or sympathetic friend of the reef. Formerly, it was a common (although completely incomprehensible and pointless) practice for divers to smash open sea urchins at the end of a dive and enjoy watching the fish gobbling up the pieces. While to many people, sea urchins are spiny and seemingly inanimate reef inhabitants, they nonetheless form a vital element in a complex chain.

Recently, the chaos theory, whereby the consequences of an event cannot be reliably predicted, has gained great credence, especially in fluid systems exemplified by meteorology and ecology. The chaos theory has been expressed simply in sentences like 'the beat of a butterfly's wing can give rise to a tornado'. It is not suggested that smashing open one sea urchin is necessarily going to destroy the reef, but why destroy it unnecessarily? Responsible SCUBA divers don't.

Deforestation and pollution

Corals have algae living symbiotically inside the cells of each polyp. The coral needs them to survive and the algae – the zooxanthellae (see page 33) – need sunlight to survive for themselves and the coral. Far-distant logging operations can destroy this delicate balance. The chain of events works like this: coral reefs occur in warm tropical zones in shallow water, but tropical regions have monsoons. Monsoons shed their water content as rain, in large quantities and in a very short time and all this sudden deluge has to go somewhere. The great advantage of forests – in this case, the aptly named rain forests – is that they, and their soil, act as both a barrier and a sponge respectively to slow down and absorb the deluge. If the forest is no longer there, the water runs straight, and savagely, to the sea. Its force in

Above: *An idyllic view of a tropical paradise where coral reefs abound. This can be preserved for future generations if we start conservation policies now.*

doing this strips off the soil and carries it to the sea. There, as the power of the water current diminishes, the silt is dropped. It makes the water turbid, cuts down light and is deposited in shallow water. Where are the coral reefs? In shallow water. What does coral need to live? Sunlight – which is precisely what coral does not get in some areas owing to deforestation a thousand miles away. A forest cleared in a few months can destroy ten thousand years of coral growth in a couple of monsoon seasons.

Some threats to reefs are global, but not natural. The sea is the final sump of the world for all waste products, and the coastal parts are the first to be affected. All over the world, industry discharges poisonous waste. Not easily detectable, yet insidious in its effect, is discharge of heavy metals, which can never be broken down. The rare metal vanadium, for example, is naturally extracted in minute quantities from seawater by molluscs, which concentrate it in parts of their body for biological reasons. They need it in the same way that we need iron to make our

blood. The problem is that the absorbing mechanism will also let in atoms of metals that are poisonous. Some fish, through the food chain, concentrate mercury and cadmium in their tissues in this way, and these two elements are very poisonous.

Conserving the reefs
As it is unlikely that in the foreseeable future these widespread threats will be curtailed, it is up to us to maintain the reef ecosystem in as healthy a condition as possible. To this end some areas have been set aside as marine nature reserves from which nothing should be taken. Those in the Red Sea and in the Caribbean are well known.

Recently, there has been an enterprise in the Philippines called *Bantay Dagat*, which means guardian of the seas. This project is working at several levels and is supported by the Philippine Tourist Department and the Philippine Commission on Sport SCUBA Diving. A media campaign to educate the public about the value of the reefs and the disadvantages of bad fishing practices (dynamiting, cyaniding etc.) has also been launched. Marine parks and sanctuaries are being designated and local people are encouraged to act as watchmen and report abuses to the authorities. Furthermore,

those whose livelihoods have been affected are being re-trained. All of this is highly laudable and will help to ensure that reefs remain a valuable resource by limiting damage to them.

With some background protection in such a reef-rich country, there is increased optimism that, at least in that part of the world, a continuing supply of fish and invertebrates will be available to the hobbyist. But the maintenance of this supply depends heavily on the sensible cropping of a renewable resource.

Conservation and the hobbyist
It is in the face of possible catastrophes that collecting activities become less subject to criticism. Yet, it is now that the aquarium trade must play its full part in maintaining reef biodiversity. It is encouraging that this is slowly starting to happen. An obvious contribution is the captive breeding of reef fish. Many species have proved recalcitrant to breeding in the aquarium. At commercially viable prices, only a few species – Neon Gobies, damsels, clownfish and some centropygids – are in trade. Others are being experimentally bred, but the cost has so far proved prohibitive. The hope is that each time another species is bred, the knowledge gained will lead towards

commercial production, thereby minimizing collection of these creatures from the wild.

As they have spectacular shapes and colours, giant clams (or, at least baby giant clams) are popular in the home aquarium, but in the wild they are under great threat. They are sought for food by many peoples of the Indo-Pacific region. As the individual clams are slow-growing and the human population is increasing alarmingly, it is not surprising that the giant clams' numbers have been decimated. Young clams were also collected for the aquarium trade which did not help their plight.

Now, however, captive breeding programmes for giant clams have been developed and help the survival of the species in the wild and also maintain their availability for the aquarist. It was about 15 years ago that various centres started the experiments necessary to find out the techniques for the captive breeding. Problems abounded; the basic biology of giant clams was known, but the detail of the larval rearing was unknown territory.

One unexpected problem was finding that the larval stages of a snail parasitize the young clams. The larvae settle between the shell and the mouth and then feed on the mantle, eventually killing the clam. The snails have to be removed manually as soon as they are noticed. Another snail with similar habits can be washed out. All these problems occur on top of the difficulties of captive breeding.

The most frequently bred species are those of the genus *Tridacna*. In this popular genus, the young do not need any feeding. Spawning in adults is stimulated while they are in large mixed groups. The larvae have their own food supply in the form of a yolk sac. So, for several days, there is no feeding problem. The still-mobile young are then inoculated with zooxanthellae algae, which enter into the cells, live symbiotically and produce the food for the growing, settled clam. One centre, in Palau in Micronesia, has usually about 10,000 growing clams

Above: *Giant clams, such as this* Tridacna maxima, *are now being bred in captivity in large numbers. Pressures on wild populations are thus relieved.*

for the aquarium trade and for future research in their biology.

The clams are sold when between one and two years old and are some 5-10cm (2-4in) in length. *Tridacna deresa* and *T. gigas* are popular for filters and *T. crocea* and *T. maxima* are kept because of the bright colours of their mantles. All in all, these ventures benefit both the aquarist and the conservationists, as well as giving employment to the local people.

Because corals grow slowly and are often difficult to keep, their removal is the most conspicuous danger to the reef ecosystem. Unfortunately, in the Third World, where there are reefs, such collecting has been a source of much-needed income to the local people. This problem has been overcome in the Philippines, where irresponsible reef collecting is discouraged. Further, an industry has now been set up to make artificial corals and sponges from fibreglass. The climate lends itself to the manufacture of these fibreglass models, and many of the former reef collectors are employed in their moulding and painting. It is a successful venture, both economically and ecologically. The false corals and sponges are very acceptable to hobbyists, look realistic in the aquarium and,

anyway, soon acquire a veneer of natural algae.

One ought to note that many of the farm-raised or captive-bred fish and invertebrates are not, at the moment, as cheap as wild-caught individuals. Because of the high cost of overheads and the expenses of the previous research into breeding techniques, the hobbyist has to pay more to support such programmes. The sensible hobbyist will realize that this is a good investment for the future so that wild stocks remain in the wild, yet are still available to the aquarium trade.

In time, when the techniques are perfected, the aquarist may have no choice but to buy farmed fish. This may happen for one or both of two reasons. Firstly, uncontrolled collecting may exterminate various species in the wild. Secondly, legislation may be introduced to prevent the sale of wild-caught animals. Either way, it is preferable to invest money into a breeding programme, and sooner rather than later. And, when the breeding techniques are perfected, the prices will fall. The future of the hobby is in our hands.

PART TWO

SETTING UP THE AQUARIUM

Planning and preparation have always been the keys to successful marine fishkeeping; long before you make any plans to bring the creatures themselves home, you will have much work to do. You will need not only to make physical preparations, but also to gain a good understanding of what makes an aquarium 'tick'. Once you, as a prospective marine aquarist, have decided what you wish to keep, practical considerations come into force, for there is no doubt that these creatures are special, coming, as they do, from the richest and possibly most varied habitat on earth.

Over recent years, many innovations have made the life of the marine aquarist very much easier, and fish and invertebrates have been enabled to live an increasingly natural and prolonged lifespan within the aquarium. Aquarium design has altered radically over the past 20 years; no more the ugly and primitive iron-framed 'waterboxes'; nowadays, people have come to expect, quite rightly, the aquarium to be an integral part of their homes; aesthetic appeal and practical application have come together in a successful and popular union.

However, a certain amount of understanding is not only desirable, but also absolutely essential; water testing is useless if the results are not clearly understood, and any amount of filtration equipment will fail to support valuable livestock if it is incorrectly installed. Setting up a new aquarium is exciting and all part of the pleasures of fishkeeping, but if mistakes are made at this stage, then the whole system will collapse, leading to unhappiness for you, and an almost certainly worse fate for the livestock! Most fish and invertebrates have made a long and difficult journey for our pleasure. We, as caring marine aquarists, owe it to them to provide the best and most natural environment possible.

Left: *Achieving a beautiful living reef aquarium such as this involves careful planning of all aspects of filtration, lighting and compatibility of stock. With some experience and common sense, the results can be breathtaking.*

Exploring the Options

The compelling colours, shapes and behaviour patterns of marine fish and invertebrates captivate most aquarists. For some, marine fishkeeping quickly becomes an all-consuming passion, regardless of cost or effort. Most people, however, must be satisfied to keep an aquarium within restricted budget and time constraints, carefully planned to suit both themselves and their livestock.

The key word here is 'planned', for taking up marine fishkeeping involves more than just parting with a seemingly large amount of money, returning home with the hardware and fishes and hoping it will turn out right; if it did not, then the hobby would be neither so challenging nor so rewarding.

If you are a newcomer to the hobby, you are urged to explore what marine fishkeeping is all about, to find out what you need to buy and to beware of what are probably incomplete assumptions. Making the right decisions at an early stage is essential for keeping these animals in the best of health for long periods. You will soon discover a wide diversity of fishes and other animals that far exceeds your original aims and ambitions.

Keeping coral reef fishes

For most aquarists, 'going marine' means only one thing: keeping fishes from tropical coral reefs, and this provides an ideal starting point for the newcomer to the hobby.

Below: *The fish-only aquarium makes a colourful introduction to the hobby for the beginner. There need be no worries about sensitive invertebrates.*

Indeed, it is recommended that you gain some experience in this section of the hobby before progressing on to the more challenging areas, should that be your ultimate goal.

Fish-only aquariums are relatively forgiving and easy to manage. Lighting is not critical, and sensible stocking ratios are easily estimated. Should disease occur, the full range of treatments are available; you won't need to refrain from using effective medication for fear of killing sensitive invertebrates.

The choice of suitable fish is wide – ranging from small, quiet fishes to large 'character fish' that can become real family pets – providing enough variation to suit everyone. All in all, fish-only aquariums are a suitable and satisfying prospect for the newcomer and experienced hobbyist alike.

Keeping tropical invertebrates

With improved collecting and shipping techniques, aquatic shops are able to offer a vast range of invertebrates from widely differing environments. Clearly, it is impossible to house them all under the same conditions. At one extreme are the burrowing Horseshoe Crabs, which live on mud and sand flats, scavenging for worms, algae and small molluscs. In contrast, many brightly coloured corals and anemones require clear, clean water and would not survive long in silty water conditions.

Compatibility is another important consideration; large lobsters and crabs can be extremely destructive, while octopi and their relations will eat any crustaceans or fish that they can capture. Other factors, such as specific food requirements or the physical disturbance of the tank decor by one species to the detriment of another, may also limit the selection of invertebrates.

Already, you will realize that keeping invertebrates is a far more challenging proposition than maintaining a fish-only aquarium. Water conditions must be kept in first-class order, and you will usually need a larger aquarium and a more sophisticated filtration system. You will also need to pay particular attention to lighting.

Above: *A tropical Octopus makes an ideal subject for an invertebrate species aquarium. Given time, they can become quite tame.*

Below: *The invertebrate-only aquarium has an aesthetic beauty all its own. In the absence of fish, and their associated waste products, many invertebrates seem to flourish more freely.*

Above: *Many marine fishkeepers have a strong desire to possess a 'piece of living reef' as illustrated here. The fascinating invertebrates and colourful fish provide a truly stunning display, but the pitfalls are numerous, especially for the inexperienced.*

As a rule, the most successful invertebrate aquariums belong to hobbyists who have served an 'apprenticeship' with fish-only systems and are familiar with most aspects of marine husbandry.

The invertebrate showpiece
The speciality invertebrate tank holds the same appeal for the marine aquarist as large cichlid aquariums have for the tropical freshwater fishkeeper. Here is a sometimes large, often aggressive, but frequently tameable animal that can make a dramatic showpiece. This sort of aquarium is often easier and cheaper to set up than one intended to house a wider range of animals because the lighting

requirements are easier to satisfy. However, in view of the large appetites of the species most often kept in such a set-up, a good filtration system is usually essential to cope with the heavy demands made on it. Again, this is not really a suitable option for the novice, as a good understanding of water conditions is usually necessary.

The living reef aquarium
Sometimes better known as the mixed aquarium, the 'living reef' tank offers the hobbyist the best of both worlds: an attractive display of tropical invertebrates, with the addition of a few well-chosen fish to add colour, movement and an increased sense of realism.

Unfortunately, the living reef aquarium holds many pitfalls for the inexperienced, and the utmost care is needed to achieve the sort of stunning display these arrangements are capable of providing.

You will need to bear in mind several important points. Firstly, the

fish will be very difficult to treat effectively for a wide variety of common diseases, as the medications contain substances (eg, copper sulphate) that would fatally damage most invertebrates. Secondly, anemone and coral invertebrates, in particular, are very sensitive to even the lowest levels of toxic fish waste; fish stocking levels must therefore be kept sensibly low, ie 2.5cm (1in) of fish for every 27.3 litres (6 Imp./7.5 US gallons) net maximum. Having said that, there is no doubt that, for the dedicated and knowledgeable hobbyist, the living reef aquarium can make an incredibly impressive centrepiece.

Keeping coldwater species
It would not be fair to assume that the above selection of tropical animal life presents the whole picture. Fishes and invertebrates found in cooler waters can also be kept in captivity. One advantage of tackling this aspect of marine fishkeeping is that, very often,

Above: *If you find the colours of the tropical marine aquarium a little too bright, then the subdued tones of coldwater marine species may be more appealing. The display can be equally impressive and the stock is often free!*

collecting specimens is not only free, but also a most enjoyable activity. These species need very similar care to tropical species, except, of course, that they don't require heating.

The only drawbacks are that most of the animals you catch lack the bright colours of many of their tropical relatives (though, of course, many enthusiasts with an appreciation for the subtler hues do not regard this aspect as a disadvantage), and that, during the summer months, the water temperature will almost certainly rise to dangerous levels. You will therefore need to buy or devise some form of cooling system.

Exploring the local seashore can be a useful lesson in conservation.

Should your specimens prove incompatible or outgrow their aquarium, you can simply release them into their natural home and, at the same time, capture a smaller or more compatible replacement to replenish your display.

The final section of this book, (see pages 362-373) looks at this aspect of the marine aquarium hobby in more detail, and features a number of suitable species.

Making the choice

There is always a tendency when visiting aquatic stores to linger and admire the spectacular display tanks that most establishments now possess, and you will be tempted to slowly convince yourself that a particular system would be eminently suitable for the family home. In reality, the case may not be further from the truth, as lack of time for proper maintenance could prove disastrous.

When making this type of important decision you will need to take into account several factors. How much time can you realistically devote to maintenance and general upkeep? Will the type of system sit comfortably within your allotted budget? Does your level of interest warrant consideration of an advanced system or should you opt for one more basic? These are questions all aquarists should attempt to address before making any decision, for an aquarium is a long-term investment in time and resources, rewards and success being in direct proportion. It is far better to start with a successful basic system that can be improved and expanded upon than to settle immediately on a large, complicated set-up requiring knowledge of advanced fishkeeping techniques and copious amounts of time that you may simply not possess.

Exploring the options can be fun, and making the right decisions at the outset can be an investment in success and trouble-free fishkeeping.

Selecting an Aquarium

Choosing a marine tank is an important decision demanding a great deal of thought and care. It will probably be the single most expensive purchase you will make as an aquarist and will form the deciding factor on what stocking options are open to you.

Tank shape

It would be fair to say that the rectangular aquarium has firmly established itself as the most popular and functional design within the hobby. Indeed, most aquarium literature and much manufactured hardware is produced with this in mind; quite rightly so, as experience has demonstrated that the cuboid provides the ideal aquarium environment for almost all livestock, marine or freshwater.

However, over recent years, aquarium design has become increasingly versatile, with its many modern intricacies catching the eye of a growing number of hobbyists. Apart from the traditional rectangular or square 'box', improvements in construction techniques have generated a wide array of hexagonal, octagonal, triangular and trapezoidal shapes, as well as irregular designs. Unfortunately, some disadvantages

Below: *The traditionally shaped rectangular tank has many advantages. It provides a natural presentation of livestock and fits most settings.*

are likely to be inherent in most of these aquariums: volume/surface area ratios may be poor, effective lighting and territory/swimming space restricted. If you do decide on such a tank, you will need to think carefully about these things before you establish stocking levels and compatibility of livestock. In addition, some designs suffer badly from display distortion as the tank panels are angled unfavourably to each other, and you may want to see a furnished example before making a final decision.

Tank construction

It is worth bearing in mind that, apart from being watertight, a marine tank must be made from

All-glass tank construction

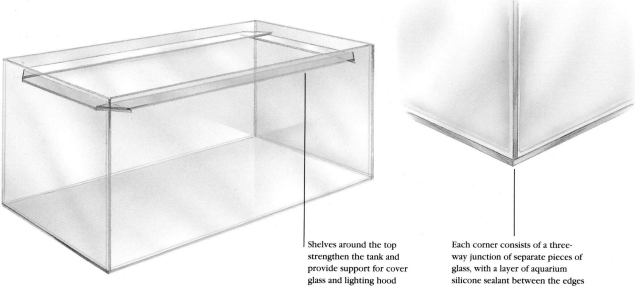

Shelves around the top strengthen the tank and provide support for cover glass and lighting hood

Each corner consists of a three-way junction of separate pieces of glass, with a layer of aquarium silicone sealant between the edges

Above: *All-glass tanks are now standard for marine applications. The panels of glass are bonded together using aquarium silicone sealant.*

materials that are neither affected by corrosion caused by sea water nor, conversely, introduce toxic materials that would harm livestock into the aquarium water. This immediately rules out most tanks that use metals in their construction, save for (perhaps) polished stainless steel, which is still occasionally used. The normal all-glass tank is ideal, although acrylic is still popular in North America, and other materials, such as polycarbonate plastic, reinforced fibreglass, and even suitably treated wood and concrete, are sometimes used. Reinforced fibreglass, concrete and acrylic tend to be favoured by public aquariums where very large tanks are likely to be the norm.

For most purposes, the ban on metals extends to the hood, light fittings and any other equipment likely to be in direct contact with the aquarium or salt spray. You can protect metal hoods and hinges with several coats of varnish, if necessary, but it is better to avoid using metals right from the start, especially if you are going to dispense with cover glasses to allow more effective light penetration (see also *Lighting the Aquarium*, pages 76-81).

Below: *This octagonal tank looks attractive enough, but may lead to problems with proper filtration, lighting and distorted viewing.*

Acrylic or all-glass?

Both acrylic and all-glass tanks have their good and bad points. All-glass types are available in a wider range of shapes and sizes and are scratch-resistant, whereas an acrylic aquarium is easily scratched but virtually crack-proof. Glass is much heavier and more easily broken, but an acrylic tank may suffer badly from display distortion, especially if it is a one-piece moulded unit.

At first glance, an all-glass tank may seem to be much more fragile and subject to greater stresses than the old fashioned and unsuitable angle-iron framed tanks. However, this is not the case if the glass tank is correctly constructed with the aid of aquarium silicone sealant glue and is installed properly.
Top-braces across the tank reduce the chance of the front or rear panels bowing outwards unduly under the pressure of water, while ledges positioned around the inside of the tank, just below the top, not only add strength, but also form a shelf on which to stand the cover glasses and hood.

The glass must be thick enough to withstand the considerable water pressure put upon it, and this must never be underestimated if cracking or shattering is to be prevented. Thickness of glass is largely governed by the overall size of the tank: the larger the tank, the thicker the glass. Reputable tank manufacturers always err on the safe side by using slightly thicker glass than is actually necessary.

It is essential that tanks of an all-glass construction are placed on a firm, level base, otherwise undue stresses are likely to occur, leading to fracturing. In many cases, you will also need to position the tank on polystyrene sheeting of a suitable thickness. It is certainly a false

economy to invest in a cheap tank, which may be incorrectly constructed from thin or second-hand glass, and thus be less reliable.

Tank design

The aesthetic appeal of different tank designs is a subjective matter, but many aquarists choose tanks that form part of a cabinet unit. These not only become a decorative feature in the home, but also have the advantage of built-in hoods that house several fluorescent tubes and eliminate the risk of corrosion.

Such a cabinet design will enable you to build a complete 'system', hiding any trace of unsightly wires or tubes. The size and complexities of such a system will be limited only by your financial constraints!

The less ambitious hobbyist can choose a basic tank and subsequently fit a complete filter system, together with the necessary heating and lighting equipment. This option has the advantage of allowing you time to develop a more complete understanding of the theoretical and practical operations involved. In the long term, opting for such an arrangement is a far more reliable alternative to having blind faith in a complex total system, which may itself be a total mystery!

Above: *This large aquarium looks impressive and is also very practical, providing a very stable environment for invertebrates, fish and algae.*

Having said that, there is much to recommend the complete 'designer system' as far as the more experienced fishkeeper is concerned. Allowing a high degree of control over water quality, such a system can be used to great advantage when keeping sensitive species, and is therefore highly recommended for the aquarist who has gained in confidence and wishes to progress further into the hobby.

Tank size

When considering the size of the tank, remember the principle that 'large is good, but bigger is better'. The larger the volume of water, the more inherently stable it is in terms of quality. The same rules applies to the filtration area, if it is to be included within the tank.

Small tanks pose problems. While it is possible to maintain fish and invertebrates in tanks with capacities of 68 litres (15 Imp./19 US gallons), adverse changes can occur very rapidly, closely followed by the decline or loss of valuable livestock. Overstocking is always a temptation – a 60cm (2ft) long fish-only tank

can safely stock a maximum of only two small fish – and a small tank severely limits the aquarist's options when faced with the vast array of fish and invertebrates on sale today.

Most beginners would be best served by a tank measuring between 90×38×30cm (36×15×12in) and 120×60×45cm (48×24×18in) with a capacity of 150-300 litres (33-66 Imp./41-82 US gallons). In such a tank you can accommodate a range of livestock *and* maintain fairly stable water conditions, whether you have opted for a fish-only, invertebrate or mixed set-up.

Stocking levels

Many elaborate formulae have been put forward over the years to calculate optimum stocking levels. These hinge around tank size, water quality, filtration systems, food inputs and waste outputs, and most fishkeepers find them confusing!

For a fish-only system, the simple rule of a maximum of 2.5cm (1in) of fish length per 9 litres (2 Imp./2.5 US gallons) of water *after one year*, still largely holds good. For the first six months no more than half this number of fish is advisable.

The situation with invertebrates is not quite so clear. As living animals, they will put some loading on the filtration system – a very small amount when compared with fish – but the range of species available is so large that it is impossible to say 'so many per litre/gallon'. As a very simple guide, it is fair to say that animals should not be forced into direct contact with one another. Many corals and anemones expand and contract and you should allow a gap of at least 5cm (2in) between two fully expanded corals.

The mixed fish/invertebrate aquarium has always presented the fishkeeper with problems. Fish produce vast amounts of toxic waste substances in comparison to invertebrates, which, in the main, find these toxins impossible to cope with. Therefore, stocking levels in the mixed reef-type aquarium must be kept very low: 2.5cm (1in) of fish length per 27 litres (6 Imp/7.5 US gallons) net after one year should be an absolute maximum if the aquarium is to be managed properly over the long-term.

Note that these are maximum stocking levels after a fair period of time, usually one year. Initial maturation of a tank allows only for the introduction of a limited biological loading without a dramatic, and deadly, rise in ammonia and nitrites. With every new introduction, the biological capacity needs time to adjust in proportion – too much, too soon, and you risk the life, not only of the new animal, but of all the existing inhabitants as well!

Siting the tank

A tank of 136 litres (30 Imp/37.5 US gallons) fully furnished with rockwork and substrate may weigh in the region of 250kg (500lb) – an immovable object! You must, therefore, consider the tank's final planned position long before you set it up. Being so heavy, the tank will obviously need to be sited on a firm base, which itself must be evenly supported on a strong floor.

Most modern houses with concrete floors, or houses with wooden floors in good condition,

Below: *Where you site an aquarium can make all the difference. This one complements the rest of the room.*

can support this sort of weight comfortably, provided the weight is evenly spread. On wooden floors, ensure the weight of the aquarium is spread across the maximum number of joists and avoid using metal stands, as these have the effect of concentrating four loading points on the floor. If you are in any doubt, especially regarding a large tank, it would be wise to consult a competent carpenter or builder.

Most tanks can be supplied with an appropriate base specially designed to support the weight safely. Think very carefully before you decide to build your own. Unless you are absolutely certain of design specifications, it can not only prove dangerous, but also generally turns out to be uneconomic. Avoid using domestic furniture to support the aquarium; it is not designed to accommodate such weight.

The aquarium will need lighting, heating (if for tropical species) and electrically operated filtration equipment, so be sure that a convenient source of electricity is available close by. Always leave sufficient space around the aquarium for access to lighting, heating and filtration systems. You must certainly allow yourself enough room to carry out feeding and regular maintenance duties, which are likely to be discouraged if access is difficult.

The invertebrate aquarium could be sited where it will receive a beneficial amount of daylight, but two to three hours of sunlight each day should be seen as a maximum to prevent overheating or growth of unsightly algae. In general, it is better to install sufficient lighting equipment, rather than rely on unpredictable natural light sources.

Many fish are susceptible to external disturbances, so avoid a site close to a constantly opening (and slamming) door.

Finally, an aquarium should be sited where it can be observed and admired comfortably. If you place it where it will receive only a cursory glance, or where it will require a back-breaking contortion to see it, you are likely to lose interest.

Heating the Aquarium

It is impossible to specify a narrow band of temperature at which all tropical fish and invertebrates will be happy. At one extreme, some animals are found in rockpools and lagoons where the water temperature may regularly exceed 27°C (80°F). On the other hand, some corals, sponges and crustaceans are found in waters more than 9-15m (30-50ft) deep that may be substantially cooler. Most commercially available species prefer temperatures in the range 24-26°C (75-79°F). Most corals and anemones are very unhappy at temperatures above 29°C (84°F) and most invertebrates are at their best above 22°C (72°F). Whatever temperature you choose within the recommended limits, it is most important that it remains stable, as all marine species resent rapid variations; accurate and reliable equipment is therefore essential.

Types of heating equipment

The easiest heating equipment to obtain is the combined heater-thermostat. This submersible unit is normally encased in a glass tube and has a built-in temperature regulator – usually a bimetallic strip that bends and straightens as the temperature changes, forming a circuit that turns the heater either on or off. Try to obtain a model that is clearly calibrated and easy to adjust to a desired setting.

The glass tube immersion heater is similar, with the exception that it lacks any form of thermostat, so a separate one must be connected. The advantage with this arrangement is that several heaters may be used in one tank utilizing only one thermostatic control. Large tanks may benefit from better heat dispersion under this system. Under-tank heating mats have been around for many years now, but they have never really gained the popularity they deserve. There are several advantages with this method

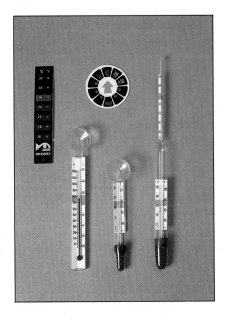

Above: *Thermometers may be of various types. Clockwise from top left: two liquid crystal external designs, one combined thermometer-hydrometer, a free-floating or captive design and a simple spirit captive type.*

of heating; not least, good heat dispersion and a lack of unsightly hardware within the aquarium itself. They cannot, however, be fitted retrospectively. As with the glass tube immersion heater, the heating mat must be controlled using a separate thermostat.

There are several types of thermostat that may be used in conjunction with either of the afore-mentioned heaters. The internal model is basically the other half of the popular heater-thermostat and utilizes the bimetallic strip principle. External models sense temperature changes either from the glass side of the tank or by extending a flexible remote probe into the water.

With the recent advancements in electronic technology, external thermostats with remote probes have become increasingly popular. Not only do they offer the capability to hold aquarium temperatures almost completely stable, but many also have built-in digital displays, dispensing with the use of a usually

less accurate thermometer. Although the heater-thermostat is likely to remain popular for many years to come, many hobbyists see the way forward in terms of the external digital thermostat for sheer accuracy and flexibility.

Controlling heat

The heating power of a heater is rated by its wattage. For example, a 200-watt unit will heat a given amount of water to a given temperature in half the time that a 100-watt unit would need. Generally speaking, most manufacturers recommend too high a wattage for a given size aquarium. These days, most marine tanks are sited in centrally heated homes or in other locations where background temperatures do not drop dramatically overnight or from one season to another.

When a tank is first filled with cold water at, perhaps, 5-10°C (40-50°F) most fishkeepers expect it to be up to working temperature 12-24 hours after switching on the heater. What they do not appreciate is that if a thermostat fails, it may very well do so in the 'on' setting. If this happens, tank temperatures can rise quickly within a very few hours, resulting in the loss of all livestock. If you are using a complete heater-thermostat unit, it is much safer to use a lower wattage unit that may take 36 to 48 hours to warm the tank initially. If the thermostat should fail in the 'on' position, more time will elapse before the tank's inhabitants are damaged, and you will stand a greater chance of discovering the failure earlier.

Most advanced electronic thermostats are designed to fail in the 'off' position and the temperature will then slowly fall to the ambient room temperature. Additionally, a 'fail' indicator light is usually built in, making this system more desirable.

In most situations, a 100-watt unit

is sufficient to heat up to 136 litres (30 Imp./37 US gallons); a 200-watt unit will heat 364 litres (80 Imp./100 US gallons); and a 300-watt unit will heat up to 682 litres (150 Imp./187 US gallons). Only if the tank is sited in a particularly cool position (which will also make for very poor viewing conditions) will you require more powerful heaters. In view of the value of marine livestock, it is well worth investing in the best and most reliable system you can afford.

A thermometer is essential to enable you to check the temperature of the aquarium, but do not opt for one of the very accurate mercury units. They are fragile and if they break in the tank, the mercury will quickly poison the water. Unfortunately, neither the liquid crystal, digital stick-on thermometers, nor the cheap alcohol thermometers are very accurate, but they do enable you to see straight away whether the temperature is stable within the correct range. A steady temperature within the recommended range is more important than one fixed precisely at, say, 26°C (79°F). For more accurate information, use a good thermometer such as the alcohol-filled units used by photographers for home printing and developing. Alternatively, some digital thermostat models can give very precise displays of temperature.

It is advisable to check the temperature of your aquarium twice a day in the initial stages, once before you switch on the lights, and again immediately before you switch them off. Some of the very powerful lighting systems now available to marine hobbyists can raise the tank temperature by several degrees and you may need to devise a fan-driven cooling/ventilation system.

It is prudent to bear in mind also that aquarium water will adapt to ambient room temperatures above that set on the thermostat. This is particularly applicable to rooms housing boilers, etc., and in very warm weather. The first instance can be avoided by siting the tank in a different location, the second is more difficult to cope with.

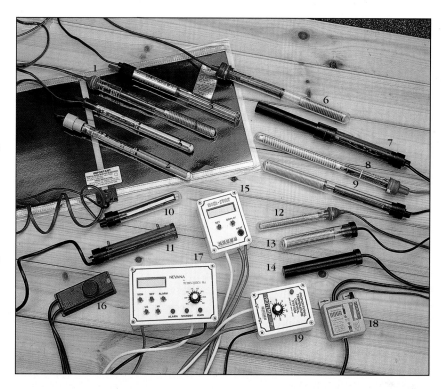

Above: *This selection of the various types of heating equipment available for the aquarium includes underwater heating pads (1), combined heater-thermostats (2-9), submersible thermostats (10-11), heaters (12-14), electronic thermostats (16,18) and temperature controllers (15, 17, 19).*

In an emergency

Cooling tank water – should heater fail 'on'

1 Switch off heating circuit.

2 Increase aeration rate to create more turbulence.

3 Immerse lower half of external filter canister in bucket of cold water.

4 Place sealed plastic bags filled with ice cubes in tank.

5 Do not feed livestock for 24 hours after temperature is normalized.

Heating tank water – should heater fail 'off'

1 Carefully warm some aquarium water in an enamelled (not aluminium) pan and *slowly* return it to tank.

2 Stand bottles of hot water in tank, taking care to avoid an overflow.

3 In the event of serious extended power failure, lag the tank with polystyrene sheets or blankets to conserve heat.

4 Reconnect spare heater as soon as possible.

5 Do not feed aquarium livestock for 24 hours after the water temperature has returned to normal.

Lighting the Aquarium

It is important to make an accurate assessment of the lighting you will need to install in your aquarium. If you are setting up a fish-only or specimen tank, you only need sufficient light to see the exhibits and provide an intensity that will be comfortable for them. In such a system, the growth of desirable green algae will be severely retarded and other 'low-light' varieties, such as red, brown or black, may occur if water quality begins to deteriorate.

Many corals and some molluscs and anemones are largely dependent for food production and waste removal on certain species of algae, known generally as zooxanthellae (see also page 33). These algae live in the fleshy tissues, where they utilize the animals' waste products and provide various foods and oxygen in return. So important are these algae that if they perish, then their host will, in many cases, also die. A typical case in point are the clownfish-type anemones, which, as their symbiotic colony of zooxanthellae algae diminishes, begin to shrink in size. Many fishkeepers are alarmed to find that, over a period of a few short months, what was a flourishing 30cm (1ft) diameter *Heteractis malu* anemone has reduced in size to 7.5cm (4in) or less, reflecting the ever-decreasing colony of zooxanthellae. The major requirement of zooxanthellae and decorative macro-algae is sufficient intensity of light at the correct wavelength.

Sunlight in the tropics is very intense compared with the more diffused light experienced in temperate regions. However, in both cases the white light is composed of various wavelengths of light, producing a spectrum of colours from violet to red. Seawater is a very efficient light filter and red light penetrates only a very short distance. Orange and yellow light are the next to be lost, and only

green and, particularly, blue light penetrates much deeper than 4.5-6m (15-20ft).

This phenomenon is clear to see when you watch one of the many undersea explorations featured on television. As the cameras venture further down, everything appears to become a deeper blue and the reds and yellows of the animals become deep brown and black, providing excellent camouflage. Many invertebrates take advantage of this; in full-spectrum sunlight many shrimps, crabs and corals are intensely red. If this colour were apparent in their natural habitat, they would pose easy targets for predators but, because of the depth at which they occur, red appears black and thus these animals are well hidden. Many fish also take advantage of this state of affairs, either for camouflage protection or to camouflage predatory activities. Many corals, anemones and similar creatures are very sensitive to blue light because successful growth of zooxanthellae is largely dependent on receiving sufficient light at the blue end of the spectrum.

A wide range of invertebrates are not particularly dependent on good lighting and these can be some of the easiest to maintain. They include the crabs, shrimps, lobsters and featherduster worms. Many species of starfish, sea cucumbers and sea urchins inhabit areas of intense light as they feed on the algae that proliferate there.

You should make a judgement on the type and intensity of lighting required with the most light-sensitive creatures occupying the aquarium in mind. Anemones and shrimps may inhabit the same tank, but high-intensity lighting should be

chosen with the anemone in mind and suitably shady areas provided for the shrimp by either varying the position of the lights, or constructing rocky overhangs.

Fluorescent lighting
Fluorescent tubes are probably the most convenient form of aquarium

Penetration of sunlight into sea water

Right: *Seawater is an efficient light filter. As sunlight penetrates the surface of the water, the various wavelengths are absorbed at different rates. Red light is lost first; blue travels the furthest.*

lighting. They give an even spread of light that some may regard as a little bland, as it lacks the natural 'rippling' effect found with spotlights. They are, however, cheap to run and cool in operation, making them ideal for smaller tanks and in situations where water spray may be a problem.

The choice of fluorescent tubes available today is both wide and often confusing; many tubes of similar colour are marketed under various trade names. There are several types of full-spectrum white tubes that mimic natural daylight and these are available in a range of sizes to suit most aquariums. White tubes with a so-called 'powertwist' give off somewhat more light than other white tubes of a similar wattage. Such 'powertwist' tubes are extremely effective, but considerably more expensive than the alternatives, and are available only in a limited range of sizes.

Grolux tubes will be familiar to many freshwater fishkeepers, but from the marine point of view, their

Below: *In deep water, red animals appear black. Blood Shrimps hide more easily from potential predators when their natural colour goes unseen.*

use is almost entirely cosmetic. Most of their output is in the red part of the spectrum and the human eye, sensitive to light in this area, finds their colour-enhancing effect on animals very appealing. However, red light is of little direct use to marine life. In some circumstances it appears to promote algal growth, but usually encourages the less attractive 'nuisance' varieties. Only introduce red light to the marine tank once you have achieved the basic white and blue balance.

All the afore-mentioned tubes were originally intended for applications other than aquarium lighting and therefore have distinct disadvantages: they are quick to lose power and have limited ability to hold their designed spectral output, so you will need to replace them every six months, whether or not they appear to need changing.

Recent advances in fluorescent tube design have improved matters considerably and a new tube has been developed specifically as an aquarium light source. This new triphosphor lamp concentrates its light output in the key areas of the spectrum essential for invertebrate well-being and lush marine algae growth, as well as being very pleasing to the human eye. Its spectral distribution takes into account the fact that various areas of the spectrum are absorbed at different rates as light passes through water. Other clear advantages with this type of tube are increased efficiency in power-to-light ratios, an extremely low power loss over a longer period of time, and an ability to retain the correct spectral qualities until the tube fails. You need only replace this type of tube when it fails to operate, and it should last a year or two. With the advent of these new triphosphor tubes, many aquarists believe that the older lamps have become redundant, with the exception, perhaps, of the actinic blue tube.

Left: *The lighting hood of this tank is mounted clear of the tank itself. Alternatively, you can dispense with the hood and use suspended spotlights.*

Actinic blue fluorescents have proved to be an economical and easy-to-use source of blue and ultraviolet light so necessary to sustain many types of invertebrates. Be sure to select the correct type; actinic tubes are available in two forms: the 03 and 05 ranges. Both give off blue light, but at different wavelengths; the 03 range is more useful to the marine aquarium. Some manufacturers have even combined the qualities of the triphosphor and actinic lamps to produce highly efficient, long-lasting lamps, which are capable of sustaining some of the most light-sensitive coral species.

Latest innovations to improve the efficiency of fluorescent tubes also include special reflectors. These may either be built into the internal structure of the tube itself as a reflective coating, or fitted externally in the form of a pressed highly polished aluminium housing; corrosion is prevented by a special coating in materials that make the product perfectly safe to use in marine aquariums.

Spotlights

As an alternative to the uniform spread of light from fluorescent tubes, the reflector hood can be removed from the tank and illumination provided by overhead individual lamps or spotlights. These lamps are ideal for creating dramatic effects and for emphasizing any surface water movement. They are particularly useful if you are keeping invertebrates and for punching light down into relatively deep tanks. Focusing a spotlight on a particular rock will benefit sea anemones, for example, which will migrate to that spot to take advantage of the brighter light. Shadier areas can be created for those species that prefer more subdued lighting.

It is important to choose the correct type of spotlights and to ensure that they are adequately

Right: *Metal halide lighting is especially desirable for the living reef aquarium, providing very intense illumination at correct wavelengths.*

Spectral output of lights used in the aquarium

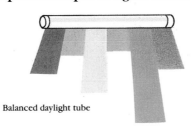

Balanced daylight tube

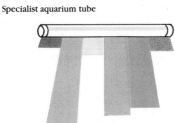

Grolux tube

Specialist aquarium tube

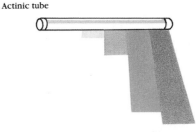

Actinic tube

Metal halide lamp

Mercury vapour lamp

Above: *Fluorescent tubes vary in terms of their spectral output at different wavelengths. With experimentation, it is possible to achieve a lighting system to suit the needs of all livestock. Metal halide and mercury vapour lamps produce the bright light essential for deep invertebrate tanks and can be very effective used with actinic blue fluorescent tubes.*

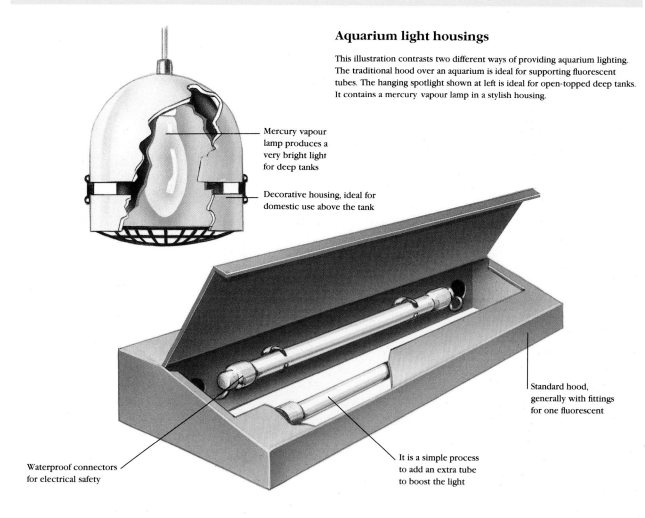

Aquarium light housings

This illustration contrasts two different ways of providing aquarium lighting. The traditional hood over an aquarium is ideal for supporting fluorescent tubes. The hanging spotlight shown at left is ideal for open-topped deep tanks. It contains a mercury vapour lamp in a stylish housing.

Mercury vapour lamp produces a very bright light for deep tanks

Decorative housing, ideal for domestic use above the tank

Standard hood, generally with fittings for one fluorescent

Waterproof connectors for electrical safety

It is a simple process to add an extra tube to boost the light

protected against water splashes. It is best to opt for those models made specifically for aquarium use, usually mercury vapour or metal-halide. Domestic tungsten, tungsten-halogen and high-pressure sodium have all proved unsuitable for marine aquarium applications, since they are expensive to run, generate a great deal of heat, are short-lived and do not produce light of a suitable spectral quality.

Mercury vapour spotlamps use mercury vapour to produce light, whereas fluorescent tubes use a mercury vapour discharge in starting, but depend on fluorescing phosphors that form a coating on the inside of the tubes for actual light output. Mercury vapour lamps have an electrical consumption of around 80-125 watts, depending on the model chosen.

Right: *This metal halide spotlight has a useful 'rise and fall' mechanism to enable correct positioning from the surface of the aquarium water.*

Two types of bulb are available: one has a built-in reflector; the other, a more traditional-type bulb, relies on an external reflector to transmit light efficiently. Most advanced aquarists regard the second type as better suited to the

needs of the marine aquarium.

Mercury vapour lamps do have several drawbacks worth noting, however. Bulb life is quite short and efficiency may drop by 50 percent over a period of less than six months; replacements are therefore

Left: The Tooth Coral (Euphyllia picteti) *requires intense illumination of the correct wavelength.*

suitable UV screening is fitted and avoid staring at active bulbs.

Although various wattages are available, a popular application is to mount a 150-watt lamp in a suitably designed reflector 30cm (12in) above the surface of the water. This will illuminate an area approximately 1.8 sq m (2 sq ft) and to a useful depth of 60cm (2ft) for light-loving invertebrates. Higher wattage units are available for larger aquariums. Owing to high-operating temperatures, coolant fans may be necessary to prevent overheating.

essential on a regular basis and may prove very expensive over a long period. Spectral analysis also shows a distinct lack of presence in the blue/green region, so you would need to supplement these lamps with additional blue lighting.

Metal halide lamps are the most intense form of lighting currently available to the fishkeeper. They are extremely powerful and usually prove very successful, especially in invertebrate aquariums. The major drawback is their initial cost. Additionally, bulbs need to be

replaced every 8-12 months to maintain maximum efficiency.

Nevertheless, remarkable results have been achieved in living reef aquariums using this source of lighting, and many aquarists consider that the high cost of such units is repaid in full by the incredible displays encouraged under metal halide units.

In common with actinic lighting, metal halides have a very high ultraviolet output, which can be damaging to the human eye. You should therefore ensure that

Combination lighting

It is now possible to obtain suspended units combining metal halide lighting with fluorescent tubes, or mercury vapour lamps with fluorescent tubes. Such units may prove useful to cover the whole of the lighting spectrum at the correct intensities. Invertebrate keepers, especially, may consider

Below: Illumination in the aquarium need not always be intense; some species, such as this Lionfish, appreciate more subdued lighting.

such units to be ideal for their purposes, although again, high cost may be a deterrent.

Light duration and intensity

There are no all-encompassing rules regarding the duration and intensity of lighting in the marine aquarium, as all applications should be considered on their merits. You will also need to balance physical necessity and personal preference.

High-intensity lighting is not strictly necessary for fish-only aquariums unless a good growth of algae is required for herbivores to graze upon. On the other hand, some species of fish, and most cnidarians (see pages 306-325) appreciate fairly subdued lighting conditions. Cardinal fishes and lionfishes may, for example, become nervous and retreat in an intensely lit aquarium. Light-loving invertebrates need high-intensity light in order to survive and, in this case, over-illumination is highly unlikely to be a problem.

Reference to the foregoing text and appropriate sections of this book will help you to make the correct choice of lighting intensity, but the photoperiod (duration) will nearly always fall between 12 and 14 hours each day. Shorter or longer photoperiods tend to result in growths of undesirable algae (although this need not be the only reason; see *Nuisance Algae*, pages 91-93), as well as being unnecessarily unnatural, leading to possible long-term stress.

Care and use of lights

Whatever lamps you use, you must protect them against water damage, either from direct spray and splashes, or from condensation. Waterproofed lamp fittings safeguard the electrics, while a cover glass fixed on the top of the aquarium between the lamps and the water surface will prevent damage and also cut down on excessive evaporation. Be sure to keep the cover glasses spotlessly clean so that none of the beneficial light is prevented from penetrating the aquarium water efficiently. In

some instances it may be feasible to dispense with cover glasses altogether and thus prevent any light loss. However, you should take great care to ensure complete safety and to prevent livestock escapes; if you are in any doubt, leave the cover glasses in place.

Many aquarists are worried about the stress caused when lights are suddenly switched on and off. Two solutions can help alleviate this problem. Firstly, switch off the aquarium lights several minutes before the main room lights so that the fish can become acclimatized to darker conditions. On dark winter mornings, switch on the room lights a few minutes before the tank's lights or adjust the photoperiod to

delay switching on the tank lights until after daylight has illuminated the room. Secondly, you could use a series of time-switches to stagger the switching on and off of lamps, simulating an artificial sunrise and sunset. Fluorescent dimmers do exist but are extremely expensive, complicated to wire up, and are not designed specifically for aquariums.

Use dim night lights sparingly, if at all. Most fish and some invertebrates require a 'sleep' period during the hours of darkness and depriving them of this could lead to stress. Most species are biologically attuned to a night period of almost total darkness and, as long as livestock is compatible, there is little cause for concern.

Types of Lighting

Type	Advantages	Disadvantages
Natural light	Correct spectral range for all animals and plants kept in the aquarium. Excellent for encouraging algal growth. Free.	Uncontrollable and unpredictable.
Artificial lighting	Controllable	Varies according to type, see below
Tungsten lamps	Cheap to install and easy to replace.	Poor and spectrally unbalanced light output. Run hot and have a short life. Expensive in use and possibly dangerous near water. Not recommended.
Fluorescent tubes	Cool running, inexpensive in use and long lasting. Even light output Available in a range of 'colours', many close to natural spectrum.	Relatively expensive to buy and install. Heavy starting gear. Performance may decline unnoticed. The ideal light source, however. Recommended.
Spotlamps, such as metal-halide and mercury vapour	High light output, making them ideal for producing dramatic effects and for use with deep tanks. Good for encouraging algal growth.	Expensive. Depending on type, light may not be totally ideal in spectral balance. Must be used at least 30cm (12in) above an open-topped tank.

Water and Testing

A shroud of mystery has often surrounded marine fishkeeping during its relatively brief existence. Many people are fearful of attempting the seemingly almost impossible challenge of maintaining the quality of the water. Gaining a good understanding of sea water and the changes it undergoes in the aquarium is, therefore, essential so that you can proceed confidently along the road to successful marine fishkeeping.

Natural sea water

Throughout the world, the chemical composition of sea water is remarkably similar. Although the total weight of salts in a given volume of water may vary slightly, particularly in comparatively land-locked waters, such as the Red Sea, the proportions are largely the same.

This being so, and the fact that 71 percent of the earth's surface is covered by sea water, you might expect it to be one of the most convenient commodities to obtain. Unfortunately, natural sea water is generally unsuitable for the aquarium, for a number of reasons.

For most aquarists it is totally impractical to make regular trips to the coast to transport large quantities of natural seawater, even if the sea is within easy reach. Secondly, it would be fair to say that most marine fishkeepers do not live in a tropical climate so their 'local' sea water is likely to be too cool for

Below: *These sea whips and sea fans are thriving in clean, unpolluted water. You are unlikely to have access to such a pure resource and may have to find a synthetic alternative.*

tropical species. If local sea water is heated for tropical livestock, there is a danger that the plankton in it will either die and cause pollution, or propagate extremely rapidly, making the water unstable and unsafe.

The third, and probably the most important, point to consider is the difficulty in finding a source of unpolluted natural water. The volume of sea-going commercial traffic, effluent from sewerage outfalls and industrial activity usually mean that coastal waters are not entirely pure, to say the least! Even if you can find a seemingly clean source of water, it cannot be guaranteed that a change in tides and currents will not introduce pollutants previously unknown in that area. As time progresses, we can only hope that we will cease using the oceans as some vast garbage

dump, but in the foreseeable future, the use of natural sea water in the aquarium is a dubious option, fraught with dangers.

Synthetic sea water

In the early years of the marine hobby, there were many 'sea salts' of varying qualities. Some were acceptable and others were certainly not. Many of the early books available on the hobby gave detailed formulae so that the aquarist could buy the various chemicals required, blend them and hopefully produce something approaching sea water. Fortunately, those days are long gone, and now there is a wide range

Composition of sea water

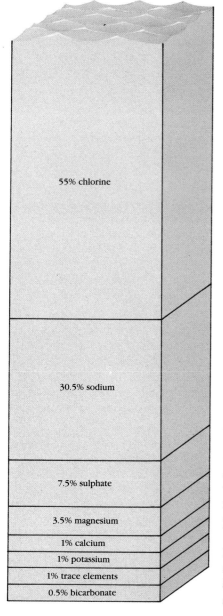

55% chlorine
30.5% sodium
7.5% sulphate
3.5% magnesium
1% calcium
1% potassium
1% trace elements
0.5% bicarbonate

of professionally produced salt mixes, the vast majority of which are perfectly suitable. As a general rule, it is a good idea to use the same salt mix as your supplier; although most salt mixes are virtually indistinguishable, there is no point in subjecting your livestock to any unnecessary environmental stress.

Generally speaking, any reputable salt mix can be dissolved in household tap water to produce a satisfactory solution, although it is important to follow manufacturers' directions exactly when using such mixes. Reseal any unused salt mix in the packet and store it in a cool dry place to prevent absorption of moisture, which may unduly affect the chemical make-up of the mix.

One word of caution, however: in some areas, particularly of high population densities, or where there is intensive arable farming, tap water may contain undesirable impurities in the form of nitrates, phosphates and sulphates etc., and you may have to purify household tap water using a de-ionizer, reverse-osmosis unit or nitrate-removing resin. Many advanced aquarists do this as a matter of course, if they are keeping delicate invertebrates and fish, and it is a practice highly recommended to all marine fishkeepers.

Additionally, it is becoming more common for local water companies to treat tap water with chloramines instead of the more traditional chlorine. Therefore, it is important to treat all water intended for the aquarium with a proprietary brand of dechlorinator coupled with vigorous aeration before adding the salt mix. It is also advisable to use a dechlorinator when using tap water to make up evaporation losses in the fully functional aquarium.

Although you can use the aquarium itself for the initial mix, be sure to make subsequent mixes of water changes in plastic or other non-metallic containers. As far as plastic is concerned, it is better to

Left: *Sodium chloride, magnesium, calcium, potassium and sodium bicarbonate are all present in sea water in significant amounts.*

use polythene or polypropylene containers made specifically for home brewing or food use, rather than coloured plastic dustbins that might release harmful toxins into the water over a period of time.

Synthetic salt water prepared in advance may be stored in a cool, dark place almost indefinitely. When required, introduce a heater/thermostat to the main aquarium temperature and aerate the water vigorously for 24-48 hours.

Maintaining water quality

The main constituents of sea water are various sodium, magnesium, calcium and potassium salts. In addition, natural sea water contains small amounts of every known naturally occurring element. Examination of a bag of prepared salt mix will show that many other chemicals are also present in very small quantities. Certain of these so-called 'trace elements' are known to play a vital role in the well-being of both marine invertebrates and fishes, in that they are extracted from the surrounding water and help to sustain the animals' metabolic processes. In the wild the concentration of trace elements is continually replenished through decomposition of dead animals and erosion of minerals from the land into the sea. Sub-oceanic volcanic activity may also be significant in this respect. Clearly, none of these forces is at work in the domestic aquarium, yet trace elements are still being abstracted by the livestock and various algae in the tank.

Freshwater tropical fishkeepers will already appreciate the importance of making regular partial water changes a means of removing accumulated waste products that filters cannot eliminate. In a marine tank, regular water changes not only remove wastes, but also ensure that trace elements are replenished. Specifically formulated trace element supplements are available and are particularly useful in a heavily stocked aquarium.

Vitamin supplements can similarly be of great benefit. In the wild,

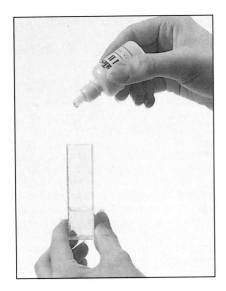

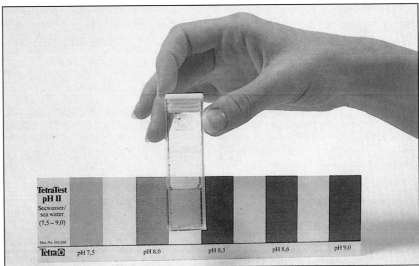

Above: *For this pH test, add the required number of drops of reagent to a sample of aquarium water and gently shake the calibrated container (with the lid on!) to ensure complete mixing.*

Above right and right: *All (non-electronic) kits for testing pH levels are based on the same principle, as shown in these two examples: a colour comparison between a mixture of water and reagent and a calibrated colour scale. It is a good idea to keep the colour chart out of strong light when not in use, as faded colours will make for inaccurate results.*

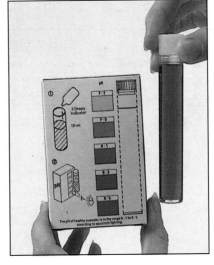

much of the food taken by invertebrates is either alive – in the form of plants and planktonic animals – or freshly killed, and therefore contains all their nutritional requirements. The marine hobbyist, however, tends to rely heavily on frozen, dried or liquid foods. While some of these are vitamin-enriched by the manufacturers, many vitamins have a comparatively short 'shelf life', so using a good vitamin supplement on a regular weekly basis will go some way towards alleviating this shortfall. Be sure to keep liquid vitamin solutions in the refrigerator.

The pH of sea water

Another feature of sea water, compared to fresh water, is its high alkalinity. You can check this using a simple pH test kit, which should produce a reading of around 8.3. A reading below 8.1 is approaching a dangerously low level, while one above 8.5 (a rarity) may occasionally result in problems with excessive ammonia if the filter system is not fully active biologically, or if you are heavy-handed with feeding. A given amount of ammonia is more toxic at higher pH levels because a greater percentage of it is in the 'free' form (NH_3), rather than as ammonium ions (NH_4+).

Synthetic sea water has a fairly high buffering capacity; that is, it is able to maintain the correct pH level in the face of influences that might alter it. However, in the marine aquarium, the animals' waste products are largely acidic and, as a result, the pH value of the water tends to fall (i.e. its alkalinity declines) over a period of time. The best way to counteract this tendency is by making regular partial water changes of, say, 20 percent every two weeks. Furthermore, many corals and molluscs, in order to produce their shells and skeletons, extract the very chemicals that maintain high pH levels. Regularly using a pH buffering liquid, which by virtue of its formulation cannot take the pH value of the water dangerously high, will maintain optimum pH levels, while at the same time ensuring there is sufficient calcium and magnesium, etc. available to the living corals, clams and other similar creatures.

Specific gravity

With the exception of areas such as the Red Sea, the salinity of sea water throughout the world is fairly constant and, within a particular small region, exceptionally stable. The salinity of water is a measure of the total amount of dissolved salts it contains; it is usually quoted as gm/litre (parts per thousand). Within the fishkeeping hobby, however, it is more usual to talk in terms of the specific gravity (S.G.) of water, since this is a good analogue, more easily tested and clearer to understand.

Specific gravity is simply a ratio of the weight of a water sample compared to the weight of an equal volume of distilled water at 4°C (39°F), which is assigned a specific gravity of 1. Since adding salts to water increases its weight as well as its salinity, the two scales are directly comparable. Thus, a salinity of 35gm/litre is equivalent to a specific gravity of 1.026 at 15°C (59°F), measured by means of the traditional floating hydrometer or the more recently introduced easy-to-use swing-needle types.

Natural sea water generally has an S.G. of between 1.023 and 1.027, depending on the location. (It can be higher in some areas; in parts of the Red Sea, for example, it can reach 1.035.) Once properly acclimatized, many fish and invertebrates seem happy to accept a constant S.G. in a fairly wide range. For most purposes, a reading between 1.021 and 1.024 is acceptable. The key factor is to keep the S.G. as constant as possible. Using a hydrometer, take regular readings and ensure that they do not vary by more than one point (between 1.022 and 1.023, for example). Although this seems a minor shift, it is still many times that which occurs in even the most extreme circumstances in any one location in the wild.

In an established tank, the main cause of changes in specific gravity are evaporation and the addition of new stock. Over a period of time, the water level in the tank tends to drop as fresh water evaporates and, thus, the salt water becomes progressively more saline. Compensate for this by regularly adding a small quantity of *fresh* water. It is not a good policy to wait until a large amount (more than 0.5 litre/approximately a pint, say) is needed, as the change in salinity will harm some delicate invertebrates and algae. Bear in mind that a small amount of salt will also be lost through emptying protein skimmers, cleaning power filters and by the accumulation of the crystalline deposits on cover glasses.

Two common, and dangerous, sequences of events can lead to a continual increase in the S.G. of the water in your tank. In the first scenario, you allow the water level in the tank to drop through evaporation and then decide to carry out a partial water change. Having drained out 10-20 percent of the water, you replace exactly the same amount with newly made up

Specific gravity/ salinity table

Specific gravity at 15°C(59°F)	Salinity (gm/litre or ppt)	Specific gravity at 25°C(77°F)
1.020	27.2	1.017
1.0203	27.6	1.018
1.021	28.5	1.019
1.022	29.8	1.020
1.023	31.1	1.021
1.024	32.4	1.022
1.025	33.7	1.023
1.026	35.0	1.0236
1.0264	35.5	1.024
1.027	36.3	1.025
1.028	37.6	1.026

sea water, plus extra sea water to make up the evaporation loss. Since the reduced volume in the tank was more concentrated, this will cause an increase in S.G. Alternatively, the water level drops and you then add more stock (and its attendant sea water), thus filling the tank and also raising the S.G. Be sure to guard against these pitfalls.

In general, it is well worth noting how much water is lost by evaporation during the first week of

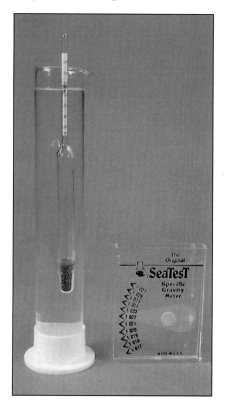

a tank's operation, while the filters are becoming biologically active and there is no livestock in the aquarium. It is then an easy task to add, say, half a cup of fresh water to the tank on a daily basis. Nevertheless, you should always check the S.G. with an accurate hydrometer at least once a week. Tanks fitted with 'total management systems' offer the ideal, if relatively expensive, solution, since they incorporate self-acting topping-up devices that automatically replace evaporation losses from a reservoir of fresh water.

Testing

It is important for the marine aquarist to understand these details about the properties of water and the importance of maintaining water quality. Marine fishes and invertebrates are highly intolerant of poor conditions and of sudden changes brought about by the good (albeit drastic, and often last-minute) intentions of the hobbyist. Any deliberate alteration to the quality and condition of the water must be made as gradually as possible to avoid stressing the fish. A case in point here is that newly bought fish introduced into an established aquarium may succumb to the levels of toxins (although relatively low) that the 'resident' fishes have become used to.

The marine aquarium is no more a slice of the ocean indoors than the freshwater aquarium is a section of river, stream or lake. The forces of nature in the aquarium may always seem to be against you, but a clear understanding of the problems, coupled with proficient management techniques, should make for success.

In addition to periodically measuring pH and specific gravity, there are other tests that you can, and should, perform to ensure that

Left: *Two methods of measuring specific gravity: a floating hydrometer, where the S.G. is read off at the water line (left), and a 'swing-needle' type, which indicates the S.G. of the water filling its container (right).*

water quality remains at its optimum. Once water has been 'lived in', the resultant effect of the waste products of fish respiration, digestion and natural decay on the water quality can be measured. The main component of waste products is ammonia, together with two other important nitrogenous compounds, nitrite and nitrate. All three compounds are toxic to fishes and invertebrates to varying degrees.

By using the relevant test kit, you can keep a check on the build-up of these unwanted by-products, and also evaluate how efficient your filtration system and regular partial water changes are at keeping them down to a minimum. (Checking the carbonate hardness of the water periodically, for example, will provide an indication of its buffering capabilities.) All these tests will indicate the state of your aquarium water, but be sure to interpret the readings correctly, otherwise you may implement the wrong remedial measure, with dire results.

Other test kits, such as those used to determine levels of copper in the water, can be of great value when treating the aquarium for disease. The usual disease treatments for marine fishes are copper based and, since copper is lethal at 'overdose' levels, it is important to administer a very accurate dose. By using a copper test kit to monitor the amount of copper given, you will avoid the 'kill or cure' method of old, in which dosing was based on correctly estimating the water content of your aquarium. This, of course, applies exclusively to the fish-only aquarium, as copper *at any level* is extremely toxic to invertebrates. It is also worth noting, if you are keeping invertebrates, that you should initially and periodically test your tap water and aquarium for traces of this toxic metal as a matter of course.

Test kits

Most test kits are made up of liquid, dry powder sachet or tablet reagents. These reagents, when mixed with the sample of water to be tested, produce a coloured solution, which is then compared against a graduated colour scale and a reading taken on the closest match. You are advised to check that each individual kit is suitable for both salt water and fresh water use, as you will need to take occasional readings from tapwater as well as regular ones from the aquarium (see page 90).

A basic selection of test kits should be acquired by all marine aquarists. Useful ones include those to test ammonia, nitrite, pH, nitrate and copper, and these will cope with most situations. However, should you wish to investigate the intricacies of marine water chemistry further, supplementary kits are available, which will enable you to measure dissolved oxygen, phosphates, carbon dioxide, iron and many more elements and compounds. Using such kits will lead you into the maze of water chemistry that many will find fascinating, and others will find a nightmare! Water chemistry is not an easy area to understand in its advanced form and may lead the unwary aquarist astray quite quickly, suggesting problems where they do not exist and, even worse, leading to remedial action where none is necessary. Most water parameters are inextricably linked, and trying to adjust one in isolation is nearly always impossible.

Electronic tests

Electronic meters are now available to measure various water parameters, some more familiar to the hobbyist than others; these include temperature, pH, conductivity, dissolved oxygen, ORP (Redox Potential), etc. Many very experienced aquarists use them regularly because of their high accuracy. However, most of these meters are extremely expensive and require a great deal of expertise to enable them to give accurate readings. Once readings have been taken, interpreting some of the less familiar parameters can be difficult and confusing owing to the prevailing conjecture over what constitutes a satisfactory result. Of these meters, only the temperature version is recommended to the average hobbyist.

Obtaining and using accurate test results

- Commercial viability has meant that some test kits (eg. nitrate tests) cannot be 100 percent accurate so a degree of allowance should be provided for.
- All reagents are highly toxic; store them safely out of the reach of children and animals.
- Liquid reagents are usually only viable for a limited period of time; discard them after their 'use-by' date.
- Always follow manufacturers' instructions very carefully to achieve the highest possible accuracy. It is particularly important to adhere strictly to standing or development times.
- Record all results in an aquarium log to help give an overall picture and to identify possible longterm problems.
- Always take tests at the same time of day to establish a base from which to compare unusual readings.
- Where possible, use the same manufacturer for all your test kits. They are far more likely to inherit the same characteristics and be easily comparable.
- Be sure that the scales of measurement are clearly marked and have broad ranges where small amounts may make all the difference.
- If a test indicates remedial action, be sure that you understand all the options available and their consequences; drastic actions usually give drastic results – but not always the desired ones!
- Prolong the life of test kits by storing them in a cool, dark place. Colour charts left exposed to light for long periods will fade, leading to inaccurate or impossible comparisons.

Nitrates

A rise in nitrate can be considered in isolation or as a general indication of worsening water quality. It matters not which, as treating for high nitrate levels nearly always improves the general water condition. 'Nuisance algae' (see pages 91-93), the demise of invertebrates and the death of sensitive fish, can all be attributed to high nitrate levels.

As we have already seen, nitrate is the last and least toxic product of the nitrogen cycle, rising slowly and almost imperceptibly in many cases over a period of time. After initial maturation, the conversion rate through the nitrogen cycle is directly proportional (i.e. 1 unit of ammonia = 1 unit of nitrite = 1 unit of nitrate) and any overfeeding or overstocking will be reflected in a correspondingly high nitrate level.

While the sudden appearance of ammonia or nitrite can make an immediate impact on the look of a previously healthy aquarium, nitrates can be *tolerated* in modest amounts (up to 20 gm/l) by many invertebrates, and some fish will survive where amounts have risen slowly to values in excess of 100 gm/l! But this is a far from satisfactory state of affairs, and the stress of continuously unfavourable nitrate levels can lead to general ill-health in livestock. You should therefore make all efforts to keep nitrate levels within safe limits; 5 gm/l or less for invertebrates, and 20gm/l or less for fish (obviously, you should observe the lower value for a mixed aquarium containing fish and invertebrates).

Some aquariums are very fortunate in that they maintain low levels of nitrate even though no special precautions are taken to prevent it. This occurs when a 'nitrate balance' has been reached. This may be due to a number of reasons, including denitrification taking place in anaerobic areas of the tank itself, e.g. within porous rocks. You should not, however, rely upon this process to take place in the natural course of events, and nitrate levels cannot be guaranteed stable for long periods of time. Test, and take measures to reduce, nitrate levels for reliable results.

Removing nitrates

Removing nitrate from aquarium water can be accomplished by four main methods, the first and easiest being the water change.

Water changes, when regularly and correctly proportioned with nitrate-free water, will immediately remove excessive quantities, with the added advantage that the aquarium will be freshened and replenished with vital minerals and trace elements. Barring emergencies, do not be tempted to change more than 33 percent of water at any one time; osmotic shock and other damaging factors associated with massive water changes can be very damaging to invertebrates, fish and decorative algae. Ideal water changes to keep nitrate levels under reasonable control fall in the area of 20-25 percent every two weeks.

The denitrifying filter employs helpful bacteria once more to convert nitrates (NO_3), firstly to nitrous oxide (N_2O), and ultimately to free nitrogen gas which is dispersed into the atmosphere. The difference with these bacteria is that they require an anaerobic (oxygen-free) environment – or one that is nearly so – in which to flourish.

Put into practice, an isolated filter compartment is filled with a suitable medium – lava rock granules, coral gravel, open-pore sintered glass etc., and the aquarium water is passed through it at a very slow rate: too quickly, and different species of bacteria will multiply, converting nitrate *back* into nitrite; too slowly, and hydrogen sulphide will be produced. Besides having a characteristic 'rotten eggs' smell, hydrogen sulphide is readily soluble in water and highly noxious. Therefore, outputs from denitrifying filters need careful adjustment and consequent monitoring. On the

Nitrate reduction system

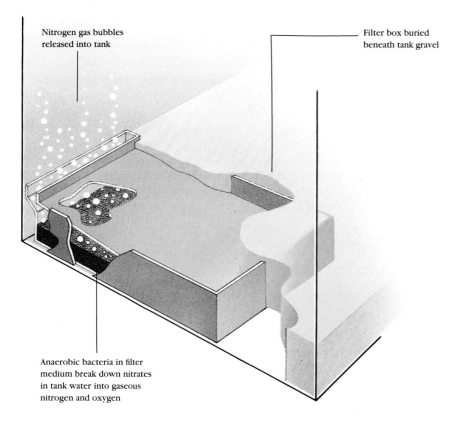

Nitrogen gas bubbles released into tank

Filter box buried beneath tank gravel

Anaerobic bacteria in filter medium break down nitrates in tank water into gaseous nitrogen and oxygen

Nitrate poisoning

Nitrate and other related toxic build-ups are generally observed in the following symptoms. (Of course, many of these symptoms could be attributable to other causes, but if these have been eliminated, consider the possiblity of nitrate – toxic – poisoning.)

In fish Loss of appetite, general deterioration in condition, susceptibility to disease, loss of colour and vitality, unusual changes in behaviour, death through no obvious cause.

In coral invertebrates Unwillingness to display full potential, inability to attach to rocks, unusual discharges, general 'shrinking', untimely death.

In crustaceans Loss of appetite, changes in behavioural patterns, inability to shed exoskeleton, sudden death.

Above: *Tooth Corals are highly sensitive to nitrate.*

whole, though, they can be extremely effective once established. Many 'systemized' aquariums incorporate a denitrifying feature, and some manufacturers have produced add-on units.

Algae can utilize nitrates, along with carbon dioxide and other potentially toxic compounds, as a source of food, effectively 'locking it up' until excessive growth can be removed, along with the offending toxins. The total effectiveness of this method is difficult to assess and a concensus of opinion may never be reached.

Even so, algae as a means of filtration is still highly regarded by many eminent marine aquarists, while some research institutes are so convinced of its useful properties that they run untreated aquarium water over a large, flat, highly illuminated tray on which hair algae are grown. This is referred to as an 'algae scrubber' and, if managed correctly, can produce superb water quality as a *sole* means of filtration! Unfortunately, 'algae scrubbers' can be highly temperamental and a collapse of the algae culture can happen suddenly and without explanation, leading to a disaster in the main showtank. This is not a method for the average hobbyist, but one that demonstrates the power of algae nonetheless.

Exchange resins have enjoyed varying fortunes as a method of nitrate removal; the disappointment has always been their reluctance to work effectively, coupled with a very high price. Resins normally take the form of small granules or impregnated pads. These are positioned where the aquarium water can flow over them at a reasonable rate, e.g. within a canister filter. Research is continuing in this area and a fully effective, rechargeable resin at a satisfactory price should be available to the hobbyist in the near future.

Mains tap water

Water changes are useless, of course, if the mains tap water already contains large quantities of nitrates (in tandem with other possible toxins). To do so merely compounds the problem, especially when topping up evaporated water,

as these harmful products become progressively more concentrated as time goes on.

All over the world, farming methods have intensified, and increasing amounts of nitrogenous fertilizers have been used to squeeze the last vestige of growing potential out of soils in many areas. As a result, one good shower of rain, and much of this simply runs off into rivers and streams ultimately to be collected by local water companies, and thence distributed to the consumer.

Fortunately for the hobbyist, several manufacturers now market filters designed to remove nitrates and, in some cases, other unwanted compounds from mains tap water. One such filter is a selective rechargeable exchange resin very similar to that designed for use directly connected to the aquarium. In operation, if tap water is passed

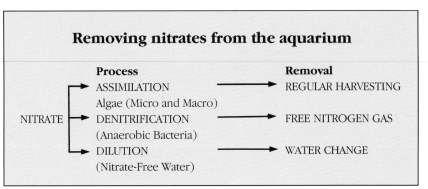

Removing nitrates from the aquarium

	Process	Removal
	ASSIMILATION Algae (Micro and Macro)	REGULAR HARVESTING
NITRATE	DENITRIFICATION (Anaerobic Bacteria)	FREE NITROGEN GAS
	DILUTION (Nitrate-Free Water)	WATER CHANGE

Left: *This nitrate filter contains a special resin to remove nitrates, phosphates and sulphates from tap water and is ideal for marine water changes.*

through the resin at a recommended rate, the emerging fluid should be free from nitrates or, at least, greatly depleted, depending on the quality of the input water.

As a bonus, many resins will remove other harmful compounds, including phosphates and sulphates, both known to encourage certain species of nuisance algae. These units are often recharged using a strong salt solution – the type of salt recommended for domestic dishwashers – and are not for direct connection to the marine aquarium.

De-ionizing units may be used to reduce nitrates in water supplies, but are more usually employed by the freshwater aquarist who needs quantities of softened water for specialized fish. Some commercial units operate on a completely different principle and are designed to be permanently connected to the household mains supply. Generally referred to as reverse-osmosis filtration units, they are becoming increasingly popular due to their ability to remove approximately 95 percent of all total dissolved solids (TDS) carried in mains tap water. Initially designed as an aid to improve the quality and taste of domestic drinking water, many fishkeepers have discovered that the water produced is ideal for the marine aquarium. Be warned, though, that many reverse-osmosis units are unsuitable for fishkeeping applications, so those destined for aquarium use should be purchased through reliable aquatic outlets.

Reverse osmosis is a fairly simple procedure. High-pressure mains tap water is forced the 'wrong way' through an ultra-fine membrane, through which only pure water molecules can pass. The waste water containing all the unwanted toxins and compounds is flushed harmlessly away. A carbon filter will be fitted to remove any dissolved gases and may be positioned either before the membrane, as a

Common causes of high nitrate levels

Cause	Remedy
Overstocking	Be realistic about aquarium capabilities.
Overfeeding	Reduce feeding to correct amounts.
Contaminated tap water	Use a proprietary tap water filter.
Wrong water change routine	Increase amount and frequency with nitrate-free water.
No denitrifying filter	Buy or make a denitrifying filter, highly recommended to maintain constantly low nitrate levels.
Insufficient primary filtration	Extract metabolic wastes at an earlier stage in the filtration process using protein skimmers and activated carbon.
Various contaminants	Reduce, or cease, the use of algae fertilizers and liquid invertebrate food.
Little or no algal growth	Encourage the growth of micro and macro algae. Harvest (remove) regularly.

protection device, or afterwards, depending on the design and sensitivity of the membrane. Settlement pre-filters may also be used to prolong the life of the membrane, which may last between one and three years – again, depending on design. The output of reverse-osmosis units may vary considerably, not only as a result of design, but also of initial water quality, temperature and pressure.

Testing for nitrates

An accurate and reliable test kit is essential; it should be easy to use and the readings clear to interpret. There are two types of nitrate test kit on the market today and it is vital that you ascertain which form you are buying to enable you to obtain accurate and understandable readings. (Be sure, also, to check that it is suitable for both fresh tap water and salt water applications.) One type measures total nitrate (NO_3) and the other measures nitrate-nitrogen (NO_3-N), which may be looked upon as a measurement of nitrogen as a relationship to total nitrates. To convert the readings from the NO_3-N kits to NO_3 multiply the result by 4.4. Conversely, an NO_3 reading may be read in terms of NO_3-N by dividing by 4.4. To give an accurate assessment of water quality, it may be wise always to record total nitrate (NO_3) values; even though the other method would appear to give a much more favourable lower figure, it may not be a true reflection of the condition of the aquarium water.

It is important not to be caught out by a slow rise in nitrates; it would also be prudent to consider testing fairly early in the life of an aquarium, say the first three months. this gives ample opportunity to adjust water changes or prepare any alternative methods for nitrate removal. Having said that, it is essential that a brand new aquarium should be filled initially with nitrate-free water. Carry out aquarium tests once every two weeks, or more often if trying to remedy a particularly high level of nitrate. Test tap water purifiers each time they

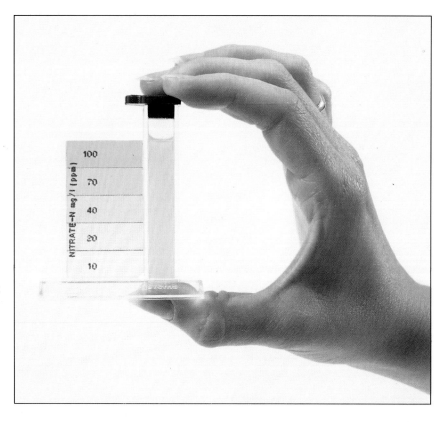

are used. Test the input water *and* the output water. If there is no reading on the input side, use of the unit may be reserved, but do consider that it may help to reduce other toxins. The quality of tap water, in many areas, can change from day to day, so it would be wise to check nitrate levels each time you collect tap water.

Above: *It is important to have accurate knowledge of the levels of both tapwater and aquarium water. Test both at every water change.*

Below: *Almost totally pure tapwater can be collected using a reverse-osmosis filter unit. The initial high cost is often outweighed by a dramatic increase in the vitality of aquarium livestock.*

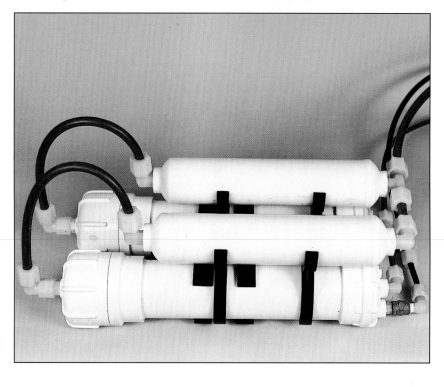

Nuisance 'Algae'

An increasing number of marine aquarists are being plagued by nuisance 'algae'. Quickly covering rocks, corals, sand and species of macro-algae in a dark red, black or brown slimy mat, they can make an attractive show tank resemble the remains of an oil tanker disaster!

In truth, the term 'algae' is used quite erroneously in this context, for this is neither an alga, nor, for that matter, a bacterium, as some have thought. In fact, the organisms usually referred to as nuisance algae form a kingdom of their own, known as Cyanophyceae. Members of this kingdom (sometimes classified as a 'class' by some authors) date back to the earliest known forms of life on the planet Earth, being particularly adaptable and resilient. It is rather unfortunate that they have been commonly termed 'algae' when their physical structure differs quite considerably, although to the unaided eye the superficial difference is indistinguishable. However, for the purpose of this book, the Cyanophyceae will be referred to as 'blue-green' or 'slime' algae in keeping with other universal but technically inaccurate references.

In common with plants, blue-green algae utilize nutrients and carbon dioxide, in this case straight from the water, and light to photosynthesize and multiply. Oxygen is produced as a by-product and, in a well-lit aquarium, bubbles of this gas may be easily visible. In particularly bad cases, whole sections of algae can be seen to rise to the surface under the support of these bubbles.

The problem with these species of blue-green algae, is that they will proliferate where other, more acceptable species, find it hard to. They are very opportunistic by nature and especially welcome deteriorating water quality.

Most aquariums, while undergoing initial maturation, will go through a brown diatom algae stage, but this usually passes quite quickly as their specific nutrients are rapidly exhausted. Green, filamentous or 'hair' algae may form if the aquarium water is rich in phosphates, nitrates and sulphates and may persist indefinitely if the source of these nutrients, usually mains tapwater, is not attended to.

Below: *An otherwise attractive marine aquarium can be spoilt by unwanted algae. Corrective action, however, may see a complete return to former glory.*

As time elapses, aquarium water becomes 'tired', especially if proper water changes are ignored, with a consequent rise in dissolved organics and other toxins, providing ideal conditions for hardy blue-green algae to take over – generally the red or red/brown species. The time period can vary from tank to tank, but if proper preventative action is not taken, it can be as little as a few months.

How nuisance algae arrive
Spores of slime algae are present in nearly all marine aquariums, carried on rock, sand, corals, macro-algae and fish waste, ready to become active when conditions are right.

However, caution should still be exercised, and no rocks, corals or other items should be acquired from a tank where the algae are plainly visible. Bare rocks can be washed in freshwater and neutralized; the same cannot, of course, be said of living items and they should be avoided if traces of the algae are attached.

Once slime algae have a firm grip on an aquarium, they rarely disappear overnight. Usually, by undertaking a suitable course of corrective solution, you should notice a gradual improvement until the whole thing is, once more, under control.

In extremely bad cases, you may be tempted to begin all over again, disposing of perfectly good rockwork and sand, not to mention the unnecessary stress put upon livestock. Alternatively, isolating the cause and making fundamental changes in the set-up or maintenance procedures can be very beneficial.

When fishkeepers set up a new aquarium, enthusiasm is high and regular maintenance is carried out without fail. Sadly, as time goes by, some aquarists lose momentum and two-weekly water changes are delayed to every three weeks, and then to once a month, sometimes even leading to the 'when-I-get-around-to-it' stage. Gradually, but not altogether surprisingly, the aquarium has transformed into a highly unattractive slimy mess. On the other hand, many hobbyists are dedicated to their tanks, yet still the problem occurs. A puzzle indeed! But there is an explanation.

As previously stated, many supplies of mains tapwater contain high levels of toxic substances (to fish and invertebrates, that is) and these start to build up in the aquarium. This is especially so when unfiltered evaporated water is replaced by unfiltered tapwater, which has an increasingly concentrated effect until the toxicity levels are so high that livestock starts to suffer. Once again, slime algae have proved very resistant.

Some hobbyists may confuse these toxic effects with the presence of nuisance algae. However, according to the Culture Collection of Algae and Protozoa based in Oban, Scotland, these species of algae are highly unlikely to prove toxic in the aquatic environment. Toxic species do exist, but the conditions under which they would flourish are extremely difficult to replicate in a closed environment.

How to prevent it
The two main causes related with blue-green algal problems are

---

Curing the nuisance algae problem

- Feed livestock very sparingly – once a day is usually sufficient for the majority of species.
- Do not overstock or add any new stock while slime algae persist.
- Reduce stock if necessary.
- Maintain good water circulation at all times, especially if reverse-flow filtration is used.
- Do not use algal fertilizers or invertebrate foods.
- Continue to use pH buffers and trace elements.
- Make sure all filters are efficient and working properly with the correctly rated pumps.
- Carry out proper regular maintenance and try to use high-quality water for water changes of the correct quantity every two weeks – more if need be.
- Do not over-estimate the nett capacity of the aquarium.
- Always operate a protein-skimmer and marine-grade activated carbon filter.
- Check lighting is of a good quality and enough variety exists to provide a correct spectral range.
- Do not alter the lighting period (photoperiod) outside recommended limits (see page 81).
- Do not add higher algae until slime algae are under control.
- Use purified water for topping up evaporated water.
- Siphon slime algae away from corals as often as possible.
- Expect improvements to take place slowly but surely once positive preventative action has been taken to remedy the situation.

Above: *A protein skimmer is a vital piece of equipment for removing toxic waste at an early stage. This is an air-operated counter-current model.*

Above: *Decorative algae, such as this* Halimeda *sp., will quickly die if smothered by nuisance algae.*

overstocking and overfeeding. So, it is essential that you are realistic about the net water capacity of the tank and the filtration capabilities.

The correct stocking levels for each type of aquarium and filtration equipment have already been discussed in detail within the relevant sections of this book and these should be closely followed. Similarly, correct feeding patterns need observing. Two sparing feeds a day are quite enough for most species; sometimes, even one would be acceptable.

Water changes Most hobbyists will carry out water changes on a regular basis, which is to be applauded, but is it often enough and in the correct quantity? A 10 percent change each month is not enough to keep seawater remaining like seawater over the long term. Most tanks require 20-25 percent every two weeks – sometimes more when heavily stocked with fish – if the environment is to be kept fresh.

Always try to use the purest water possible for water changes and topping up evaporated water. Areas plagued with high levels of nitrates, phosphates, sulphates and other assorted toxins will require the use of a tapwater filtration unit. Even if tapwater seems suitable, it is often amazing to see the improvement in livestock health once filtered mains water is used on a regular basis.

Lighting Contrary to previous popular belief, it now appears that lighting has little effect on slime algae. Low- or high-intensity lighting appears to make very little difference, as the algae will adapt to either. It is a mistake to be tempted to turn the lights off for long periods, especially in a coral invertebrate aquarium, for although the algae will diminish, this is only a temporary state and, on returning to normal lighting patterns, the algae will reappear in the same quantity.

Refrain from adding higher forms of algae (such as *Caulerpa* spp) while there is a carpet of slime algae. Far from 'starving-out' the offending algae, it usually falls victim itself by being covered and choked until it eventually dies, causing further pollution and aggravating the whole problem.

However, if all other solutions have been attempted with little success, it may be worth trying a new combination of bulbs or tubes. This has shown to be effective in certain situations, although the reasons still remain unclear.

Other methods Maintaining a constantly high redox potential in the region 400-450 mv has been shown to have had some success in clearing fish-only aquariums of slime algae. This requires running an ozonizer at very high levels and monitoring carefully with a reliable ORP probe/dosing timer; it cannot be done by guesswork.

The short-term effects on fish are observed to be little, but it would be unwise to try this sort of treatment in a coral invertebrate aquarium, as invertebrates have been observed to fare very badly. The theory behind this practice is that the aquarium water is kept in a very high state of cleanliness as the ozone oxidizes many of the metabolic wastes, while also destroying algal spores. It is certainly not a method to be recommended to the inexperienced marine aquarist.

Natural predators Although natural predators do exist in the wild, there are none that can be practically kept by the hobbyist.

Chemical controls As yet, there are no chemical controls for this condition. Some herbicides were used at one time, but were hastily removed from the market when found to contain substances dangerous to humans! Be wary of any such chemicals, as you may put at risk your livestock's health, as well as your own!

Beneficial algae
It is important to bear in mind that micro-algae do have a positive side to them and an invaluable part to play in the biological equilibrium of the aquarium. To discourage every species to the point of extinction would be a mistake.

The back and side of the aquarium should, at least, be allowed colonization by non-invasive algae. Not only do they provide a useful filtration function in themselves, for reasons already stated, but they also provide shelter for a multitude of micro-organisms, essential foods for success with Mandarinfish and various Blennies.

Not all species of micro-algae – and there are many of them – are rampant and to be regarded as undesirable. Slow-growing species are common and may be left undisturbed, as they rarely interfere with other forms of life. These species require good water quality for survival. Therefore, if there is a slime algae problem, these other more beneficial species are unlikely to be found in any great quantity.

Filtration

The most critical and, to the beginner, most confusing aspect of the marine aquarium is how to provide adequate filtration. The function of filtration is to keep the water clean, visibly and chemically, so that the animals can continue to live, hopefully grow, and possibly reproduce within the aquarium.

The purpose of the aquarium filtration system is to remove as much unwanted 'dirt' from the water as possible, thus delaying the need to replace water for a reasonable period of time. As well as visible dirt and sediment stirred up by the actions of the fishes and other tank occupants, the water also contains invisible dissolved organic compounds (waste products and the result of their decomposition) and these, too, must be removed.

Many types of filters are used in the aquarium hobby, ranging from simple air-operated box filters to highly sophisticated water purification systems. Filters vary not only in design, but also in their mode of operation, some providing simple mechanical straining, while others exert a chemical or biological influence on the water flowing through them. Many filters perform all three functions.

In addition to these 'standard' filter types, many of which find a ready application in both freshwater and marine aquariums, there are a number of water treatment systems that are used more or less exclusively for marine creatures. These systems include algal filters, protein skimmers, ozonizers and ultraviolet sterilizers. We shall discuss the operation and merits of these filtration methods later in this section. First, we will look at the options open to the marine aquarist among the more conventional type of aquarium filters.

Mechanical filtration
Mechanical filtration serves the simple function of removing visible particulate matter from the water and delivering it to a point where it is easily removed and/or where it can be broken down into less toxic end products.

The simplest type of mechanical filter suitable for a marine aquarium is the external sealed canister power filter. Its basic advantage over air-operated designs widely used in freshwater fishkeeping is that it delivers a satisfactorily high throughput of water. An electric motor mounted in the top of the canister powers a pump that draws water through the filter media placed inside the body of the unit. By varying the media, it is possible to 'fine tune' the effect of the filter.

The most commonly used 'mechanical' filter medium is some form of floss made from spun nylon, dacron or some other man-made fibre. In order to prevent rapid clogging of the fibres, it is normal practice to have some 'prefilter' medium ahead of the floss in the water flow. This can be small pieces of ceramic pipe, but many fishkeepers use plastic pot-scourers.

External power filter

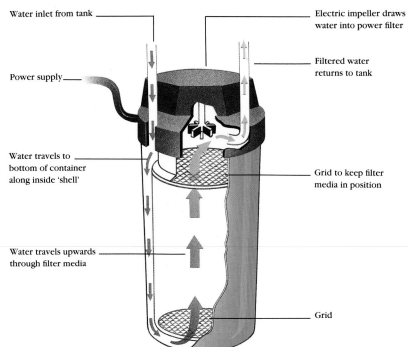

Water inlet from tank

Power supply

Water travels to bottom of container along inside 'shell'

Water travels upwards through filter media

Electric impeller draws water into power filter

Filtered water returns to tank

Grid to keep filter media in position

Grid

Below: *There are now an excellent range of high-quality external canister filters available for every aquarium situation. They vary in filter media volume and pumping capacity.*

The maturing process of a biological filter

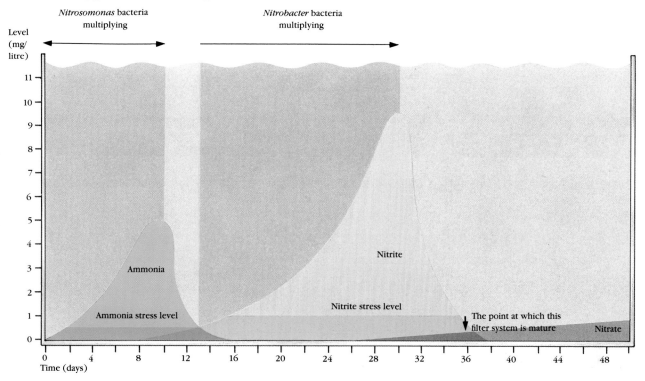

Level (mg/litre)

Nitrosomonas bacteria multiplying

Nitrobacter bacteria multiplying

Ammonia

Ammonia stress level

Nitrite

Nitrite stress level

The point at which this filter system is mature

Nitrate

Time (days)

Above: *In a newly established tank, the levels of ammonia and nitrite form overlapping peaks as the bacteria in the biological filter multiply and begin to break down these substances.*

Preformed foam pads that fit snugly into the filter body not only perform a simple straining function but, if not renewed on a frequent basis, also act as biological filters once bacteria have colonized the foam. This may be considered undesirable, as every time the medium is changed, there will be unnecessary interference to the biological balance of the filtration system. Therefore, it is essential that you regularly renew mechanical filter floss and pads, preferably every two weeks.

Chemical filtration

The most common chemical medium used in canister filters is activated carbon. This is, basically,

Right: *In the natural world, harmful ammonia and nitrite are continually converted into less dangerous nitrates, which are in turn used as food by plants and algae. This process is replicated in the aquarium environment with efficient biological filtration.*

wood or bone charcoal that is 'activated' by being baked at a high temperature to open up tiny pores in the particles and to make the surface of the carbon more 'attractive' to certain substances. As aquarium water passes through the carbon pieces, dissolved waste

products are adsorbed over the very large surface area available. (This also makes activated carbon an ideal site for bacteria to flourish, but holds the same disadvantage as mechanical media with regard to biological imbalance each time the carbon is replaced.)

The nitrogen cycle

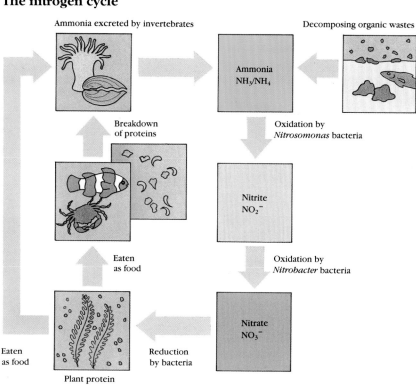

Ammonia excreted by invertebrates

Decomposing organic wastes

Breakdown of proteins

Ammonia
NH_3/NH_4

Oxidation by *Nitrosomonas* bacteria

Eaten as food

Nitrite
NO_2^-

Oxidation by *Nitrobacter* bacteria

Plant protein

Eaten as food

Reduction by bacteria

Nitrate
NO_3^-

The carbon is usually sandwiched between two layers of floss in the filter body to prevent it from being drawn into the aquarium. Alternatively, the charcoal can be contained in a nylon bag, which serves the same purpose. The effective life of activated carbon is not easy to estimate and is greatly dependent on the stocking levels within the aquarium. However, a good-quality carbon, used in the correct quantity, will need renewing every two months on average. If this is not done, the adsorbed chemicals are likely to be released back into the aquarium, thus negating all the filter's previous beneficial work.

In addition to activated carbon, other 'convenience' chemical filters are commonly available. These usually take the form of resins in granules or pads, and are designed to do a specific task. They may act to remove ammonia, nitrates, phosphates or other pollutants. Some require a high throughput of water and need to be situated within a canister filter, while others may only require 'hiding' within the aquarium itself. It is important that you use these types of filters in conjunction with related test kits so that you can monitor their continued efficiency and discard the media as soon as it is exhausted. This may, again, prevent any adsorbed pollutants from discharging back into the aquarium.

Sea water is a very complex chemical solution containing many trace elements and organic substances, both beneficial and detrimental to its inhabitants. It is very difficult to manufacture a chemical filter that removes only the detrimental molecules, while leaving the others available to the marine animals and plants. This has led to much, sometimes heated, debate regarding the value of chemical filtration in the marine aquarium. Generally speaking, if supplementary trace elements, buffer solutions and vitamin additives are used regularly (i.e. to replace those filtered off), then the benefits of chemical filtration outweigh its limitations.

Biological filtration

In the wild, the cleansing action of the sea converts and disperses the waste substances produced by the animals therein. In the closed confines of an aquarium, the livestock depend on the fishkeeper to provide an efficient system to purify the water. Biological (bacterial) filtration is a natural means of removing toxic nitrogen-based wastes (i.e. ammonia, nitrites and nitrates) from the aquarium.

Fishes and invertebrates excrete ammonia as a waste product. This is

Downflow biological filtration system

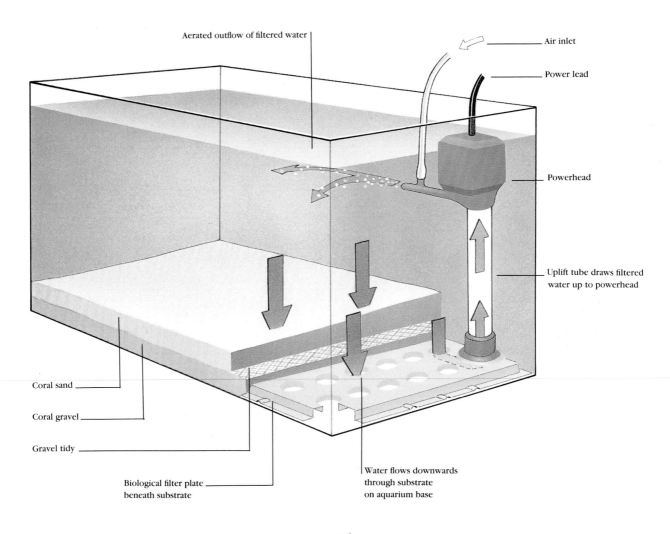

Aerated outflow of filtered water

Air inlet

Power lead

Powerhead

Uplift tube draws filtered water up to powerhead

Coral sand

Coral gravel

Gravel tidy

Biological filter plate beneath substrate

Water flows downwards through substrate on aquarium base

added to the ammonia produced by bacteria working on other waste materials in the aquarium, such as dead plants and uneaten food. Ammonia is toxic to fishes and invertebrates and if it is not removed, or converted into other less harmful substances, then aquatic life will soon perish. Fortunately, a substance that is poison to one living organism is food for another and thus, there is a natural way of dealing with ammonia disposal.

Aerobic (oxygen-loving) bacteria of the genus *Nitrosomonas*, convert ammonia to nitrite, a slightly less toxic substance, but one that is still dangerous to fishes and invertebrates. A second group of bacteria, belonging to the genus *Nitrobacter*, transform the nitrite to nitrate, a much safer substance, but one that can still cause problems if it builds up in the aquarium. Different

marine creatures vary in their sensitivity to these nitrogenous compounds. Nitrate levels, for example, should be maintained below 20 gm/l for fishes, and below 5 gm/l for invertebrates (even lower for some species).

If further anaerobic ('non-oxygen-requiring') bacteria are allowed to get to work, then the nitrate can be converted back to atmospheric nitrogen (by so-called denitrification) and vented from the aquarium. Herein lies a paradox, however, since, if we are to provide ideal situations for the complete disposal of all nitrogenous substances, it appears that we must provide two sets of conditions for opposite types of bacteria.

In the basic marine set-up, it is usual to provide the conditions required by the first group of aerobic bacteria, i.e. those that thrive on oxygen. The inevitable

subsequent build-up of nitrate is kept under control by carrying out regular partial water changes – always assuming that the 'new' water has a lower level of nitrates than the tank water. Here, treating tap water with a nitrate-reducing resin can be particularly useful. (In most modern 'total system' aquariums, the filtration system provides both aerobic and anaerobic conditions at different stages in the filtration process.) In its simplest form, providing the necessary oxygen-rich environment for the *Nitrosomonas* and *Nitrobacter* bacteria to flourish may be found in the traditional undergravel form of filtration.

Undergravel filters
In essence, an undergravel filter is a perforated plate, covered by a 7-10cm (2.5-4in) layer of suitable substrate, ideally, equal quantities of coral gravel and coral sand

Reverse-flow biological filtration system

Airstone

Dirty water intake

Dirty water drawn out by power filter

Filtered water expelled by power filter

Air tube

Air pump

Power lead

Clean water rises from undergravel filter

Filtered water drawn into undergravel filter

Filter media in external power filter remove solid and toxic waste

(separated by a plastic mesh sheet, known as a gravel tidy). In a traditional 'down-flow' filtration set-up, polluted, but oxygenated, water is drawn downwards through the substrate, where colonies of nitrifying bacteria living on the surface of the substrate particles break down nitrogenous wastes in two stages.

It is vital that this conversion process occurs efficiently because ammonia and nitrites are extremely dangerous, even at very low levels. The nitrifying bacteria are aerobic (oxygen-loving) and so it is important to keep the water moving and well oxygenated. Water movement can be achieved in two

Below: *A commercially produced 'above the tank' trickle filter – simple, but nevertheless very effective.*

basic ways: by using air pumps or water pumps. Many newcomers to the marine hobby will have arrived via tropical freshwater fishkeeping and will be familiar with air pumps. These can be used to drive undergravel filters, but they are by no means completely satisfactory. The main drawback is that the flow from an air-operated undergravel filter is not strong enough to produce the powerful water currents required by many of the most commonly kept marine fish and invertebrates. Furthermore, dry salts tend to accumulate at the tip of the air tube connection to the undergravel filter, whether it is left open or has an airstone fitted. This reduces the flow of air, which in turn reduces the rate at which water is pulled through the filter. In extreme cases, it may cause the complete failure of the biological filter system.

A much better alternative is afforded by submersible water pumps, known as powerheads, which draw water through the substrate. These pumps simply sit on top of the undergravel filter uplifts. Even the smallest units move a great deal of water and generate

beneficial currents. Most models also have a venturi facility, which allows additional air to be drawn into the aquarium to improve the aeration still further. Not least of the benefits of powerheads is their comparative silence in operation, compared to air-pump-operated filters (provided the air inlet is valved to regulate the influx of air).

In down-flow undergravel filters there is a tendency for debris to build up within the substrate and effectively clog the system. Using the alternative 'reverse-flow' approach helps to keep the substrate clean by first straining the water through a power filter packed with mainly mechanical media and then pumping it under the filter plate and thus upwards through the substrate into the body of the tank. This has the added advantage of keeping food particles in suspension, which will benefit many invertebrates. When carrying out a partial water change, use a gravel cleaner to remove debris and agitate the gravel, thereby preventing the development of channels in the filter bed, which may lead to an uneven water passage.

Owing to their design, 'reverse-

Trickle filter

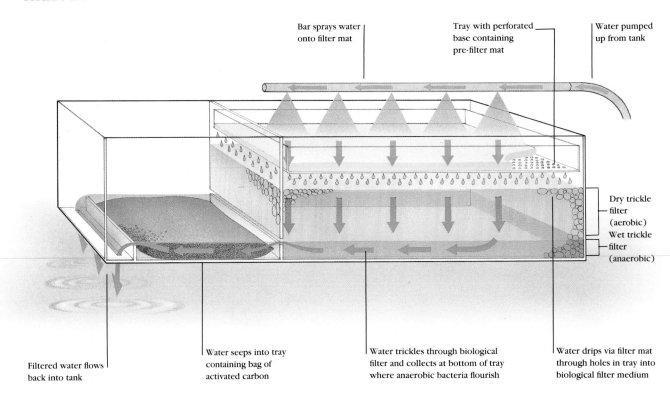

Bar sprays water onto filter mat

Tray with perforated base containing pre-filter mat

Water pumped up from tank

Dry trickle filter (aerobic)

Wet trickle filter (anaerobic)

Filtered water flows back into tank

Water seeps into tray containing bag of activated carbon

Water trickles through biological filter and collects at bottom of tray where anaerobic bacteria flourish

Water drips via filter mat through holes in tray into biological filter medium

Counter-current protein skimmer

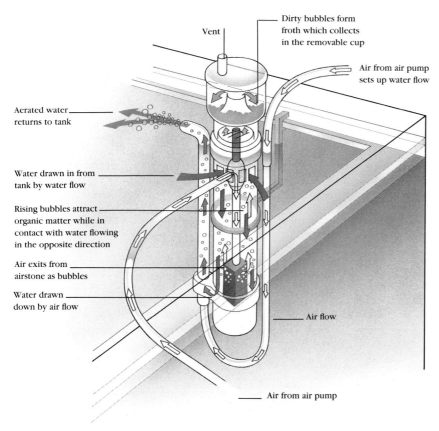

Vent

Dirty bubbles form
froth which collects
in the removable cup

Air from air pump
sets up water flow

Aerated water
returns to tank

Water drawn in from
tank by water flow

Rising bubbles attract
organic matter while in
contact with water flowing
in the opposite direction

Air exits from
airstone as bubbles

Water drawn
down by air flow

Air flow

Air from air pump

flow' filters are unable to provide sufficient water circulation within the aquarium and suitable supplementary internal water pumps should be provided as a matter of course.

With both the 'down-flow' and 'reverse-flow' methods of filtration, the total aquarium water volume should be passed through the filter bed at least three times every hour.

Advanced filtration
Traditional undergravel filtration has been substantially improved upon over the last few years. Although the basic bacterial processes remain the same, significant advancements have been made on filter design to make for more efficient, convenient and attractive units.

These units are usually referred to as 'trickle filters' or 'wet and dry filters'. Being remote from the main aquarium, design can be very flexible and filters may range from small, basic units, to extremely large and complicated 'total water management systems'.

Trickle filters
The development of nitrifying bacteria is limited by three main factors: their supply of food, the surface area for them to colonize and, most critically, the amount of oxygen available to them. Under-gravel filters meet the first two demands well, but the oxygen-carrying capacity of water is limited.

'Dry' trickle filters overcome this problem by removing the filter bed from the tank into a separate container, either above or below the tank. Water is pumped, or taken from the tank by gravity, and sprayed over the substrate. The beauty of this system is that the nitrifying bacteria have a virtually unlimited supply of oxygen and therefore develop at a prodigious rate. Correctly managed, a dry trickle filter substrate (porous clay granules or similarly inert material) has a biological efficiency at least 20 times that of a similar volume of substrate in a submerged undergravel filter.

Trickle filter systems are available

commercially, but keen hobbyists will find it a simple task to set up their own. Remember to pass the water through an outside power filter first, before it flows into the trickle filter. Such mechanical straining will help to keep the filter medium clean and effective for a much longer time.

So-called 'wet' trickle filters provide a way of removing varying quantities of nitrates from the aquarium. At their simplest, these consist of a very porous filter medium kept submerged in a box or canister into which a very slow trickle of water enters from the aquarium. The water is ultimately pumped back into the tank. Because of the slow flow of water through this unit, oxygen is in short supply.

Anaerobic denitrifying bacteria develop here and obtain their necessary oxygen by breaking down nitrates (NO_3), first to nitrous oxide (N_2O) and, ultimately, to free nitrogen gas. Very few 'wet' trickle filters are available commercially, other than those included as part of complete tank and filter system packages. Very simple units that consist of a box filled with a suitable filtration medium and buried beneath the tank gravel perform a function along similar lines at a very low cost, although results from them can be very variable.

Algal filters
In a marine aquarium devoid of plant life, an algal filter can be especially useful. A considerable amount of algae is required, so the filter will need to be a correspondingly large affair, which may not be practical in most domestic environments.

Protein skimmers
There are other water purification systems that are especially useful in the marine aquarium. Among these are air-stripping devices, known as protein skimmers, which are now widely used and appreciated by marine hobbyists the world over. Many harmful waste products, particularly proteins, phenols and albumens, are easily and efficiently

Ultraviolet sterilizer

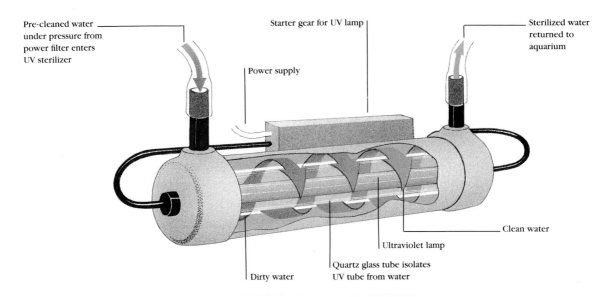

Pre-cleaned water under pressure from power filter enters UV sterilizer

Starter gear for UV lamp

Sterilized water returned to aquarium

Power supply

Clean water

Ultraviolet lamp

Quartz glass tube isolates UV tube from water

Dirty water

removed by these skimmers. In the simplest air-operated units, a column of enclosed water is fiercely aerated with very fine bubbles. The proteins and other wastes 'stick' to the bubbles (a natural tendency of so-called 'surface-active' dissolved organic molecules) and are carried to the top of the column in a stiff foam, familiar to anyone who has seen a polluted stream bubbling over a weir. This foam spills over into a plastic cup, where it collapses into a murky brown liquid, which can then be discarded. Motor-driven venturi types are expensive, but even more powerful and efficient.

The efficiency of protein skimmers is governed by their vertical height, the amount of air supplied to the unit, and the water flow rate. The last two parameters determine the amount of contact between the stream of air bubbles and the water flowing through the unit. In practical terms, this means using the tallest unit that will fit in the tank, a strong air pump and a skimmer of the counter-current type, in which the water flows against the current of the air bubbles, thus increasing the beneficial contact between the two.

Protein skimmers are rather clumsy looking units, but once in position they can be hidden behind rockwork and you can carry out any servicing from above. The amount

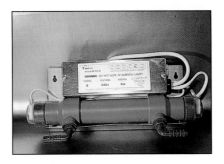

Above: *A UV sterilizer with ballast unit attached above. A proper through-flow of water is essential.*

of waste removed by these units can be considerable, and their contribution to longterm success with invertebrates and fish is second only in importance to good biological filtration.

Ozonizers

The popularity of ozonizers in the domestic aquarium has waned in recent years, but they are still widely used commercially. Here, a supply of dry air is pumped over a high voltage electrical discharge, where a proportion of the oxygen (O_2) in the air is converted to ozone (O_3). Ozone is chemically very reactive, a 'fierce' oxidizing and disinfecting agent, which was thought to be a useful background aid to disease prevention, since the gas kills bacteria and any other free-swimming micro-organisms that come into direct contact with it. The

main commercial use of ozonizers, however, is to increase the efficiency of protein skimmers. By injecting ozone into the air supply to a skimmer, the efficiency rate of the skimmer is vastly improved.

If you buy an ozonizer, *never* feed the ozone directly into the tank, but couple it to a skimmer and regulate the generation rate so that you cannot detect the sharp tang of ozone around the tank. Any unwanted ozone should be ducted from the room or passed through a carbon filter to neutralize it, especially if you are using high levels. Always use plastic airline when dealing with ozone; it will damage rubber tubing and pump diaphragms, and excessive levels will even make plastic brittle. WARNING: USE OZONE WITH EXTREME CARE; IT CAN CAUSE HEADACHES, NAUSEA AND OTHER UNPLEASANT EFFECTS.

Ultraviolet sterlilizers

Ultraviolet light has long been known to have a strong sterilizing effect, and UV-penetrating units have many industrial uses as bactericides. Such units, which pass precleaned water from a power filter close by an enclosed UV light, are now widely available to the fishkeeper. The light they emit can kill algal spores and bacteria and may affect some small parasitic organisms. The efficiency of UV sterilizing units is governed by the internal cleanliness of the unit, the clarity and flow rate of the water going through it and the intensity and precise wavelength of the UV light. If too much water is pumped through too small a unit, then many organisms will not be killed. If the correct size unit is balanced with the right water flow, though, UV units can be beneficial in reducing algal blooms. Their main use, however, is to reduce the incidence of disease among fishes kept in invertebrate systems, where chemical treatments are inadvisable.

The output of a UV unit is rated by its wattage, 15 watts being the smallest available. For most purposes, a rating of 15 watts per 136 litres (30 Imp./36 US gallons) tank capacity is adequate. The water in the aquarium should pass through the unit approximately twice per hour. The combination of a power filter and UV light can be very effective, if relatively expensive. Replace the UV tube about every six months for the best results.
WARNING: ULTRAVIOLET LIGHT CAN BE VERY DAMAGING TO THE HUMAN EYE;

Below: *External water management systems, such as this, are capable of keeping water quality extremely high.*

NEVER LOOK AT A UV LAMP WITHOUT PROPER EYE PROTECTION.

Complete systems

Nowadays, it is possible to purchase a total water management system that not only employs trickle filtration, with all its advantages, but also incorporates mechanical and chemical filters, heaters, pumps, protein skimmers, oxygen reactors, nitrate removers and practically every form of device designed for the improvement of water quality.

These units may be situated to the side, rear or below the aquarium and may come as standard units or be custom designed to suit the particular needs of the aquarist.

Many experienced aquarists

Above: *Ozonizers are very compact units and easy to adjust. Use in conjunction with a protein skimmer.*

researching into advanced filtration techniques see external trickle filters and 'total' systems as the way forward for all hobbyists. They point to the glaring inadequacies and limitations of undergravel methods that make them both wasteful of natural products ('total' systems need use only plastic media), and inefficient, as water flow is difficult to regulate. Additionally, the pores of coral sand and gravel become clogged and unusable by beneficial bacteria within a relatively short space of time. Given the convincing arguments – and there are many more – it could be that the days are numbered for the undergravel filter and that more and more aquarists will either convert to or begin with a more advanced trickle filter.

Using filters with medications

The treatment of disease in the marine aquarium can be complicated by the presence of sophisticated filtration systems. Activated carbon will adsorb many medications, for example, so filters containing carbon should be disconnected for the duration of any treatments. Certain resins may be affected in a similar way, so read instructions carefully.

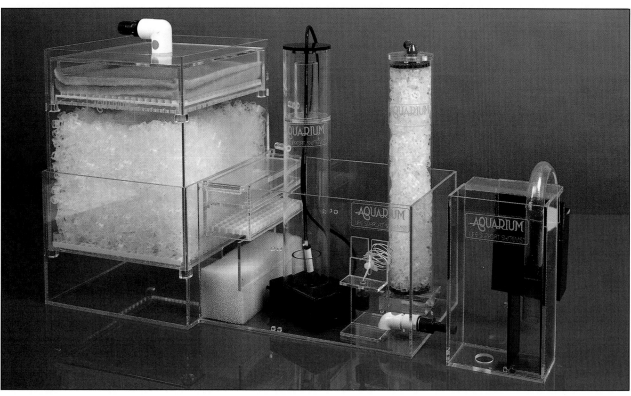

Rockwork and Decoration

The presentation of an impressive aquatic display is largely due to the imaginative use of the background decoration. Bare tanks may be practical for maintenance purposes, but most people would agree that they provide a boring presentation, to say the very least.

Carefully arranged background structures not only show off the marine creatures to their best advantage, but they can also prove essential to the well-being of the aquarium. Many fish and invertebrates rely on the cover of rockwork to give them a sense of security, providing 'bolt-holes' and retreats that are essential for a stress-free life. Sedentary and sessile invertebrates will make full use of decorative structures as secure attachment points, as will many species of algae.

Calcareous rocks can also act as a useful aid to water quality; being of a very alkaline constituency, they help to counteract the natural tendency of the aquarium water to acidify. This action is commonly referred to as 'buffering' and can, in certain circumstances, help to prevent rapid drops in pH. Where no calcareous materials are used at all, proprietary brands of liquid buffering agents may need to be applied. (However, if you carry out correct regular water changes, you may be able to dispense with the need for buffering media.)

Rockwork and contaminants
One phenomenon peculiar to calcareous rocks and substrates is their ability to absorb copper commonly used for medication purposes. This can then be leached back into the aquarium water at an uncontrolled rate, varying with changes of pH and specific gravity. As all invertebrates are highly sensitive to even the smallest quantities of copper, calcareous materials that have been exposed to copper medication must not be used in the invertebrate aquarium.

You should also avoid rocks containing any sort of metal, or toxic poisoning is sure to result. Similarly, various decorative woods, usually sold for the freshwater aquarium, are totally unsuitable for marine applications. (The only exception to this may be driftwood that is well-worn and denuded of any substances likely to be of harm to marine life.)

Contaminants can also arrive as concealed pieces of metal within bags of coral sand and gravel. Collection areas for such materials are often sited close to World War Two battle areas, where sunken shipping has broken up and spread various metals over a wide area in vast amounts. The deteriorating sides of transport barges can also impart rust flakes. Most responsible companies use powerful magnets, as well as sieves and visual checks, to eliminate the problem, but some often seems to get through to the hobbyist. Make a thorough visual check of all coral gravel and sand. In addition, you can pass a magnet wrapped in a polythene bag over it; any pieces of metal drawn to the magnet can then be thrown away with the plastic bag.

Be careful not to strip the copper wiring of various pieces of aquarium equipment above the aquarium, as any small strands of copper falling into the sand will prove toxic and almost impossible to locate.

The substrate
Whether you use an undergravel or external filter in the aquarium, the bottom of the tank will still be seen as part of the decor. Undergravel filters look natural owing to their use of coral sand and you can sprinkle the same sand to a maximum depth of 12mm (½in) over the base of a tank using external filters. Even in this shallow depth of sand, however, detritus may collect and cause pollution problems, and even a thin layer of sand will be unsuitable for burrowing invertebrates and fish, such as wrasses. In these cases you will need to filter any thicker layer of substrate to prevent the build-up of anaerobic conditions, which would constitute a subsequent pollution hazard.

If substrate is not a necessity, it is quite possible to leave the base of the aquarium bare. In a short time it will take on a covering of beneficial and decorative algae, as well as having the advantage of remaining cleaner and less prone to pollution.

Building safe rockwork
Unsafe rockwork has been the downfall – quite literally – of many an aquarium. Tumbling rocks have resulted in shattered bases and sides due to precarious positioning. As a rule, any rockwork structure should be as broad in the base as it is high. Place the largest rocks at the bottom to provide a firm foundation, and ensure that the rest of the rocks decrease in size as the structure rises. Tufa makes an excellent foundation rock, being very easy to shape with the most basic of tools. Make one face as flat as possible and place it directly on to the floor of the aquarium. Treat the rest of the base rocks in this way to create a safe rockwork structure and to prevent sliding and rolling.

Keep rockwork structures as simple as possible; complicated arrangements using copious amounts of small rocks and coral skeletons not only look messy, but are also unstable and trap large amounts of detritus. In addition, should you ever have to extricate a specimen, the whole process could prove to be a nightmare, as you will need to dismantle most, if not all, of the rockwork structure. This is another good reason why only totally compatible fish should be housed in a reef aquarium.

Bear in mind that fish require

Finding the true volume of water

Calculating the exact net volume of water in an aquarium is essential for the correct use of medications, trace elements, vitamin supplements, stocking levels, etc. Most people over-estimate the net volume of their tank by some considerable amount. This is usually due to underestimating just how much water decorative rockwork can displace.

A simple solution that need not waste valuable sea water is to fill the aquarium accurately, using fresh water. (Keeping count of buckets of a known volume is practical for smaller tanks, but you may need to use a flowmeter in line with the filling hose to accurately calculate the volume of water in a large aquarium.) Once you have filled the tank to the desired level, place the rocks into the water and remove and measure the displaced water. Once you have introduced all the rocks, you can deduct the displaced water total from the original tank-full figure to calculate the correct net volume.

Cleaning corals for the aquarium

Soak corals in a solution of household bleach until clean. Rinse in fresh water to remove all traces of bleach

Use gloves as a potection against spiky corals and the effects of the bleach

You can clean small pieces by boiling them for at least an hour

An effective way of rinsing corals is with a hose

Avoid using high water pressure that might damage delicate corals

adequate swimming space and you will need to make suitable allowances for each species. Rocks, sand and gravel will all displace large amounts of valuable water, something that is not to be recommended. The 'trade-off' between decoration and water is not an equal one – water will always prove to be the more valuable.

Artificial coral

While most decorative coral skeletons and sea fans are collected in a responsible and self-sustaining manner, the marine hobby has always tried hard to find alternatives, and synthetic replicas can now be found in increasing numbers. These are manufactured using moulds and safe resins and are usually virtually indistinguishable from the real thing. One of the exciting things about this idea is that many synthetic corals are produced in the countries where real corals would normally be collected. In this way, not only are jobs preserved, but additional ones are also created in regions of the world where foreign income is desperately needed.

Preparing corals

Decorative corals are the skeletal remains of millions of polyps that make up parts of the coral reef. Since a piece of coral is dead when you buy it, you must ensure that it no longer contains the remains of the original animals (or other creatures that have made their home in it) before you put it into the aquarium. As long as the corals are reasonably clean, do not smell of bleach, and you rinse them beforehand, you can safely add them to the tank during the maturation period. Bacteria will then break down any organic material and

Left: *All traces of animal remains must be removed from corals and shells intended for use in the aquarium. Deeply convoluted and indented corals are the hardest to clean. You can clean coral by soaking it in bleach or boiling it. Rinse it thoroughly in clean water to avoid pollution, particularly from bleach residue.*

there need be no risk to future livestock, once the full maturation cycle has been completed.

Adding further decorative corals and shells to an already established aquarium is a different matter, though, and you should take every care to eliminate possible sources of pollution. It is true that most decorative corals available these days have already been cleaned to some extent by pre-bleaching or being allowed to dry and bleach in a hot tropical sun, or sometimes both. However, to be on the safe side, consider the following cleaning options. The first, and safest, method is to boil the corals for at least half an hour, preferably one hour. After cooling, they are then ready for immediate use. The second method involves the use of household bleach and is usually reserved for

items too large to be boiled. Soak the coral (and other shells) in a solution of household bleach (2 cups per 4.5 litres/1 Imp./1.25 US gallons of water) for one to two weeks. Then flush it thoroughly under fresh water. During the next week, soak it in several changes of fresh water until every trace of the bleach smell has disappeared. Only then will it be safe to introduce it into the marine aquarium.

Decorative corals are quickly overrun by algae and the point of cleaning to a highly white finish is often defeated. Aquarists who insist on having only clean, white corals generally possess two sets of decoration: one in the tank and one being cleaned. After the initial cleaning, the corals may be given a secondary light cleaning by bleaching, but they will only require

a relatively short soaking of two or three days and a thorough rinsing before being serviceable again.

Treat bleach solutions with the utmost care; not only is bleach highly toxic in the aquarium at trace levels, but it can also pose a hazard to humans and animals. Make sure that you cover treatment containers that are in use and keep them in a completely safe place at all times.

A coral diorama
A totally safe and effective way of using all forms of decorative rocks, corals and shells, etc., is the diorama. This involves placing a well-lit box behind the main

Below: *A dry, well-lit diorama of corals, rocks and dried sea fans behind the aquarium provides a natural-looking background to the underwater scene.*

Creating a coral diorama

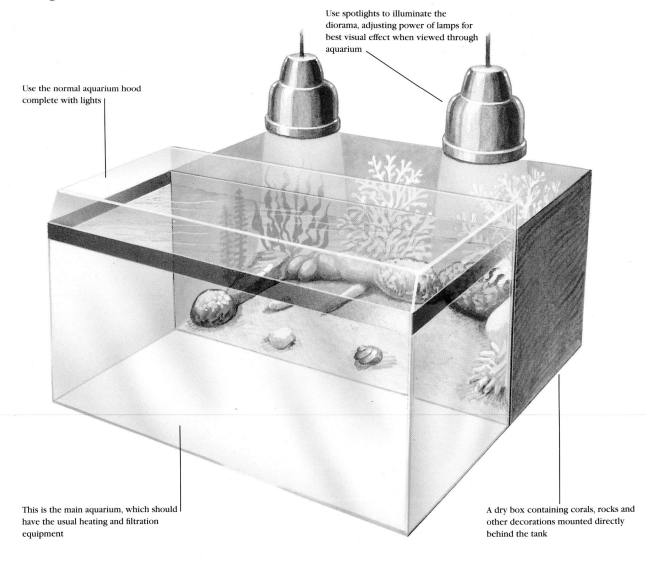

Use spotlights to illuminate the diorama, adjusting power of lamps for best visual effect when viewed through aquarium

Use the normal aquarium hood complete with lights

This is the main aquarium, which should have the usual heating and filtration equipment

A dry box containing corals, rocks and other decorations mounted directly behind the tank

Above: Dead, cured coral
A brittle and sharp-edged decoration.

Above: Dead, cured coral
The blue colour will fade with age.

Above: Tufa rock
A soft rock, easily carved and shaped.

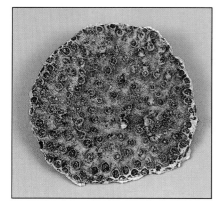

Above: Artifical coral
A natural-looking synthetic replica.

Above: Artificial coral
Man-made using moulds and safe resins.

Above: Man-made lava rock
Light, porous and quite expensive.

Above: Dead, cured sea fan
An ideal decoration for a diorama.

Above: 'Rainbow rock'
A carved and sand-blasted sandstone.

Above: Petrified wood
An usual decoration from Nevada, USA.

aquarium and arranging the various corals, rocks etc. in a pleasing manner within it. When viewed through the main tank, the effect can be quite remarkable, giving a tremendous depth to the whole scene. You will need, however, to keep the rear glass of the aquarium scrupulously clean at all times, as the whole effect will be lost if any trace of algae obscures the view.

Suitable tank decorations
Let us then consider the main decorative options available to the marine hobbyist.

Calcareous stone (limestone) was the only rock in regular use in the early days of the marine hobby. Although it is perfectly acceptable, it is extremely heavy and tends to displace unacceptable amounts of

water. Some conservationist-minded countries use this material to produce 'living rock' for the aquarium trade (see below).

Corals are readily available in many different forms. They tend to be quite expensive and very brittle. Since the corals you buy from an aquatic store are only the skeletal remains of living polyp colonies, their bleached white appearance tends to look rather incongruous in a natural living reef situation. Artificial corals, while still expensive, are likely to provide the longterm alternative to the natural resource.

Slate is totally safe to use in saltwater and makes an ideal spawning surface for species such as anemonefishes.

Shells and barnacle clusters are justifiably popular, providing valuable shelter for many smaller fish and invertebrates. They should be treated by boiling, or – only if boiling is impractical – by bleaching.

Lava rock and tufa are the best rocks to use in the marine aquarium. Tufa is a naturally occurring calcareous rock with a similar chemical composition to that of coral sand. It is also very soft and easy to shape. Being almost totally constituted of calcium, it has a highly beneficial buffering effect on the pH of seawater, keeping it well within acceptable parameters if used in sufficient quantities. The displacement of water by tufa rock is moderate owing to its porous nature, which may not be immediately apparent, as it is usually concealed by a thick layer of aggregated dust; this should be rinsed away before placing the rock in the aquarium. Many encrusting corals and algae quickly colonize tufa and its initial stark whitish colour quickly tones down.

Lava rock is a man-made product and not to be confused with natural lava, which must never be used in marine tanks since it is often rich in sulphur and metals that can cause severe pollution in seawater. Although it is the most expensive stone on the market, it is increasingly popular, and rightly so. Its advantages are numerous. Varying in colour between dark coffee and purple-brown, it looks very attractive. It is also exceptionally light for its volume, displaces acceptable amounts of water and is very easy to handle

(although care should be taken on the sharp edges). It stacks together well and is easily colonized by worms, sponges and algae. This is a first-class product for any type of marine aquarium.

Sea fans and sea whips present a different and pleasing texture to the aquatic scene, but it is important that they are first properly cleaned. Before cleaning, they have a chalky appearance and are often pink, yellow or buff in colour. When washed thoroughly in several changes of water (bleaching is unnecessary), all that remains is a black 'wire' skeleton. Unless they are in this condition, do not use them. Unlike corals, the sea fans and sea whips do not have a calcareous skeleton, but are organic in nature and subject to biological degradation. This process may take some time, but at the first signs of deterioration, they should be removed and discarded.

Living rock is the accepted name for pieces of calcareous rock taken from tropical seas, usually from a site close to the shoreline. When first removed from the sea, this rock is liberally clothed with various algae, sponges and miniature anemones, and usually houses a considerable population of small shrimps, crabs, worms and many other 'mini-beasts'. Some years ago, large quantities of living rock were shipped from Saudi Arabia and the short flight times to Europe meant that this rock usually arrived in very good condition and justified the name 'living rock'. As a result of political constraints, this source of supply is no longer available and most living rock now comes from the Caribbean or Singapore.

In view of its weight, living rock is shipped dry, i.e. not submerged in water but packed in sealed plastic bags that retain a high level of humidity. As much of the rock comes from the tideline, many of the animals and plants living upon it are well able to cope with a period of several hours out of water, but invariably some of the more delicate

organisms may succumb before reaching the dealer's tanks. Caribbean rock usually arrives in good condition, but it is always expensive and in limited supply. Rock from Singapore is cheaper but, because of the longer flight time, much of the life which makes it so appealing is either dead or dying.

Some authorities recommend maturing marine tanks with living rock, but generally speaking it is not a wise practice and newcomers to the hobby should never attempt it. The theory is that if you introduce approximately 0.45 kg (1 lb) of living rock per 4.5 litres (1 Imp./1.25 US gallons) of water to a newly set up aquarium, you provide the tank with a good initial 'inoculation' of nitrifying bacteria, plus a thriving ecosystem of 'lower' animals and plants. For the next two weeks, carry out regular tests for ammonia and nitrite and add more expensive livestock only when tests are clear.

This method can work well, but is more often a failure for a number of reasons. As we have seen, much of the life on the rock may be dead or dying when introduced to the tank and, as it decomposes, it will produce unacceptably high levels of ammonia and nitrite. Although many of the animals on the rock are fairly resistant, other species are unable to cope with such high levels of pollution, will die and further exacerbate the problem. In the worst cases, the tank's water may turn milky white and foul smelling, and a complete change of water and strip-down of the tank are the only remedy. In less extreme cases, the high levels of toxins can delay the maturation of the tank and so put back the day when you can begin stocking it. If you do give serious consideration to the living rock method of maturing a tank, make a point of smelling each piece of rock before buying it. Be sure to refuse rock that smells foul, or pieces that have a white bubbly coating.

Good-quality living rock does, however, have an important part to play in establishing a thriving living reef system. When introduced to a bacterially mature system, it will encourage a population of micro-organisms, sponges and small shrimps, etc., all of which play an important role in the ecology of the tank. Trying to produce a living reef that contains a good cross-section of invertebrate types without some living rock in the tank, is like gardening without earthworms, soil bacteria and beneficial fungi.

As a general rule, do not introduce more than 2.2-4.5kg (5-10lb) of living rock to the tank at any one time. When you buy the rock, it will be packed in much the same way as it was when originally imported, i.e. sealed in plastic bags. Most of the water will have drained out of the rock, so be sure to tumble the rock over in the tank to release any trapped air before placing it in its final position. Burrowing molluscs, sponges and many worms – the very animals you want in the tank and have paid for – will die if trapped in an air bubble. If they do, ammonia/nitrite pollution may result and the value of introducing the rock is diminished.

Remember, too, that all living rock has a top and a bottom, i.e. one area will have been exposed to light and another will have been shaded. If you put the algae-covered lit surface upside-down in the tank and expose the shade-loving animals to light, neither will survive.

Living rock is often graded to indicate quality and the profusion of life forms that may be expected. Owing to its very high cost, it is well worth seeking some confirmation as to what has been ordered – some low grades of living rock are no more than rocks that have spent time in the sea!

Always buy the best possible quality available and collect it from the aquatic dealer, unopened, as soon as it is delivered if at all possible. In this way, unnecessary changes of water are avoided and there will be no fear that animals will migrate into other rocks that

Below: *An air diffuser has been hidden in this sunken jug to provide a practical function and pleasing effect.*

you may not be buying.

In some parts of the world, living rock is actually cultured. Quantities of limestone rubble are positioned below the low water mark and, within three to ten years, these have been colonized and are ready for sale as living rock. As this process is repeated on a rotational basis, there is a constant and fully sustainable supply of living rock without harm to the environment in any way. Schemes such as this are to be applauded and encouraged in every way by the responsible hobbyist.

Unusual decorations
When you visit an aquatic store, you will see many decorative items, from plastic plants to shipwrecks, divers and bridges. While these may be practical or amusing in a freshwater context, they serve little or no purpose in the marine aquarium. However, provided that they prove non-toxic in saltwater, there is no harm in using them in any way you desire – just don't expect the scene to look *too* convincing!

Setting up the Aquarium

Siting and setting up the aquarium is important for practical reasons and, although it is an exciting task, it is one that should only have to be done once! Plan the procedure well in advance and make sure everything needed is close at hand. Electricity and water make a lethal combination, so if any doubt exists, consult a professional electrician.

Siting the tank

The aquarium should be sited in a prominent position that can be comfortably viewed when seated. Some direct sunlight – a maximum of two hours each day – would be beneficial, especially for a coral invertebrate aquarium, but certainly no more, if problems with unwanted algae and overheating are to be avoided. Keep the tank away from doors in constant use, as draughts and sudden loud noises are most undesirable.

Do not forget that aquariums are very heavy, even when empty. Enlist the help of as many people as are necessary to move the aquarium easily. The alternative could be a permanently injured back and a broken tank!

Choose a site for your tank that is near an electric power outlet and totally accessible, not only during the setting up stage, but also to allow for easy maintenance. Place a slab of expanded polystyrene on the stand so that the tank sits on a firm, level cushioned surface.

Clear away any obstacles in your path that might cause you to fall – especially when carrying buckets of water or valuable corals. Make sure that you have enough airline and electric cable, together with connecting blocks, air valves and electrical plugs before you start. A shelf below the tank will provide space on which to place some of the external equipment; you can always box this in afterwards.

New tanks should not leak, but if you have any doubts, test yours (outdoors!) before you use it. Leaks in used tanks can be simply repaired with aquarium silicone sealant; this is another outdoor job, as the sealant gives off strong fumes. Once any necessary repairs have 'cured', remove all traces of excess sealant and wash the tank out thoroughly with plain fresh water. You can also use this opportunity to clean all the glass panels.

At this stage you can attach a cable tidy and a bank of ganged air valves to a convenient spot on the outside of the tank or stand (but you will not be making use of them just yet). If the tank is to stand against a wall, this is the time to paint the outside rear glass panel or tape on background 'picturescape' paper so that the room decor does not show through and spoil an artistic underwater picture. Alternatively, if there is room behind the tank, build up a dry 'aquascape' or diorama of corals and other suitable decorations. Illuminating this with a spotlight will appear to extend the visible 'depth' of the aquarium when viewed from the front.

Fitting a biological filter

If you opt for one of the very sophisticated total system filters, you will find that comprehensive details for assembling the system are provided with the equipment. In this book we are assuming that the vast majority of aquarists intend using undergravel biological filtration with ancillary chemical/mechanical filtration.

Always fit the biological filtration system into the aquarium before you add the substrate. The filter plates must cover the whole of the aquarium base to provide the maximum area of filtration and to ensure that there are no anaerobic

Left: *Place the biological filter plate in the tank first. With large tanks you may need several separate plates or a number of small lock-together types.*

pockets. It is vital that the water flows evenly through the substrate. To ensure that there are no gaps around the edges of the filter plate, you can seal it in position with aquarium sealant. If you do this, allow a further 24 hours for the sealant to cure before continuing the setting up sequence.

Fit the airlift tubes and make sure that they are firmly attached. On some air-operated filters the airline should be connected right at the base of the airlift, so now is the time to fit it. Later on, the airline may well be under the surface of the substrate and much harder to get at. Connect the other end of the airline to an outlet on the ganged air valve block. Uplifts may need cutting when used in conjunction with powerheads so that the top of the water pump is roughly level with the water surface. Most powerheads are water cooled and, submerged, they avoid the risk of overheating.

If you are using powerheads do not fit these to the top of the return tube yet. Fitting a powerhead at this stage may inconvenience you and there is always a risk that it may topple from the unsupported tubes

and damage the tank. Similarly, if you are using a 'reverse-flow' system, connect the flow pipe from the external power filter to the top of the tube at a later stage. Although you may not be fitting these items together at this point, be sure to check that they *do* fit together; you may need to ask your dealer for different pipes.

Adding the substrate
With the filter plates in position, the next step is to cover them with a layer of coral gravel or similar coarse material designed for this purpose (cockle shell, dolomite gravel, etc.). Use approximately 4.5kg (10lb) per 0.1sqm (approx 1sqft). Rinse the material thoroughly before use to remove dust and other small particles and take this

Below left: *As an alternative to an air-operated uplift, a powerhead pump may be fitted. An adaptor is usually provided to fit any bore of uplift.*

Below right: *You can then add coral gravel. Cover this with a plastic mesh gravel tidy, carefully cutting it to shape, especially around the uplift.*

opportunity to check for metals and other foreign bodies.

Cover the coarse material with a perforated plastic mesh, known as a 'gravel tidy', and then add a further layer of coral sand, treated in the same way as detailed above, to a total substrate depth of about 7.5cm (3in). The mesh will help to prevent small grains of substrate from getting under the filter plate and possibly being sucked up into the powerhead, thus causing it to fail.

There is very little point in creating a filter bed more than about 7.5cm (3in) deep unless you plan to keep those few animals that habitually burrow deeper than this. The theory that very deep filter beds will benefit filtration has been largely discredited. We now know that most of the beneficial biological activity in the aquarium takes place only in the first few centimetres of substrate that the water enters. Nitrifying bacteria demand very high levels of oxygen and little is available in the water once it has passed through 7.5cm (3in) of substrate. In down-flow undergravel filtration systems, the top layers of substrate will be the most

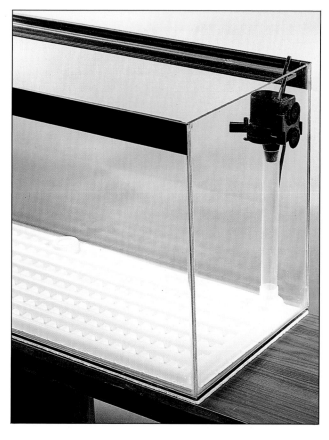

biologically active; in reverse-flow systems, the opposite applies, since the lower levels of substrate receive the well-oxygenated water first.

If you wish, you can vary the level of the substrate by sloping it from the back of the tank to the front.

Do not add the final decorative rocks until you have put all the 'technical equipment' into the tank.

Heating and lighting

At this stage you can install the heater-thermostat unit in the tank. Choose a position that is out of the way, but still allows you to see the neon 'working' indicator light clearly. Almost all models are fully submersible, but they are best not laid horizontally on the sand; mount them diagonally or almost vertically. Set the temperature at about 24°C (75°F) and then position a thermometer at the opposite end of the aquarium.

In view of the high water flow rates used in marine tanks, there is very little risk of hot spots developing. However, bear in mind that some invertebrates, particularly anemones, have a habit of moving slowly about the tank and, in the process, may adhere to a heater unit. When the heater next switches

Above left: *Once you have added the final layer of coral sand, the whole substrate depth should be about 7.5cm (3in). A deeper layer is unnecessary.*

Above right: *Bank up the substrate slightly to the rear to give a better effect and to encourage debris to fall forward. Fit the heater at this stage.*

itself on, the anemone will be unable to move away quickly and may be severely burned. Such burns usually prove fatal but can be easily avoided by the simple precaution of sliding a length of wide-bore, perforated, non-toxic plastic tubing over the heater. Alternatively, fit undertank heating mats to eliminate the problem altogether (see *Heating*, pages 74-75).

Next, install the lighting equipment. If you are using fluorescent tubes, make sure that there is no risk of the ballast units being splashed with sea water. The end caps that fit on the tubes are normally designed to be water-resistant, but do bear in mind that sea water is an extremely good conductor of electricity, so use splash trays or cover glasses.

Suspend mercury vapour and metal halide spotlights no less than

23cm (9in) above the tank. If you are using a supplementary actinic tube with these lights, house it within its own reflector; a length of half-round or 'U'-section white guttering fits the bill perfectly (but there are also special reflectors available now).

Foundation rockwork

Now bed the foundation stones of the background rockwork into the gravel. Given the chance, many invertebrates will burrow under stones, so it is much easier to anchor these securely when the tank is empty. Bear in mind the effect of refraction, which will make the tank look narrower once it is filled with water. When adding rocks to a dry tank, do not place them too near the front glass or it may look cramped once filled.

Filling the tank

Work out the volume of the tank and add enough of the dry sea salt mix to the tank to make up between 85 and 90 percent of the tank's volume. (To calculate the volume of an empty tank, multiply the length × width × depth – all in cm – and divide the result by 1000 to arrive at the volume in litres. Then multiply

Left: *Tufa rock can now be placed carefully in position to create a visually pleasing background as well as masking unsightly aquarium equipment.*

Below left: *Tufa need not be used to the exclusion of all other rocks. Here, lava rock has been added to a base of tufa to provide a mixture of textures.*

long it takes to fill a container of known volume and use this to calculate the volume of the tank. Thus, if a 22 litre (5 Imp./6 US gallon) bucket takes a minute to fill and the tank takes 20 minutes, then the tank must contain 20 × 22 = 440 litres (97 Imp./121 US gallons).

Switching on the power

Having wired all the electrical equipment into the relevant cable tidy facilities, switch on the electrical supply and check that the pumps are working correctly, that water is being circulated through the undergravel filter, that the heater has started to warm the water and that the lighting is operational.

After 24 hours, the tank should be up to its working temperature and all the salt will have dissolved, leaving the water crystal clear. Use a hydrometer to check the specific gravity; if the reading is too low, add a little more salt; if it is too high, remove some water from the tank and replace it with cold tap water. When the specific gravity is correct, you can begin to add any non-living decoration in the form of rocks, shells, etc. to the tank.

Completing the decoration

To achieve a reef-wall appearance, build up plenty of rockwork in a honeycomb pattern with a few jutting platforms. Cover the back of the tank to within about 10cm (4in) of the final water level. A wide variety of sessile (non-moving) invertebrates will make their home in the many crevices and by adding the rocks, the tank should be filled to the correct level – about 1.25cm (0.5in) from the top.

This type of rock arrangement is fine for corals, anemones, shrimps and, indeed, the vast majority of

by 0.22 for Imperial gallons or 0.27 for US gallons.) For example, if the tank is nominally 181 litres (40 Imp./50 US gallons) put in enough salt for 159 litres (35 Imp./44 US gallons). Allow the cold tap to run to waste for a few minutes and then fill the tank to within about 5cm (2in) of the final water level. Alternatively,

mix the salt and water in a non-metallic container, using a plastic or wooden spoon, before adding it to the tank. It is a good idea to fill the tank using a clean bucket of known volume so that you can establish the precise total capacity.

When filling large tanks with a running hosepipe, first time how

invertebrates but, as we have seen, it is impossible to satisfy all their requirements within one tank. A large lobster, for example, would probably knock over most of the rock and then hide behind the pile!

In a fish-only aquarium it is important to strike a balance between enough rockwork to be decorative and functional to the fishes' lives, while leaving plenty of space for swimming. Remember also, that rockwork displaces large quantities of useful water.

Maturing the filter

Once the rockwork is in place, the process of maturing the filter system can begin. The gradual development of bacterial activity in the gravel generally takes place before the addition of any livestock. Since all animals' waste products are poisonous to them, it is essential to produce a situation where the bacteria are 'ready and waiting' to go to work on these wastes before you add any valuable animals.

There are a number of ways of achieving this goal, the simplest of which is to use a maturing fluid or similar bacterial culture. Add these to the tank on a daily basis in accordance with the manufacturers' recommended dose until a nitrite test produces a reading of about 15 mg/litre (ppm). Discontinue the dosage, but carry on with daily tests until the nitrite reading is nil. This may take from two weeks to over a month, so patience will be needed. At this stage, there will be a moderate bacterial population and, providing the pH test kit and hydrometer both give satisfactory readings, the tank is ready to receive its first few animals.

Living bacterial cultures are now available that can speed up the initial maturation process considerably by innoculating the filter with huge quantities of beneficial *Nitrosomonas* and *Nitrobacter* bacteria, and are well worth considering.

Below: *A canister filter can be fitted at this stage. Is should be packed with high-quality marine-grade activated carbon and filter floss.*

Other methods of maturation may be suggested to you, but they are best avoided. The most common of these is to introduce a few 'hardy' fish – usually damsels – into the 'raw' aquarium. If they survive this very stressful procedure, it is likely that they will have formed such a strong territorial bond that any new inhabitants will be attacked and perhaps killed. This method risks the lives of perfectly good fish unnecessarily, and the practice is discouraged by caring aquarists.

The second method is to introduce a few handfuls of sand from an established tank to 'speed' the maturation process along. Unless the established tank can be guaranteed free from parasites, pests and unwanted algae spores, this could prove very undesirable.

Always remember that this initial maturation can only support a very few lifeforms and that it will take another 6-12 months before the aquarium can be regarded as 'established'. In the meantime, follow the guidelines to stocking levels detailed on page 72.

Installing ancillary filtration

Before adding any livestock, you should install ancillary filtration, i.e. the protein skimmer and external power filter. Place the protein skimmer in a rear corner of the tank, hidden behind rockwork. Here, any maintenance can be performed from above without disturbing the main body of the skimmer. Ensure that the slots in the top of the main column are at, or below, water level to allow water to enter the unit. When the two connections on the skimmer are attached to an air supply, foam is generated within the plastic column, while water is pumped back to the tank through the skimmer return pipe. If both air supplies are valved, you can regulate the foam so that it rises slowly until it is above water level. The foam thickens and then collapses as it falls into the

collection cup, leaving a brownish fluid that is discarded. You can drill a small hole into the cup, insert a length of thin plastic tube and direct this into a container that will only need emptying occasionally.

Position the external power filter either alongside or underneath the aquarium; most cabinet aquariums have base units specifically designed to house such a filter. Load the canister from the bottom upwards with ceramic rings or coarse mesh filter fibre, marine grade activitated carbon and, finally, a thin layer of filter wool. Place the inlet tubing, with its strainer, in the tank about 2.5cm (1in) above the sand. A quick suck on the outflow pipe will start water siphoning down into the canister, which will begin to fill up. Water then rises up the outflow pipe to the level of water in the tank. Shake the canister gently to remove

Above: *The fully established tropical marine aquarium soon becomes the ambition of most hobbyists. Brilliantly coloured fishes share the tank with the delicate hues of hard and soft corals, which, in turn, contrast beautifully with the various green and red algae. Such an aquarium will need both sound basic theory and experience to succeed. It is likely, and recommended, that filtration for such a set-up, having graduated from a basic undergravel system, will be of a more sophisticated (systemized) variety, while lighting will almost certainly be provided by metal halide spotlamps.*

trapped air and prevent cavitation (i.e. the formation of gas bubbles in the water flow). Switch on the power filter and water will be drawn through the filter and then fed back into the tank. The system is now operational . . .

PART THREE

MARINE FISH AND INVERTEBRATE CARE

Once suitable 'life-support' arrangements have been made, you can look forward to the next and, arguably, most enjoyable task: that of selecting and looking after suitable livestock. Although a pleasurable pastime, it is not always easy, and you will need to exercise great care, especially if you are just embarking on the hobby.

Buying on impulse is a common route to disaster, so always plan your purchases in advance. Try to see fish, in particular, feeding before you take them home and always make sure that all livestock is 100 percent healthy; decline offers of sick or ailing stock, which may be for sale at cheaper rates.

The key to long-term success is correct, regular maintenance. Neglect this at your – and your livestock's – peril! Maintenance should be seen as an enjoyable task and a chance to learn about and improve the aquarium environment. If you have a sensibly stocked aquarium with a good, well-balanced feeding regime, you should rarely encounter problems with diseases. It is wise to be prepared, though, and a basic range of medicines, as well as a quarantine tank, is strongly recommended.

With any luck, and having provided the optimum conditions, you may witness the greatest compliment the creatures can pay you. The excitement of seeing fishes spawn for the first time in your tank can be tremendous, and the fry of some species, such as anemonefishes, are not exceptionally difficult to raise. Coral may divide and multiply or even colonize spontaneously.

Remember that your charges rely entirely on you for their well-being and deserve the best possible care, and in return you will be rewarded with a colourful, fascinating and trouble-free display for many years.

Left: *The close relationship between clownfish and anemone is one of the most endearing scenes of the marine aquarium, but such a harmonious arrangement will not last long without your dedicated care and attention.*

Choosing and Introducing Fishes

The most vital choices of a practical nature that you have to make start here. Always allow the aquarium to mature before you introduce any livestock, although no doubt you will be keen to get going. Make haste slowly is the best advice; do not spoil your careful preparations by making an inconsidered choice of fishes. Always buy healthy stock from a reputable dealer and, using Part Four of this book (see pages 160-301) as a guide – do read it before you visit a dealer – choose fish that are suitable for the aquarium and compatible with other species. Remember to add new fishes to your collection gradually.

Introduce the fish into the tank with care; a frightened, shocked fish will only decline in health in the days and weeks that follow. Having taken all possible care to set up the tank and provide the ideal conditions, do not ruin your pain-staking preparations by choosing unsuitable fish to put in the aquarium. Several factors must influence your choice of fishes: their compatibility, both with fishes of the same species and members of other families; their feeding requirements; their tolerance of artificial surroundings; their appearance, and even their cost.

Buying healthy stock

Observe the fishes carefully in the dealer's tank before you buy them. They should not be thin, nor have a 'pinched-in' appearance. Freshwater fishkeepers are used to seeing their healthy fishes swim with erect fins, but remember that some marine fish naturally swim with their fins clamped shut, so this is not necessarily a sign of ill-health. Furthermore, some species of marines lack ventral, or pelvic, fins.

Since you are probably buying marine fishes for their dazzling colours, be sure to avoid any that look dull or have poorly defined markings for their species. It should

not be necessary to warn against buying fish with skin ulcerations, excessive skin mucus, protruding or clouded eyes. Look out for abnormal swimming or breathing actions; persistent scratching against rocks or corals; a failure to maintain a steady position in the water; a rapid respiration rate or a fish deliberately hanging in the bubbles from an airstone. Any of these signs may indicate sickness or distress.

One of the major hurdles faced by the marine fishkeeper is how to accustom new fishes to feeding under aquarium conditions. Always ensure that fishes are feeding before you buy them. It may not be enough simply to see a fish take food; it is not unknown for fishes to regurgitate hastily eaten food on the journey home and die because of raised ammonia levels in the bag. Ideally, try to see the fish feeding and then say that you will leave a deposit and return the next day or

Below: *Choosing livestock for a marine aquarium can be a pleasurable pastime, but needs to be done carefully.*

in a few days' time to pick up the fish. In the end, you must trust your dealer; good dealers will ensure that the fishes are feeding properly.

Before you buy fish for the first time, it may be a good idea to establish the specific gravity of the water in which the fish are currently held, so that you can prepare their future tank accordingly. Subsequently, unless you intend to buy all your future stock from the same dealer, you cannot be certain that the specific gravity will be the same as that in your aquarium once it is established. In most cases, however, any slight difference in specific gravity should not cause any problem, provided that you acclimatize new fishes.

Shop around before you buy any fish. This will help you to judge the quality of fish in different stores. A store with a large turnover of fishes may be doing fantastic business, or it may be merely replacing losses of fish it cannot keep. Beware of unrealistically low prices. These might suggest that the fish have not been quarantined or that they have

been collected by cheap and undesirable methods involving the use of cyanide or even explosives. Either way you risk buying unhealthy stock doomed to a short life in your aquarium.

Ideally, buy your aquarium and fishes from the same reputable dealer, who will then be familiar with your set-up, its operation and likely fish-holding capacity. Such dealers will not mind advising you when you come back for stocks of fish and will discuss any problems that may arise. If your fish come from several different dealers, it may be difficult for you to pinpoint the source of any trouble with accuracy, and unfair to expect one dealer from a number to accept responsibility for any problem you might encounter.

The journey home

Once you have bought them, the fishes begin the final leg of their long journey from coral reef to home aquarium. They have survived all the traumas of capture, air travel and several changes of aquarium conditions, so you must ensure that their introduction into your aquarium is as stress free as possible. (See page 146 for stress and ill-health.)

Always tell your dealer how long you will take to get home. If you have a journey of several hours ahead of you, your dealer can use oxygen instead of air in the plastic bag or, if oxygen is not available, provide you with a larger bag. The plastic bag in which the fish are being transported should be enclosed in a dark paper bag to reduce the light intensity for the fish. Even though you may be keen to inspect your purchase, do not take the plastic bag out in bright sunlight immediately outside the shop. This will inevitably stress the fish. The best option for the journey home is to use a darkened, well-insulated container. This will not only reduce stress from bright light, but also cut down heat loss.

Introducing the fishes

Always unpack fishes in dimly lit conditions (preferably with the tank lights switched off). Before releasing each fish, float the bag in the unlit aquarium for about 15 minutes. Then open the bag and add a small quantity of water from the aquarium. Wait for five minutes and repeat the process. Doing this four times over a 20 or so minute period will not only bring the temperature of the water inside the bag in line

with that in the tank, but will also gradually acclimatize the fish to any slight variations in pH and specific gravity levels. If necessary, use an airstone to provide gentle aeration in the plastic bag. Since the fishes may already have been confined for several hours, it is not advisable to prolong this transfer process beyond about 45 minutes.

If you quarantine new fishes in a separate aquarium, ensure that this is maintained at the same standard as the main tank. Of course, any fishes in quarantine tanks will need to be acclimatized to the main tank in the same gradual way.

Despite the best of intentions, it may be argued (with some degree of logic) that any extra transference of fishes from one situation to another causes stress, no matter how slight. The extra quarantining stage may be an example of this. You must balance up all the pros and cons as they exist in your particular situation. Careful handling of healthy stock from a reliable commercial source will generally prove to be successful, but if any doubts arise, then take any

Below: *Always choose healthy fish. Reject any with cloudy eyes, wounds, hollow bodies or swimming difficulties.*

Compatibility guide for tropical marine fishes and invertebrates

	1	2	3	4	5	6	7	8	9	10	11	12	13	14	15	16	17	18	19	20	21	22	23
1 Anemonefishes																							
2 Angelfishes																							
3 Batfishes																							
4 Blennies																							
5 Boxfishes/Trunkfishes																							
6 Butterfishes/Scats																							
7 Butterflyfishes																							
8 Cardinalfishes																							
9 Catfishes																							
10 Croakers/Drums																							
11 Damselfishes																							
12 Dottybacks																							
13 Dragonets/Mandarinfishes																							
14 Filefishes																							
15 Fingerfishes																							
16 Gobies																							
17 Grunts																							
18 Hawkfishes																							
19 Jawfishes																							
20 Lionfishes																							
21 Moray Eels																							
22 Pine-cone Fishes																							
23 Porcupinefishes																							
24 Puffers																							
25 Rabbitfishes																							
26 Sea Basses/Groupers																							
27 Seahorses/Pipefishes																							
28 Snappers																							
29 Squirrelfishes																							
30 Surgeons/Tangs																							
31 Sweetlips																							
32 Triggerfishes																							
33 Wormfishes																							
34 Wrasses																							
35 Invertebrates																							

Legend:

- Compatible
- Incompatible
- With some caution
- Best kept in a species tank
- Young specimens best kept in brackish water

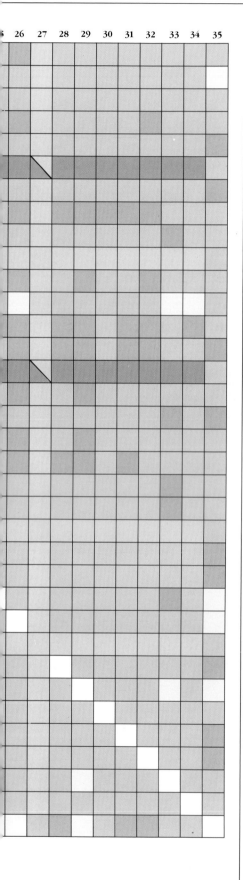

necessary precautions to protect your existing aquarium stock from health risks likely to be introduced by new additions. Quarantining is fine in principle – everyone agrees that is the thing to do – but in practice, things are often different, especially where impatience tempers even the highest ideals!

Do not attempt to feed newly introduced fish on the same day, as they will need time to settle in and food will, more than likely, be ignored. The day after will be soon enough, when a light feed should be offered. Needless to say, all uneaten food should be removed to prevent pollution in the tank.

When adding new fishes to an established aquarium, you can reduce the likelihood of territorial squabbles by employing some 'fish psychology'. Feed the inhabitants of the main aquarium to take their minds off the newcomers, or even consider rearranging the rockwork layout in the aquarium immediately before introducing new fishes so that all the fishes, old and new, are more preoccupied with home-finding than with quarrelling.

Compatibility

The aggressive behaviour of some fishes may be suppressed in a crowded display tank. Only when a fish has room to 'flex its muscles' in the comparative spaciousness of a home aquarium does it reveal its true nature. This is another good reason for providing enough swimming space and a choice of refuges for all the fishes in the tank.

The chart opposite gives an indication of the compatibility between each family of fishes featured in this book. Using the table, the advice and the guide to feeding requirements (see page 130), you can select those fishes that will suit your needs. A 'fish plan' is essential. Purchasing in an ad hoc manner usually leads to disaster and a tank full of incompatible fish!

Always introduce the least territorial species first and the most territorial last. That way more sensitive species should avoid a battering by established tankmates. (This is an admirable case for *not* introducing aggressively territorial damselfish as first fish – no matter how hardy they may be!)

Lifespans

Our knowledge of marine fishkeeping is increasing all the time, but it is still difficult to give any accurate guidance to the projected lifespan of fishes kept in aquarium conditions. Nor has there been much research on the lifespan of species in the wild. Usually, lifespan is proportionate to size and the figures given below for lifespans in the home aquarium can only be approximate in the context of an aquarium environment under the very best of conditions. The lifespan of fishes in conditions approximating more nearly to their natural environment, i.e. those in very large public aquariums, is understandably longer.

Lifespans

The following is an *approximate* guide *only* to the lifespan of tropical marine fishes in home and (public) aquariums.

Anemonefishes and damselfishes	12+ (20+) years
Angelfishes	12+ (28+) years
Batfishes, groupers and lionfishes	12+ (15+) years
Butterflyfishes	12+ (28+) years
Pipefishes and seahorses	1-3 (1-5) years
Puffers	5-8 (11-15) years
Snappers	5-8 (12+) years
Surgeonfishes	10+ (10-17) years
Triggerfishes	10+ (20+) years

Choosing and Maintaining Invertebrates

All animals and plants are subject to a range of diseases, be they induced by chemicals, bacteria, viruses or parasites. Unfortunately, because most invertebrates are of very little commercial value compared with other animals, and because such a huge range of animals is involved, almost nothing has been discovered about diseases of invertebrates (see also pages 158-159). Some good work has been carried out in the United States, where bivalve molluscs and crustaceans are farmed for the food market. Although several causative agents have been described, until recently there was

little progress in the development of remedies suitable for the aquarium. Indeed, the commonest response to invertebrate disease was to destroy afflicted stock, sterilize the tank and restock. Now, the situation is beginning to improve. With increasing interest in marine fishkeeping generally, new products are beginning to appear on the market that do not contain

Below: *In order to achieve an impressive marine invertebrate display, such as this, it is vitally important that you choose only specimens that are 100 percent healthy.*

chemicals that will disrupt a biological filter system and are safe and effective in an invertebrate aquarium when used at the recommended strength.

Remember that the invertebrate keeper does have one advantage over the fishkeeper; not being involved in a monoculture (i.e. keeping just one species, genus or family), the risk of losing entire stocks of animals to disease is negligible. Given good environmental conditions and an adequate source of food, vitamins and trace elements, most species appear very resistant to disease.

Many of the problems associated with invertebrates are either environmental or the result of incompatability with tankmates. With care and consideration on the part of the aquarist, many of these can be resolved quite easily. However, bacterial infections are a different matter. The majority of aquarium bactericides available commercially are primarily designed for treating freshwater or marine fishes. Unless you are sure that they are safe in an invertebrate aquarium, refrain from using any chemical medication because the potential risks are usually far greater than the possible advantages. However, the appropriate use of UV sterilization and ozone as methods of reducing the background bacterial count can prove very valuable, with a UV system being preferable if a choice must be made between the two.

It has often been said that prevention is better than cure, and it is certainly true that you can do a great deal to reduce the chances of disease. Start by selecting only good-quality specimens, with no signs of physical damage that could lead to bacterial infections or, in the case of starfish, for example, to the rapid and complete collapse of the animal.

Owing to their sensitive nature, invertebrates need special attention from the dealer. In general, invertebrate tanks should be well-lit and sessile invertebrates arranged so that they are not touching. Tanks should be free from 'nuisance algae' and 'plague flatworms' (see pages 91-93 and 383); if they are not, do not buy, otherwise you will invest in a great deal of potential trouble. Look at the general condition of all the invertebrates. If there is a preponderance of dead or dying stock, then find another dealer. Above all, try to know what you are looking for; it is all too easy to purchase a deteriorating specimen if you do not know what a healthy one looks like!

Invertebrates should be introduced into the aquarium in much the same way as fish, but take a little longer so as to avoid osmotic shock caused by any differences in specific gravity. This can be crucial to long-term survival, as delicate tissues must not be damaged. Float the bag in the aquarium for 15 minutes to equalize temperatures. Open the bag and add a small quantity of water from the aquarium. Repeat this every five minutes for the next 30-40 minutes. Rushing this process will cause the rupture of vital tissues and lead to deterioration and, finally, death.

The following guidelines will show you what to look for when selecting and maintaining stock. (See also pages 158-159.)

Below: Goniopora *is very sensitive; choose specimens carefully. It should have a full complement of light brown polyps. Avoid animals that are shedding polyps or have a mucous coating.*

Sponges
Ensure that the base is attached to a small piece of rock and not buried in the substrate. There should be no discoloured patches and no overgrowth of film or hair algae.

Maintenance Many aquarists encounter problems with sponges, yet these are intrinsically among the most resilient of animals. Much of the difficulty lies in not appreciating the natural lifestyle of these animals. They are normally found in dimly lit areas, where they do not become coated with algae, and where there is a moderately strong water flow to bring them oxygen and food and take away waste products. Few species can withstand exposure to the air, as this is very easily trapped within the body, thus preventing

various metabolic processes from taking place.

Sponges are food for many animals and they are easily damaged by contact with the stinging cells of cnidarians. Damaged areas are very pale in colour and commonly occur at the tips of branches and at the base, where the animal has been removed from the substrate. Place sponges on rocks – not on the sand substrate of the aquarium – to allow good water movement around them. In good conditions, sponges grow rapidly, anchoring themselves in position in the process.

Cnidarians

Anemones should have expanded tentacles and the central mouth should be closed. A healthy anemone will show a reflex reaction when disturbed, either by withdrawing its tentacles and rapidly shrinking in size, or by becoming very tense and firm. In *Cerianthus* species, the anemone should withdraw into its tube. Look for evidence of damage. Animals may have heater burns, or the flesh – particularly the foot – may have been torn during capture.

Polyp colonies and mushroom polyps should be expanded, but should contract or stiffen when disturbed. There should be no visible damage, or any smell of decomposition when the colony is brought to the surface of the water.

Corals should be firm and plump and intact. Pay particular attention to the base; leather corals are often damaged during collection and rapid decomposition spreads quickly, often destroying the animal within a few days. *Sarcophyton* species regularly shed a skinlike mucus, but this is of no concern.

When selecting soft corals, the main precaution is to check for areas of almost powdery decomposition. Regularly siphon away any debris that collects in the dishlike structure of many species.

The thin, colourful flesh that covers the horny skeletons of sea whips and sea fans is easily damaged. In perfect conditions, these lesions may heal and the

animal recover but, more commonly, the bald areas spread and the animal eventually dies. The most common points of damage are the extreme tips of the branches and the base, where the animal was fixed to the seabed. The best specimens are those that remain attached to a small piece of rock.

Stony corals, such as *Goniopora* and *Euphyllia* species, should be expanded and erect. The coral head, as it is called, should have no bald patches, or areas of grey, bubbly decomposition.

Maintenance Although an ailing specimen will generally die quite rapidly, healthy anemones are some of the hardiest invertebrates. However, air or gases resulting from decomposition show up as pale areas within the body which, because of the buoyancy involved, seems to stretch upwards. Remove any anemone with these symptoms from the tank, as it will decompose very rapidly and can cause a major pollution problem. From time to time, most anemones will contract quite dramatically to void the contaminated fluid that accumulates within the body. A string of thick brown mucus may also appear. This is quite normal and the animal should reinflate within a few hours.

Many aquarists do not appreciate that interspecific fighting among corals and anemones is a common cause of death among these animals.

To allow for this, leave a gap of at least the diameter of a given coral between that one and the next. This also allows for the fact that many corals and anemones do not expand to their fullest extent until after dark. Anemones and corals may also shed undischarged nematocysts and can therefore sting each other at a distance. An efficient protein skimmer or power filter will reduce this risk, but it is not a good idea to keep anemones of different types within the same tank.

With very few exceptions, all corals and anemones appreciate a strong flow of water. If the flow is too slow, a 'skin' of almost still water forms around the animal, which is then unable either to catch food or rid itself of its own waste products.

Stony corals require very good water and lighting conditions. Provide a high pH level (but not exceeding 8.4), and a good reserve of calcium in the water. When lighting the aquarium, bear in mind that corals and anemones can be loosely divided into two groups. Those within the colour range of purple through red to orange and yellow are lacking in zooxanthellae and generally do not require intense lighting. In contrast, those with beige, brown, green or blue

Below: *Flatworms are colourful and very appealing, but they are notoriously difficult to maintain because of the specialized diets they require.*

colouring require strong lighting of the correct spectrum if the algae are to function correctly. Many of the problems encountered with these species can be traced back to poor environmental conditions. The corals fail to expand and feed, and the flesh gradually shrinks until only the rocky skeleton remains.

Another common problem, particularly with *Goniopora* species, is physical damage leading to bacterial infection. If their soft tissue is punctured, possibly by a tumble within the tank, the action of predators or through inadvertent contact with sharp-footed tankmates, then bacterial infection often follows. This commonly takes the form of a greyish brown disc of slime that rapidly enlarges, while the remaining flesh lifts away from the skeleton. Gently siphoning the decomposing tissue away may slow the process, but the long-term prognosis is generally poor.

Corals should react in a similar way to sea anemones when handled. Do not remove them from the water unless absolutely necessary, and never expose them while expanded, as the weight of water inside them will almost invariably cause the delicate flesh to tear and allow bacterial infections to set in.

Platyhelminthes
Flatworms should be active, with no tears in the body, and should seek shelter when disturbed.

Maintenance Do not house these delicate animals with potential predators, such as crabs, lobsters and large shrimps. Very little is known about the diet of these creatures, but they may accept finely chopped shellfish and similarly sized particles. Scavenging species may benefit from a good background growth of algae.

Annelids
The main species of interest to aquarists are the various tubeworms that live within parchment-like or stony tubes. These should show evenly shaped heads of feathered tentacles and the worms should

withdraw rapidly into their tubes when disturbed. Avoid specimens that hang limply from their tubes or do not retract when touched.

Maintenance These animals do not demand the bright lighting required by many other invertebrate species, but they do enjoy a moderate to good flow of water. In the wild, they are efficient filters of both zooplankton and phytoplankton (microscopic algae). In the aquarium, this dietary requirement is easily met with liquid foods.

Crustaceans
As a general guide to the health of crustaceans, check the mouthparts. All crustaceans have a complicated system of mouthparts that are

Above: *The crowns of these serpulids may reach 4cm (1.6in) in diameter. Check that they retract when disturbed and that the tube is undamaged.*

forever in motion. If there is no sign of movement around the mouth, and the animal does not respond to disturbance within the tank, then it is ailing. Crabs, shrimps and lobsters normally make determined efforts to avoid capture – hermit crabs should retract rapidly into their shells. Do not buy lethargic animals.

Crustaceans should have their full complement of limbs, but provided

Below: *Bear in mind when choosing invertebrates that many species, like this Red Reef Lobster, are nocturnal and so rarely seen during the day.*

the animal is not subject to predation and can still feed, the loss of one or more legs or claws is neither unusual nor cause for concern; they will be replaced at the next change of exoskeleton. Deformed or broken antennae are of little significance. Check shrimps and crabs for fungal growths.

Maintenance If a shrimp or crab is injured in the tank to the extent that it may be at risk from one of its tankmates, isolate it in a perforated plastic container within the tank (floating breeding traps are ideal) until the missing limbs are replaced. Keep a check on the pH level of the water; if it is too low, the new shell may not harden properly and the animal will become deformed.

The internal fungal infections, to which shrimps and crabs are susceptible, typically show up as white patches within the body or as discolorations on the surface. Some crabs suffer from a fungus that invades the entire body and is visible externally as a tufty growth on the underside of the carapace (the main 'body shell'). At the moment, there is no readily available medication for this problem. Generally, it is best to remove infected animals in the hope of preventing transmission to related species.

Molluscs

In clams, scallops and mussels, the mantle should be expanded and the breathing siphons extended. These bivalves should close quickly if disturbed; clams with mantles falling inwards, away from the shell, are dead or dying. Univalves (eg cowries) should be active and, where applicable, the gill tufts on the dorsal surface should withdraw or contract if the animal is touched. Check the shells of bivalves and univalves for major physical damage.

The shell-less snails, or sea slugs, should be active and, where applicable, the gill tufts on the dorsal surface should withdraw or contract if the animal is touched.

Octopus, squid, and Nautilus, etc., all belong to the class Cephalopoda. They are mobile animals that should actively try to evade capture. Tentacles should be firm and mobile, with no major physical unhealed damage. The complete loss of an arm, provided the site has healed, is not critical.

Maintenance Most of the problems encountered with molluscs tend to be associated with their environment and, of course, many species are predated upon by other tank inhabitants. Most molluscs are continually extracting calcium from the water to enlarge their shells and any shortfall in this essential mineral can cause severe problems, particularly among the bivalves.

Octopi and their close relatives

Below: *Unfortunately, the colourful and desirable Spanish Dancer is difficult to maintain, as there is still much to learn about its dietary needs.*

require perfect water conditions and very slow, careful acclimatization to a new environment, preferably in semi-darkness. They are also very sensitive, nervous animals and losses can often be ascribed to shock. When disturbed, most species are able to eject sepia, an inklike substance that provides a smokescreen to cover their escape from predators in the wild. In captivity, 'inking' can cause a pollution problem, particularly in a small aquarium.

Echinoderms

Echinoderms generally deteriorate very rapidly if infected or damaged, so you can be fairly confident of selecting a healthy specimen.

Starfishes should have plump bodies and be firm to the touch, with their tube feet expanded. Any flaccidity in the arms may be a sign of failure of the internal water vascular system, which is almost invariably fatal. Introducing specimens too rapidly into a new tank may also produce this symptom, with a similar end result. Damaged arms often show white eruptions; reject specimens in this condition. It is worth checking the underside of the arms for damage and parasitic molluscs. Healed scars can be ignored because, in the wild, starfishes show remarkable regenerative abilities, but these are largely lost in captivity. Obvious mechanical damage is clearly a serious disadvantage.

Sea urchins are quite easy to choose. The spines – and the tube feet among the spines – should be erect and mobile; indeed many species will actively point their spines towards questing fingers. If the spines are depressed, giving a 'thatched' appearance, the urchin is almost certain dying. You may notice small bald patches of missing spines, but as long as the tube feet are still active, the animal will generally regrow new spines and recover. If both spines and tube feet are absent, reject the urchin.

Sea cucumbers should be plump and, if the feeding tentacles are expanded, these should be largely

intact and should retract rapidly when the animal is disturbed.

Crinoids, or feather stars, are very brittle animals and easily damaged. The loss of one or two arms may not be too harmful; in good conditions, they will slowly regrow, but any further damage is not acceptable. Check that the small, clawlike, gripping appendages are present on the underside of the animal.

Maintenance Some species of sea urchin seem particularly prone to getting air trapped within the body, which invariably proves fatal. In view of this, never remove urchins from the water when transferring them from tank to tank. Place them in a small, clean, plastic tub or thick plastic bag and allow them to become slowly acclimatized to the conditions in the new tank. All echinoderms require this period of gradual readjustment.

Sea cucumbers are very resilient animals, but one fairly common problem is damage caused by heat burns. These very slow-moving animals occasionally settle on heater units and when the thermostat

Above: *Starfishes should have all their arms intact and be firm to the touch. Introduce them slowly to a new tank.*

switches on, the cucumber is rapidly and severely burnt. Avoid this risk by hiding the heater behind rockwork, protecting it with a perforated plastic tube or using an outside heat source. The feathery tentacles of the more commonly kept species are often targets for predators. As the tentacles are used to trap food, their loss is obviously serious, but they usually regenerate over a period of several weeks, provided the predator is removed.

If you see threads protruding from the anus of a sea cucumber, the animal has probably been incorrectly acclimatized and has voided part, if not all, of its intestines. In the wild, this deters predators and the cucumber normally regenerates its internal organs. In captivity, such evisceration is usually followed by death.

The commonest problem in crinoids is major limb loss, with numerous arms being broken. The arms are very flexible in a vertical

plane, but have very little lateral flexibility. If you place the animal in a fierce current, such as is produced by some of the submersible powerhead water pumps, many arms may snap off in a short period. Predatory or heavy, clumsy animals, such as cowries, can cause severe damage to brittle crinoids.

Chordates
While most commonly introduced by accident to the tank, a few of the larger, more colourful species of sea squirts occasionally appear on the market. Their breathing and feeding siphons should be open, but should quickly close when the animal is disturbed. Check the base and sides carefully for tears.

Maintenance These animals require the same care in captivity as featherduster worms (see page 123).

Living rock
Although we have covered living rock in greater detail elsewhere in this book (see pages 106-107), it is worth outlining the best method of purchasing this particulary useful and attractive aquarium addition.

There are three grades of living rock, decreasing in quality and cost. Grade 1 should carry an abundance of life, whereas Grade 3 can be no more than rocks that have been in the sea! Always buy the best grade you can afford and pool a larger order with friends for the most advantageous price.

A reliable source is essential and a guarantee of where the rock originates from should be given. Caribbean sources are generally the best for UK aquarists as they have a shorter journey time, when compared to some Pacific locations. For the best possible success make arrangements with your dealer to collect the rock on the delivery day still bagged – there is no point in exposing the animals to more stress than necessary, nor in giving them time to migrate into rocks that you may not be buying!

Introduce all living rock into a pre-matured aquarium by the normal routine as outlined above.

Feeding in the Aquarium

Providing the correct food for your fishes is one thing, getting them to accept it in captivity may prove a little more difficult. Try to present the food to them in a manner that is as natural as possible; carnivorous fishes may begin with livefoods then be gradually tempted to eat dead food. Supply greenfoods for herbivorous species and be prepared to keep a supply of cultured livefoods for finicky feeders. Study your fishes' feeding actions and habits and don't neglect the nocturnal feeders. It is a good idea to use feeding time as an opportunity to count the inmates of your tank. The marine fishkeeper faces many challenges in the quest to maintain an aquarium of healthy contented fishes. Second only to the problem of controlling water quality is the question of how to feed the fishes. This means not only providing suitable foods, but also, in some cases, teaching or persuading the fish to recognize and take them.

In nature, every animal is part of a food chain, living on, or being eaten by, the animals adjacent to it in the chain. Many marine fishes, for example, eat polyps from the living coral heads, take crustaceans, and scrape algae from surfaces of rocks. Small fishes are, in turn, the prey of larger fishes. This food chain is fine in the open ocean, where there is a continuous supply to keep it going. In the aquarium, the fishkeeper must provide suitable substitutes for the fishes and other tank occupants.

Marine fishes can be grouped into four feeding categories, depending upon their requirements and how they feed. These groups can be summarized as open-water feeders, reef grazers, shy feeders and specialist feeders. Open-water feeders feed confidently, taking food

Feeding types (in nature)

Open-water feeders
Blennies, clownfishes, damselfishes, fingerfishes, some gobies, lionfishes, rabbitfishes, snappers, squirrelfishes, triggerfishes.

Reef grazers
Angelfishes, butterflyfishes, filefishes, parrotfishes, surgeons, wrasses.

Shy feeders
Basslets, cardinalfishes, hawkfishes, jawfishes.

Specialist feeders
Pipefishes, seahorses.

in midwater or scouring the seabed debris. Reef grazers confine their feeding attentions to polyp and algal life on the coral reef and associated rocks. Shy feeders may hide in caves or burrow entrances waiting for feeding opportunities to present themselves. Another reason for feeding shyness can be normal nocturnal habits. Such species do not feel confident enough to compete for food in the brightly lit aquarium. Specialist feeders, such as seahorses and some pipefishes, require food especially selected for them, mainly tiny live foods.

In the relative emptiness of the coral reef, each fish can feed in its own way, occupying its natural place in the food chain. In the aquarium, the fishes are forced to rub shoulders with each other; shy, hesitant feeders must take their

Right: *Angelfishes, such as this Blue-faced Angelfish, have very strong mouthparts, which are ideal for grazing on encrusted sponges and algae.*

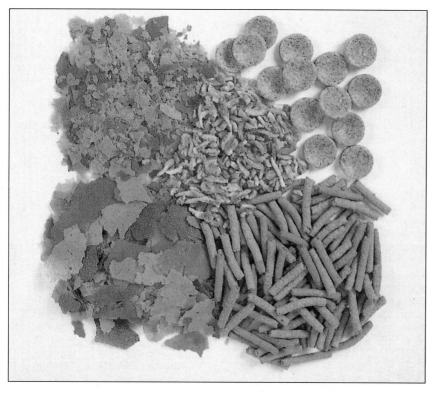

Above: *Dried foods will help provide a balanced diet, especially when used in conjunction with live and frozen foods.*

Clockwise from top left: small flakes, tablet foods, floating foodsticks, large flakes and freeze-dried krill.

chances alongside the more boisterous fish. All these factors make up a complicated feeding puzzle for the fishkeeper to unravel.

Sources of food

Here we examine the advantages and disadvantages of a variety of foods available for marine fishes.

Prepared foods Many freshwater flake fish foods have been modified by their manufacturers to suit marine fishes. Research into foods specially formulated for marine fishes continues unabated and will result in a continually wider range of acceptable foods that will simplify the job of the marine fishkeeper.

Prepared foods can be used in different ways. Flaked foods, marketed as 'marine staple diet' are easy to use. Large fish will accept them as they are, or they can be crumbled up for smaller species. Marine tablet foods can be offered in a variety of ways depending on the feeding manner of the recipient. Sometimes they are snapped up as they 'free-fall' through the water, or

else they may be stuck to the aquarium glass to be pecked by grazing species.

Fresh meat foods As the staple diet of fishes in nature comes from the sea, it should be possible to buy supplies from the fishmonger for your fish. Shellfish meat, such as mussels and crab meat, etc. are usually accepted by bold feeders, although there is always a strong risk of introducing disease from these sources and also fouling the water. The main advantage of feeding 'like to like' is that the natural foods will contain valuable trace elements. Specially formulated prepared foods and gamma-irradiated foods are safer.

Frozen foods Many types of frozen foods are produced commercially specifically for marine fishkeepers and these provide the bulk of the diet for many marine fishes. These foods will keep for several months in a suitable freezer and 'meal-size' portions can be thawed out as required. Ideal frozen foods for

marine fishes include brine shrimps, *Mysis* and other shrimps, krill, squid, clam, crab and lobster eggs. The commercial suppliers of frozen food for aquarium use treat it with gamma rays to ensure that it is free from disease organisms. Use the wide range of frozen foods with complete confidence.

Freeze-dried foods The freeze-drying process has made it possible to preserve many natural foods for fish in captivity. Small aquatic animals, such as brine shrimps and krill (marine crustaceans), are available in freeze-dried form, and provide an excellent alternative to flaked food. They help to vary the diet and prevent the fishes from becoming bored with a monotonous feeding regime.

Live foods Brine shrimps (*Artemia* spp) are an excellent disease-free and nutritious living food for marine fishes. The eggs of these saltwater shrimps are readily available from aquatic dealers and are easy to hatch in a well-aerated saline solution. The newly hatched nauplii stages are ideal for filter-feeding invertebrates and are relished by butterflyfishes, seahorses and pipefishes in particular. Both newly hatched brine shrimps and the larger adult forms provide food for butterflyfishes, but obviously there is more nutritional value in the larger forms. Be sure to organize a continuous supply, or else wean such fishes off brine shrimps and onto a more practicable diet. Raising the tiny brine shrimps with yeast-based foods over several weeks will produce a stock of fully grown shrimps suitable for larger fishes. Live *Mysis* shrimp, often available from aquatic stores, provide yet another tasty snack. You can also culture marine rotifers. These are tiny planktonic animals, half to one-third the size of newly hatched brine shrimps, whose dormant eggs can be hatched and raised in a micro-algae liquid to provide food for the larval stages of fishes.

Live freshwater fry and adults, particulary guppies, mollies and

goldfish, are used by some hobbyists keeping such fish as Lionfish to the exclusion of all other foods. The mistaken belief is that Lionfish will *only* accept live foods. This is not true; in reality the fish has become 'fixed' on live fish. With patience, 99 percent can be weaned on to dead foods such as frozen lancefish. There are three good reasons for avoiding the use of feeder fish: it is not a 100 percent reliable source; diseases may easily be passed on; the vast majority of people have come to realize that this is an unaccceptable and unnecessary practice.

Fresh vegetable foods It is essential to provide vegetable materials in the marine diet, since many marine fishes are naturally herbivorous. Algae provide the most natural form of vegetable food for many marine fishes. You can grow algae in the main aquarium or, should the number of herbivorous fishes exceed the supply, you can set up a spare tank as an 'algal farm'. Simply furnish the tank with a few rocks, fill it with sea water and stand it in a sunny position. You can scrape algae from the rocks, or transfer the entire algae-covered rocks in to the main aquarium for the benefit of grazing fishes. Lettuce and spinach leaves are effective sources of vegetable matter but should be blanched before being offered to break down any cellulose content, which would otherwise make them indigestible.

A balanced diet
Understandably, most of the research on fish nutrition has been centred upon species raised in captivity for conservation purposes

Above: *A selection of natural fish foods (shown here unfrozen) that are suitable for use in the marine aquarium. From left to right: cockles, lancefish, large shrimp,* Mysis *shrimp, lobster eggs.*

or for food – trout bred on fish farms, for example. It seems logical, however, that these findings also apply to other fishes maintained in captivity, including ornamental marine fishes.

Such research has shown how vital it is to provide a diet containing the right balance of carbohydrates, proteins and fats to provide energy and build body tissue. This has led

to the production of high-quality prepared foods for a wide range of fishes, as we have seen, and should underly all our decisions about suitable diets. It is clear, for example, that fishes need a ready supply of so-called 'essential' amino acids (the building blocks of proteins) for normal growth and healthy development. These requirements are species dependant and vary with the age of individual fishes. Deficiency in any particular essential amino acid produces characteristic symptoms, ranging from a reduced growth rate to death. Although it is possible to add these amino acids to the diet in a 'free' form, some fishes cannot make use of them in this form; hence the need to ensure that all species receive a balanced and varied diet.

Nutritional research also shows that fishes need highly unsaturated fatty acids for good health. One vital reason for this is that such fats enable the delicate membranes throughout the body of a fish to remain flexible and fluid at relatively low temperatures. It is particularly important to protect fats in fish foods from becoming rancid, i.e. oxidized, since rancid fats in the diet can cause specific health problems in fishes.

It is also important to remember the vital role that minerals and vitamins play in the diet. Although needed in only very small amounts, vitamins may act as 'catalysts' to activate the nutritional processes. Some vitamins are manufactured in the fish's body; others must be

Left: *Live brineshrimp – a nutritious and disease-free food for marine fishes – are easy to hatch in the aquarium using this device and an air pump.*

Setting up an algae farm

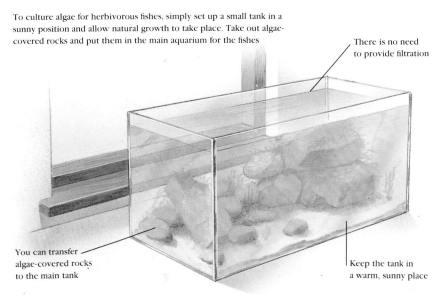

To culture algae for herbivorous fishes, simply set up a small tank in a sunny position and allow natural growth to take place. Take out algae-covered rocks and put them in the main aquarium for the fishes

There is no need to provide filtration

You can transfer algae-covered rocks to the main tank

Keep the tank in a warm, sunny place

Feeding methods

The well-tried formula of feeding 'a little and often' applies equally well to marine fishes, with one very important proviso: to prevent pollution of the aquarium, always clear away any remains of food once the fish have lost interest in it. (A check on the nitrite level will indicate any evidence of overfeeding, or perhaps the presence of a decaying fish body, or an inefficient biological filter.)

Bear in mind that merely casting flakes on the water surface will not satisfy every fish in the aquarium. Some fishes are not physically adapted for surface feeding, and others may not care to feed in the bright glare of a fully lit aquarium.

constantly available in the food. No single food provides all the essential vitamins, but all foods contain some vitamins. Thus, part of the bonus of providing a varied and balanced diet is that the fish receives all the vitamins from a variety of sources.

Vitamin groups may be fat soluble (A, D, E and K) or water soluble (B and C). Vitamin B is a collective name for a group of vitamins known by individual names and/or numbers, such as B_1 (Thiamine), B_2 (Riboflavin), B_6 and B_{12}.

Below: *Tangs and many angelfishes appreciate a green supplement, such as blanched lettuce or spinach. Trap the leaf between two halves of an algae magnet and let it sink to the bottom. This will stop it floating away and allow chunks to be easily torn off.*

Right: *With patience nearly all fish that normally prey on livefoods (this is a Fu Man Chu Lionfish) can be taught to accept 'dead' foods. Many of the larger fishes will become tame and take food from your hand, but as many have very strong jaws, sharp teeth or venomous spines, it is safer to offer food on a stick or using tongs, as here.*

Grazing fishes can be accustomed to the taste of dried foods by painting a liquidized emulsion of meat and algae foods on to a rock; when dry, place the rock into the tank for browsing fish to peck at.

Fishes that normally prey on livefoods can be taught to accept dead food by attaching it to a short thread and jiggling it in front of their noses until they learn to take it. Lionfish are a good example of this.

Observe the habits of the fishes and note your own successes and failures when feeding them. By collating the results in a notebook (see *Regular Maintenance*, page 136), you should be able to gradually work out a feeding routine that suits both you and the fish. Some of the larger fish become hand tame after a time, but it is still a wise precaution to impale food on a cocktail stick before offering it, in order to avoid receiving a painful bite from sharp teeth.

Feeding fish during holiday periods may cause some concern, but it is not a difficult problem to overcome. You cannot leave marine fishes to their own devices for as long as freshwater species, so the best arangement is probably for someone to feed the fish in your absence. A fellow hobbyist or reliable neighbour may be willing to help out, but to avoid any risk of overfeeding, make up small packets containing the exact amount of food for each meal, stipulating that on no account should additional supplies be given. This is much safer than allowing a well-intentioned but inexperienced enthusiast to look after your fish, since they always overestimate the amount of food required. It is better to come home to a tank of hungry living fishes than to a tank of overfed dead ones.

Ease of feeding in the aquarium

Group	Easy	Reasonable once feeding	Difficult	Difficult, need live foods
Angelfishes		●	●	
Basses	●	●		
Blennies	●	●		
Butterflyfishes (depending on species	●	●	●	●
Cardinals		●		
Catfishes	●			
Clownfishes	●	●		
Damselfishes	●			
Eels	●			
Filefishes		●	●	
Gobies	●			
Hawkfishes	●	●		
Jawfishes		●		
Lionfishes		●		●
Seahorses			●	●
Squirrelfishes		●		●
Tangs	●			
Triggerfishes	●	●		
Wrasses	●			

Feeding Invertebrates

The whole question of what, when and how to feed invertebrates causes confusion among beginners and experienced hobbyists alike. A number of useful 'invertebrate foods' are available commercially, both frozen and in liquid form. Certain brands are very good, and the manufacturers are continually improving their products, but none should be considered as a complete diet for all invertebrates. Their main value is as a substitute planktonic diet for filter feeders, such as corals, featherduster worms, some molluscs and sea cucumbers. You have only to consider the varying roles that invertebrates fill in the ecology of the seas to realize that no single type of food can meet the dietary requirements of all species.

Invertebrates appear at every level of the classic food pyramid, from primary food organisms in the form of corals, micro-crustaceans, etc., to top predators such as octopi, squid and some large crustaceans. Clearly, a suspension of fine particulate food will not satisfy an octopus, while the frozen shrimps that they appreciate are way beyond the capabilities of some of the attractive filter-feeding worms that extract very fine particles of food from the water around them.

WARNING: Overfeeding, especially with liquid foods, is one of the most common causes of invertebrate mortality, owing to the enormous water-polluting potential of these foods. If you do use liquid food, you should monitor water conditions daily and adjust amounts of food accordingly. Once you have established a feeding pattern that is successful, stick rigidly to it. If you are keeping fish with invertebrates, it may be totally unnecessary to add liquid feeds (see *Feeding in the Mixed Aquarium*).

For practical purposes, it is convenient to divide invertebrates into three feeding groups: the hunters; the opportunists and scavengers; and the filter feeders and self-sustainers.

Hunters

A number of invertebrate groups are active predators, which means that they are capable of catching active prey animals. While most of these are of little interest if you plan to keep a 'living reef' aquarium, you may see them for sale and you may even introduce them into the aquarium by accident.

The cephalopods – octopus, squid, cuttlefish and Nautilus species – are becoming increasingly popular, but only octopus species are regularly available, although generally quite expensive. All are fascinating creatures that quickly overcome their initial shyness when introduced into the aquarium. Unfortunately, they will eat almost any fish, shrimps, crabs or lobsters with which they are housed. Indeed, these animals should be fed a diet of defrosted whole shrimps, prawns and small fishes. The amount will vary with the size of the cephalopod, but an octopus with a 30cm (12in) span would need 2 or 3 shrimps, or a 5-6cm (2-2.4in) fish, daily.

Equally predatory mantis shrimps and swimming crabs are often introduced accidentally with newly imported living stock. Both are highly active predators known to kill and eat shrimps and quite sizeable fishes. Never introduce either of them into an aquarium containing a mixed collection of animals, as expensive losses will undoubtedly soon occur.

Opportunists and scavengers

Many of the most attractive, desirable and easily maintained invertebrates are included within

Right: *The octopus is an efficient hunter. It relishes live shrimp, but you may be able to train it to accept frozen shrimp or lancefish.*

Right: *This starfish is quite capable of prising out the soft fleshy body of a periwinkle. In captivity, starfishes will settle for similar frozen foods.*

this grouping. The vast majority of shrimps, crabs and starfishes (and some molluscs) will eat almost anything edible that comes their way. Their needs are easily met by feeding them with small pieces of fish, or various shellfish. These animals can subsist on surprisingly small amounts of food and you must take care that the tank does not become polluted with excess uneaten food. A certain amount of trial and error regarding quantity and frequency of feeding will be necessary but, as a guide, offer as much as the animal can eat in 10 minutes every other day.

Bear in mind that with their wide and catholic tastes, many species will quite happily eat the other sessile (non-moving) invertebrates that you hope to culture. Some of the starfish are well-known offenders in this respect – for example, the colourful red and white *Fromia* eats sponges. As a general rule, those species with large knobs on the dorsal (top) surface will eat almost anything upon which they can settle. Some of the small, brightly coloured dwarf

Feeding invertebrates

FEEDING TYPES	SUITABLE FOODS	QUANTITIES AND FREQUENCY OF FEEDING
Hunters Large crabs, lobsters Large univalve molluscs, e.g. Queen conch octopus, squid, cuttlefish, Nautilus	Small whole fish, shellfish meat, defrosted whole shrimps, prawns. Flaked food tucked into fish meat to supply essential trace elements.	Feed once a day. Remove uneaten food after one hour. Quantities depend on the size of the animal, e.g. 12in diameter octopus needs 2–3 large shrimps, or one 5–6cm (2–2.4in) fish.
Opportunists and scavengers Small crabs, shrimps Horseshoe crabs Cowries, sea slugs Starfishes, sea urchins, sea cucumbers	Small pieces of fish, shellfish meat, shredded prawn, *Mysis* shrimp, brine shrimp, cockles, mussels. Algae	Feed once a day at most. Remove uneaten food after one hour. These animals require little food and may subsist on the leavings from any fish in the tank.
Filter feeders and self sustainers Sponges Sea pens, soft corals, sea whips, sea fans, anemones*, jellyfish, hard corals Flatworms Fanworms, tubeworms, featherduster worms Barnacles Clams, scallops, oysters Crinoids Sea squirts *Anemones are classified here as filter feeds, but will accept small pieces of food once a fortnight.	Liquidized fish, shellfish meat or mussel blended with salt water from the aquarium. Newly hatched brine shrimp nauplii. Algae, pulverized or as a suspension. Live rotifers. Prepared plankton food. Liquid feed. Vitamin supplement.	Feed once a day at most; filter feeders are better underfed than overfed. Under intense lighting many corals and clams need little if any supplementary food. One drop per day per animal of a good liquid feed is sufficient, then switch filtration off for half an hour to prevent premature removal of food before it has been consumed.

lobsters may also be destructive.

Some invertebrates, such as several molluscs (*Sarcoglossan* sea slugs, for example) and sea urchins will browse on algae in the tank. Feed these largely herbivorous species with vegetable matter such as lettuce, spinach (in moderation) and seaweed preparations from health food stores.

Filter feeders and self sustainers

As their name suggests, filter feeders have evolved feeding mechanisms that enable them to trap small particles of detritus and planktonic organisms. Many worms, sponges and corals fall within this group. Anemones can also be considered as filter feeders, but they trap much larger particles of food than other species. Offer them one or two pieces of fish, shrimp, squid etc., every week or two. Gently push the food into the tentacles; do not force it into the mouth, and remove any pieces not enfolded by the tentacles.

The self sustainers include many of the shallow water coral species and sea anemones. Nearly all corals and anemones house algae, known collectively as zooxanthellae, within their tissue. As we have seen, zooxanthellae rely heavily on lighting of the correct spectrum in order to survive. In good conditions, where the host animals are given sufficiently intense lighting, the algae will prosper and the animals need very little supplementary feeding. However, under poor lighting conditions the algae die, followed fairly shortly by their host.

Below: *The extended polyps of* Dendrophyllia *sp. are much like tiny anemones. These corals are best fed live foods in the aquarium.*

Types of food

Most of the supplementary feeding requirements of these and other invertebrates can be met quite easily with readily available prepared foods. Most of the frozen foods are useful, but the dried foods familiar to freshwater fishkeepers have little place in the invertebrate aquarium. The only possible exception is the finely powdered dry food, sold for newly hatched fish larvae, which is acceptable to some filter feeders.

Liquid feeds are a convenient food source for filter feeders, but a freshly prepared solution is preferable to one that may have been lying on a shop shelf for weeks or even months. To prepare a solution, simply force a piece of shrimp, fish, or mussel, etc., through a fine meshed net and allow the resultant liquid to drip into the tank. You can prepare a larger supply of food by liquidizing a 10mm (0.4in) cube of fish, mussel, squid or shrimp, straining the liquid through a net to remove any larger particles, and then diluting it with 0.28 litre (0.5 pint) of sea water. Freeze this in ice-cube trays ready for use.

It is generally safer to offer only gamma-irradiated sterilized frozen foods. Although invertebrates appear less susceptible than fishes to disease introduced via non-sterilized foods, it is unwise to take any avoidable risks.

Other valuable foods for filter feeding invertebrates are the newly-hatched nauplii stages of brine shrimps, which are easily produced from eggs available from aquatic dealers, and the rotifers and algae used by marine fishkeepers in their efforts to breed and rear stock. These latter foods are fairly simple to produce if you follow the detailed instructions provided by suppliers of the initial cultures. Unfortunately, however, the space and time required for the process is more than most hobbyists are prepared to give. This seems a great pity given their high nutritional value.

The table opposite provides a summary of suitable foods for the different invertebrate groups and a guide to frequency of feeding.

Feeding in the Mixed Aquarium

Ask most hobbyists why they keep fish in the same aquarium as invertebrates and the usual reply is, 'because the aquarium looks unnatural without them.' That is a very fair response, but it is also strange when you think that by far the biggest threat to an invertebrate's well-being is *fish* waste and not their own waste products, which are minimal.

Waste is produced in direct relation to the amount of food input and it stands to reason that, the more food you put in, the more waste will result. The biggest challenge the aquarist is faced with is finding a balance between the two. As long as the mixed (sometimes referred to as reef) aquarium is not overstocked (remember 2.5cm/1in of fish to every 27 litres/6 Imp./7.5 US gallons net absolute maximum) then an equilibrium is not difficult to achieve. However, overstocked aquariums are very troublesome and, in the main, a correct balance is impossible to maintain.

The right fish

Having decided that a very low fish stocking level is essential for a mixed tank, it is crucially important to make the right choice of fish. Messy, wasteful feeders are first on the list of exclusions. Small fish with good appetites are ideal. Anemonefishes (*Amphiprion* spp.

Below: *The mixed aquarium is truly a thing of great beauty and appeal. Correct and careful feeding is one of the keys to success.*

and *Premnas* sp), damselfishes (*Pomacentrids*) and basslets (*Gramma* and *Pseudochromis* spp.) are good choices for all areas of the water column, whereas many gobies and blennies will cover the bottom and rocky areas efficiently.

Essentially, in a well-established, healthy aquarium, many fish will find masses of organisms inhabiting the areas of algae and coral growths and require little, if any, feeding whatever; Mandarinfishes fall into this category and confirm the point by generally faring very poorly in bare, fish-only aquariums.

Feeding in harmony

While discussing the feeding of invertebrates, we noted that many of the filter feeders will thrive on liquidized shrimp, etc. Frozen fish

foods contain these particles in abundance and, being too small for fish to pursue, they are captured by filtering invertebrates and thus the latter do not need any extra liquid food additions.

Many anemones and their relations seem to thrive under these conditions and rarely need gross feeding with lumps of shrimp, squid or lancefish. In fact, many anemones do very badly when being gross fed. Their digestive systems are designed to operate on a 'little and often' basis and if fed too often with large pieces, will reject the morsel, which, in turn, leads to pollution. Watching anemones consume large portions of food can be fascinating, but this must be kept to an absolute minimum if anemones are to thrive; about twice or three times a month is sufficient.

Crustaceans are mainly gross feeders and appreciate lumps of food, suitable to their size. To prevent fish from stealing their meal, it would be wise to train crustaceans to accept the appropriate food from aquarium tongs. By offering it in the same place at the same time every day, crustaceans quickly learn to accept their food in this manner – particularly shrimps and lobsters.

Never feed?

In the above manner, the mixed or reef aquarium need never become polluted due to overfeeding. In fact, several authorities have had great success with not feeding at all! The correct fish are chosen – mainly micro-organism feeders – and strong lighting used to encourage the symbiotic zooxanthellae algae, which sustains many of the coral and anemone species. As expected,

filter feeders and crustaceans do not really have a place in this environment and should, on the whole, be omitted. Large, regular water changes supply all the useful nutrients necessary to maintain an on-going food supply and, as a bonus, keep the quality of water at a very high level of cleanliness.

The key to successful feeding in a mixed fish/invertebrate environment is to understand how your livestock gains its nutrition. You may never need to feed many invertebrates and some fish. However, if you feel you do need to make regular feedings, *always* err on the light side – no aquarium should *need* feeding more than twice a day. Feed a variety of foods such as brine shrimp, *Mysis*, squid, lobster eggs and even a little marine flake. Generally, live foods are preferable to frozen; brine shrimp and rotifers are favourites with fish and invertebrates but are likely to remain occasional treats.

Overfeeding will encourage plagues of slimy algae, bristleworms, copepods and some flatworm species; time taken to decide on the correct feeding regime for your situation will be well spent.

Below: *Tiny rotifers and brineshrimp nauplii can be easily raised from commercial kits. They are an important form of nutrition for many fish and invertebrates. Liquid foods, while convenient, have a high pollutant value and should be used very sparingly.*

Regular Maintenance

To keep an aquarium sparkling with vitality, regular maintenance is absolutely essential. Indeed, the cornerstone of long-term success is firmly embedded in this. The tasks involved should always be seen as part and parcel of the joys of fishkeeping and a chance to renew and revitalize the aquarium.

In the hustle and bustle of today's modern living, tank maintenance can often be overlooked, delayed or even believed to be unnecessary. What an opportunity missed! It is far easier and less time-consuming to maintain a tank on a regular basis than it is to revive an aquarium that has sunk into the depths of neglect.

The siting of an aquarium can play an important role in whether it receives the correct maintenance or not. Site an aquarium where it is going to be seen, preferably in the main relaxation area of the house; a place where it will be noticed and admired. What a wonderful reason to keep it looking bright and healthy! A tank hidden away in a dark hallway or alcove rarely gets the attention it deserves and the incentive to maintain it correctly is quickly lost.

Know your aquarium

As the initial flush of enthusiasm with a new aquarium wanes a little, you will need to get into a more organized routine. An aquarium diary (log, notebook – call it what you will) is indispensable as a memory jogger. As the phrase implies, 'regular maintenance tasks' are performed on a predictable basis and can be listed in advance.

Although maintenance times can be kept to a minimum, in many cases just a few minutes each day, and an hour or two every few weeks, it is important to emphasize that you should always be looking out for the unusual or signs of impending trouble, so that you can

Below: *A marine aquarium makes a beautiful addition to a main 'living' area and a tank in a prominent place is more likely to be regularly maintained.*

Above: *It is important that you observe your aquarium closely, and regularly, for signs of trouble. You can then take remedial action before a crisis occurs.*

take remedial action before the situation becomes critical. In this respect 'getting to know your tank' can become a useful exercise. If you know what is normal for your situation, spotting the abnormal should present no difficulties.

Routine checks

There are a number of checks that you should make on a daily and weekly basis. First and foremost is a daily temperature check, together with a 'head count' of the fishes and crustaceans. Inspect sessile invertebrates for any signs of deterioration and, if bad enough, remove them immediately. Counting fishes and crustaceans is best done at feeding time, when an appearance can usually be guaranteed. Locate any absentees and search carefully for missing fish; they may simply not be feeling hungry or sociable, but if a particular specimen continues to behave this way, or seems off-colour, then you must assume something is wrong and begin further checks (water conditions, disease and bullying). If the fish or crustacean is dead, remove the body before it decomposes and pollutes the water. Of course, this may be easier said than done, especially if you have a reef aquarium and the animal has died right at the back of the tank. If this has happened, you can only try to remove the body in the best way possible without disturbing the rest of the livestock too much. If you do not find a missing animal in the tank, look on the floor, since some marine species are expert escape artists!

Check pH, ammonia, nitrite, specific gravity and nitrate levels in established tanks at fortnightly

---

Regular maintenance schedule

Daily
- Check temperature
- Check livestock
- Remove protein-skimmer waste
- Check all equipment is functioning properly
- Remove any uneaten food

Every other day
- Top up evaporated water
- Remove algae from front glass

Weekly
- Check residual current breaker (RCB) operation for electrical safety in the aquarium
- Clean cover glasses
- Remove any 'salt creep'
- Add trace elements, pH buffers and vitamin supplements

Every two weeks
- Partial water changes – 20 percent of net tank content ideally
- Test for ammonia, nitrite, nitrate and specific gravity in an established aquarium (this needs doing more often in a new tank)
- Remove any detritus from tank base or filters
- Replace filter floss in canister filters

Monthly
- Rake through coral sand

Bi-monthly
- Replace any tank airstones, including those in a protein-skimmer
- Check all electrical connections
- Harvest unwanted algae (more often, if necessary)

- Change carbon
- Clean protein-skimmer
- Rinse biological foam filters in tank water

Quarterly
- Clean out all canister filter hoses with hose brushes
- Clean pumps and check for wear
- Change air filters
- Clean out all water courses on a 'systemized' aquarium

As required
- Replace lighting tubes and bulbs according to manufacturers' instructions
- Clean lighting reflectors
- Record all events, test results, etc. in a diary for later reference.

intervals. The pH level should remain between 7.9 and 8.3. In a newly set-up aquarium it is a good idea to check all levels every two or three days.

It should go without saying, but removing uneaten food comes firmly under the heading of regular maintenance and is often overlooked by the novice fishkeeper. Excess food is a major potential source of water pollution and should be removed as soon as livestock lose interest in it – after a few minutes or up to half an hour, depending on the species.

Water changes

This is, arguably, the single most important maintenance task you can undertake. Without regular water changes in the correct proportion, the chemical composition of tank water can change dramatically, causing stress and disease, not to mention encouraging unsightly algae and pests.

The water change performs many useful tasks. As previously mentioned, it helps to keep the true composition of sea water within tolerable limits, reduces nitrate, phosphate and sulphate levels by dilution (as well as many other accumulating harmful substances), maintains the buffering power of the aquarium water so that the pH remains stable, and removes metabolic waste products that may otherwise remain in solution. (Medications can also be removed in this way.) In an emergency, a large water change will dilute high levels of ammonia and nitrite and preserve precious livestock until faults can be rectified.

Ideally, change 20 percent net content of the aquarium water every two weeks, or three weeks maximum. Some people may choose to vary this to 10 percent each week or even to arrange for a permanent drip feed from a reservoir and a suitable overflow.

The specific gravity should remain stable but in the region 1.020-1.024

by replacing any evaporated water with *fresh* water (salt cannot evaporate). The more frequently this is done, the better, but every other day is ideal for most aquariums. Always choose high-quality water for mixing up water changes. It should possess no nitrates or other impurities and, if necessary, tapwater should be pre-filtered using correct resins, de-ionizers or reverse-osmosis units.

Rising nitrate (NO_3) levels can be a useful indicator of deteriorating water quality and should be kept as low as possible, whether the aquarium is a fish-only or an invertebrate arrangement. As a guide, levels of NO_3 should not

Below: *Some of the equipment that will become essential in the aquarist's maintenance 'armoury'. 1 tongs; 2 algae scraper with nylon pad; 3 algae scraper with fitted blade; 4 coarse nylon cleaning pad; 5 algae magnets; 6 soft nylon net; 7 self-starting siphon tube; 8 hose brushes; 9 substrate cleaner.*

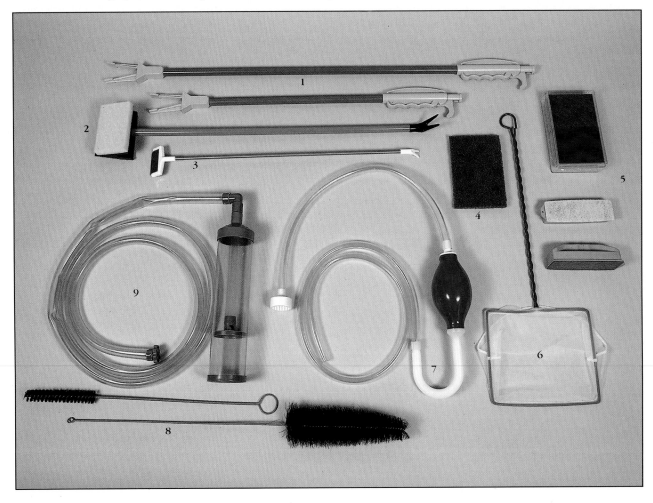

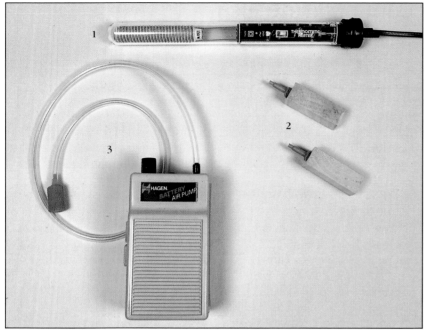

exceed 25 mg/litre for fish-only and 5 mg/litre if *any* invertebrates are kept. If levels increase before a regular water change is due, do not wait, carry one out immediately.

Some publicity information suggests that certain systems require *no* water changes. Experience has shown that this is a very foolish route to follow and usually leads to the dreaded 'wipe-out'.

Above: *Items to keep spare: 1 heater/ thermostat; 2 battery-power air pump; 3 wooden air diffusers for the protein skimmer or for general aeration.*

Dealing with emergencies

These usually fall into four categories: major power failure; overheating due to weather conditions; major pollution, and leaking aquarium.

Major power failure Short power failures usually present little problem in a sensibly stocked tank. Most can cope with up to 24 hours by using a battery air pump and, during the cooler months, lagging the tank with blankets and coats. Do not feed during power failures or for 24 hours after a major one. If the power is expected to be off for days, ask a friendly aquatic dealer if they will board your stock, or hire a portable generator.

Overheating Most fish and inevertebrates will cope with temperatures close to 86°F (30°C); above that, not many methods – apart from fitting a chiller – can be considered successful. Floating bags of ice cubes or pouring in refrigerated tank water may bring the temperature down a degree or so, but it soon rises again and the effort may not be equal to the results. Try to reduce all external sources of heat, such as lighting and keep the room as cool as possible. A large fan aimed at the aquarium may help a little. In warm climates, a chiller will probably be needed as standard equipment.

Major pollution A fish or invertebrate dying overnight can cause the whole tank to go cloudy and may raise ammonia and nitrite levels dangerously high. After removing the source of pollution, mix up as much new water as possible and exchange it. Repeat as many times as necessary. Renew any carbon and add more if possible. Do not feed the tank inhabitants until all levels are back to normal. Keep a close eye on livestock for the next two weeks for any signs of disease.

Leaking aquarium Gather as much tankwater as possible into buckets and put the surviving animals in them. Aerate and keep them warm. Ask a friendly dealer to board the surviving stock. Thankfully, this is a rare event nowadays, but you should only purchase a good-quality tank made of first-generation glass of the correct thickness.

Maintenance hints and tips

● Algae magnets left submerged for long periods can contaminate the aquarium. When not in use, dry and remove.

● Yellowing, cloudy, frothing or smelly water are all signs of deteriorating water quality requiring immediate attention.

● If cover glasses are fitted, keep them clean to enable the full intensity of lighting to reach the bottom of the aquarium. Stubborn calcium deposits can be removed by rubbing with a solution of citric acid or lemon juice. Rinse thoroughly.

● Always wash biological media in tank water to avoid killing the useful bacterial colonies.

● After a while, undergravel filter beds begin to compact and the through-flow of water is reduced. Every month, rake your fingers through the sand and siphon off any debris. Push a siphon tube down the filter uplifts to repeat the procedure below the filter plates.

● Within two years, the pores of coral sand become clogged and full efficiency is lost. Begin a gradual replacement programme as follows: divide the sand into five strips and change one a month for fresh sand. That way, the least damage will be done to the biological filter.

● Keep a spare heater. Heaters do not last forever and will eventually fail.

● Keep a battery-powered air pump handy. If the electricity fails, some aeration will be vital. (Make sure the batteries are new and not drained!)

Breeding Marine Fishes

There was a time, not many years ago, when it was thought useless to try to breed marines in captivity. Since then, and owing mainly to the hard work of a few dedicated hobbyists, advances have been made that early aquatic pioneers would have thought impossible. However, there is still much work to be done in this area and what we believe to be impossible today will, no doubt, become commonplace in years to come. This is especially likely as the hobby exchanges ideas with the increasing number of salmon and sea trout farms.

When compared with freshwater, the number of marine species successfully reproduced in captivity is very small indeed. The reasons for such comparative lack of success are easy to understand. Firstly, marine fishkeeping is a young hobby, and much of the original work concerned itself with just keeping the animals alive, let alone breeding them. Anyway, livestock was relatively easy to obtain from the wild and, thus, the incentive to devise captive breeding programmes was lacking. Thankfully, a recent reversal in emphasis has taken place. Secondly, and more importantly, the life cycles of many marine creatures are complex and not easy to replicate in captive environments.

Life cycles

Many fish and invertebrates are involved in webs of life that we are only now beginning to untangle. A great many species, although beginning their life as a familiar egg, do not hatch out into fry as we understand the term by freshwater standards; they are much less well developed, and commonly referred to as larvae. These eggs and larvae make up much of the vast bulk of the planktonic layers in the oceans, where they may remain for some time, dispersing the species.

While in the plankton, larvae feed on smaller organisms and are, in turn, fed upon by larger ones until the strongest and best developed migrate back to their natural niche in the oceanic world to begin the process all over again.

Before you can hope to have any success breeding marines, you must be aware of the life cycle of the species in question and try to replicate or substitute each requirement in turn. Unfortunately, little is known about the life cycles of many families, but where information is available, it has been given under Family Characteristics headings and individual species entries (see pages 160-301).

Spawning methods

Marine fishes employ various methods of spawning, in much the same way as their freshwater cousins; these include egg-scattering, egg depositing, mouth-brooding and pouch-brooding.

The egg-depositors feature anemonefishes, damselfishes and many gobies, whose spawning behaviour and parental care is similar to that found in freshwater cichlids. The jawfishes and cardinalfishes incubate eggs in their mouths, while the male seahorse is famed for adopting a similar role, with eggs deposited in his abdominal pouch by the female.

Below: *Both the male and female Common Clownfish clean the site upon which the batch of eggs will be laid – a practice shared by all clownfish species.*

Within the framework of the different spawning methods, there is also a diversity of breeding behaviour. Fishes may, for instance, form longterm partnerships; a male may set up an attendant harem of females; shoals of fishes may seasonally congregate for a mass spawning; or a male and female may spontaneously spawn as and when the opportunity occurs.

Seasonal conditions can also affect spawning activity. Many species are also heavily influenced by the lunar phases, water temperature and photoperiod (length of daylight).

Basics for breeding

There are some basic requirements for successful spawning: compatible pairs (groups) of fishes being well conditioned and of sexual maturity; the right environment, including optimum water conditions; and the correct amount of space to encourage spawning. Space, or lack of it, is probably the most restricting factor as far as the average hobbyist is concerned.

However, once fulfilled, these requirements should form a basic framework on which to work.

Identifying a likely pair of parent fish is not an easy task. So little work has been done on identifying the differences between sexes that many successful pairings have relied

solely on chance, an experienced intuitive eye, and not a small amount of luck! Having said that, most fishkeepers could begin by making some commonsense assumptions. For example, you could expect a female to become stockier when full of eggs, while a male might have intensified colours or elongated fins.

The best clues to sexual differences, apart from the obvious physical ones, are to be found by close observation of behaviour patterns. This may be a long and time-consuming task but, in many cases, the only reliable way of pairing two fish.

Space is important when attempting to breed fish, for two reasons: they require space both to establish spawning territories and also for the spawning act itself. The amount required varies depending on the species, but anemonefishes and Neon Gobies need less room than larger wrasses or angelfishes.

'Accidental' or 'planned'

Most aquarium spawnings, it would be fair to say, are chance occurrences in the company of other fish. Left to their own devices, eggs or larvae have virtually no chance of surviving in the aquarium. If breeding is to be taken seriously, then eggs must be removed to a

friendlier environment at the appropriate moment.

An even better arrangement would be to house the breeding fishes separately – although the eggs would still, in all probability, have to be removed and the larvae cared for in a 'safe' aquarium. Exceptions to this would possibly occur among the mouth and pouch brooders, where young fish are well developed and may be left in the parents' company.

However, you may be happy to witness the spawning activity with no intention of rearing larvae. This, too, can be useful, especially with fish that are known to be difficult. Notes on all aspects of aquarium care and breeding behaviour would be eagerly received by any marine society, thus adding to the greater fund of knowledge.

Even so, our collected information could clearly be made the subject of a separate volume, well beyond the scope of this book. If sufficiently interested, the dedicated fishkeeper could certainly make it a lifetime's study. However, for the purposes of this book, an outline of the spawning and rearing of anemonefishes is appropriate. Much of their information applies to other species, in particular Neon Gobies and damselfishes, which breed in much the same fashion.

Breeding anemonefishes

By chance and good fortune, some of the easiest fish to spawn and raise in captivity just happen to be the most easily obtainable and most popular. All anemonefish species are readily spawned in the aquarium given the correct conditions.

There are three main ways of making sure you have a breeding pair. Firstly, you could buy two very small anemonefishes; at this stage they would both be males (or juveniles), but eventually one would grow considerably larger than the

Left: *After the female has laid a line of eggs, the male passes over them to immediately fertilize each new row. The process is repeated and practically all the eggs are successfully fertilized.*

other, having changed gender to a female. Secondly, you could purchase several small specimens; a dominant female will become apparent and she will pair with the next dominant male. (The excess fish will have to be removed for their own health.) Finally, you could acquire a proven spawning pair.

Whichever method you choose, make sure that the fish are fit and highly coloured – pale-coloured fish give rise to pale-coloured offspring.

Conditioning

Whether growing-on or conditioning adult stock, the food provided should be of the best quality, varied and abundant; a female cannot produce eggs on a poor diet. Feed live brine shrimp, as well as high-protein frozen foods, three times each day. A good quality marine flake is also recommended to provide added minerals. In addition, vitamins may be added straight onto the food.

Because of this rich diet, the aquarium should either be for the exclusive use of the anemonefishes, or be sparsely stocked with compatible species. No more than 2.5cm (1in) of fish to every 27 litres (6 Imp./7.5 US gallons) net.

Spawning

Although it is possible to spawn anemonefishes in small tanks, chances are greatly increased if capacities of around 135 litres (30 Imp./37.5 US gallons) plus are used.

Good filtration, coupled with a protein skimmer and activated carbon, along with fortnightly 20 percent water changes, will keep both fish and anemone happy. Anemonefishes will spawn in the absence of an anemone but, again, the chances of regular spawnings are much increased if an anemone is available, especially where wild stock is concerned. One anemone and one pair of anemonefishes in the aquarium will cement the pair bond and the fish can concentrate their efforts on breeding. The presence of other anemones and anemonefishes will divert attention and may even cause the pair to split.

At about 3.8-5cm (1.5-2in) the female will become sexually mature and will become interested in spawning. The much smaller male may well be cleaning prospective sites with his mouth long before this, but the female will only spawn when ready. Help the process along by placing convenient spawning receptacles, such as pieces of slate or clam shells, around the base of the anemone. If the tentacles of the anemone brush over the spawning sites, so much the better.

Have patience and refrain from disturbing the tank's occupants unduly. The anemonefishes will spawn when they are ready. In temperate latitudes, the longer days of spring and summer are often the spur needed to spawn, and artificial lighting may not always be able to substitute for the quality of daylight.

At last, the pair will pick out a spawning site and begin to clean it until they are satisfied. The sticky eggs are usually laid in the late afternoon and the clutch may contain several hundred yellow/orange eggs. These are then vigorously guarded by the male, and, to a lesser extent, by the female. All intruders, big or small, are attacked, sometimes with great ferocity, although the pair rarely do any damage. Even your hand will be treated with contempt – anemonefishes are quite capable of drawing blood! Nearby algal clumps are often torn to shreds because they 'refuse to move'!

The yellow/orange eggs gradually darken as they mature and, in about six days, the silvery eyes of the

Above: *Although it is nearly always the male that attends the eggs, the female takes an occasional interest, as shown here. At six days old, the silvery eyes of the unhatched larvae can be seen.*

developing larvae can be clearly seen. Depending on tank conditions – temperature, pH and specific gravity – hatching takes between 7 and 10 days and always occurs in pitch darkness. Anemonefishes use lunar activity to time the hatching of their eggs on a moonless night.

Keep careful notes throughout, especially on the length of the hatching period; these will be invaluable references for the future. It is a good idea to just observe the first few spawnings; don't make any attempt to rear the larvae at this stage. Once started, anemonefishes are very difficult to *stop* spawning – usually on a monthly cycle, but sometimes more often.

Hatching

In the wild, the larvae become pelagic and swim to the upper planktonic layers of the sea to feed and develop. In the aquarium, they will not survive until morning unless other arrangements are made.

From the notes you have already made, you should know precisely the date of hatching. Two days in advance of this date, prepare a bare 45-55 litre (10-12 Imp./12.5-15 US gallon) tank with heater, airstone and fresh seawater mixed to the same specific gravity, temperature and (if possible) pH, as the water the eggs are already in. No filtration or lighting will be needed at this stage, apart from a small pygmy bulb above the tank, but it should be possible to black out the tank.

On the evening of hatching, take out 50 percent of the rearing tank water and exchange this for water taken from the main tank. Two hours before normal lights out remove the eggs and the rock on which they were laid to the rearing tank. Never expose the eggs to air or remove them from the rock they are attached to. Place the rock near the airstone and aerate it to give good, but not powerful, circulation.

Now, black out the tank completely for two hours and do not be tempted to look. Any light source at this time will delay a complete hatch for 24 hours, giving fungus a chance to form. After two hours, remove the blackout and you should have an aquarium full of hungry larvae.

Raising
For the average hobbyist, it is not commercially viable to raise all the fish and a batch of between five and 20 is usually enough. Many of the larvae will have failed to survive the first 12-24 hours, and it is often unnecessary to cull the batch drastically, although weak and deformed individuals should be siphoned off immediately using airline tubing or a fine pipette.

Brine shrimp nauplii are too large for a first food, so rotifers of the species *Brachionus plicatilis* should be fed. Commercial kits are now readily available to raise these useful animals and enough should be cultured to enable the young fish to feed constantly, 24 hours a day, for at least four days (hence, the need for a pygmy bulb).

On day four, newly hatched brine shrimp should be fed, in addition to the rotifers, which you should continue to use for at least a further nine days. From then on, brine shrimp and pulverized flake food will become the staple diet of the young fish.

Growing on the fry
As with freshwater fry, regular water changes are needed to promote

good growth rates, and 20 percent every two days will be required, as will a larger growing-on tank. A 90 litre (20 Imp./25 US gallon) tank, properly filtered, will be needed for up to 20 fry, which can begin to be transferred on an individual basis from 10 days onwards, as adult colours start to develop. Transfer the largest and strongest first to give the less well developed a chance to catch up, until all are converted.

Gradually, the young fish can be introduced to other frozen foods, which may have to be finely chopped. As growth rates accelerate, larger living quarters will have to be found to prevent stunting.

All marine larvae and fry are very delicate and should only be moved

by a gentle siphon or container. Nets do untold damage to fins and gills and should not be used.

Breeding other species
Like anemonefishes, demersal (bottom-dwelling) spawners are reared in much the same way, although incubation and raising periods may alter. For instance, Neon Gobies require rotifers exclusively for up to six days, at which time they begin to get some colour, achieving full adult coloration after 21 days. Adult colours in Yellow-Tailed Damsels may take 21-28 days, while the Electric Blue Damsel can be slow at up to 50 days.

Pelagic (surface or midwater) spawners scatter their eggs and take no further part in the rearing process. As the eggs are usually very small, at about 1mm (0.04in) in diameter, and floating, the prospective breeder must work very hard to retrieve them. Even when successfully hatched, many of the larvae are too small for rotifers and alternative microscopic cultures must be reared. Techniques are being developed to raise species of angelfish, mandarinfish and several types of wrasse.

The popular but unpredictable seahorses quite often give birth to live young in the aquarium and this is one of the very few species to take newly hatched brine shrimp as a first food. This should make rearing easy, but these are slow growers and require almost continuous and copious feeding. Failing to do so usually results in sudden deaths, which can happen at any stage in development, including adult.

Meet the challenge!
As you can see, the list of marine species bred in captivity is quite short, but it represents no mean achievement by those fortunate enough to be involved. It also holds out more than just a ray of hope for future marine fishkeepers. If you feel you are up to the challenge, the rest of the marine fishkeeping world awaits your finds and results with bated breath.

Fishes spawned in captivity

The following species have been spawned in captivity and are commonly available. (Many have never been raised to adulthood.)

Egg-scatterers

Angelfishes	*Centropyge* spp.
	Holacanthus spp.
	Pomacanthus spp.
Blennies	*Aspidontus* spp.
	Petroscirtes spp.
Butterflyfishes	*Chaetodon* spp.
Mandarinfishes	*Synchiropus* spp.
Wrasses	*Pseudocheilinus* spp.
	Thalassoma spp.

Egg-depositors

Anemonefishes	*Amphiprion* spp.
Damselfishes	*Abudefduf* spp.
	Dascyllus spp.
Gobies	*Gobiosoma oceanops*
	Lythrypnus dalli
Grammas	*Gramma loreto*
Hawkfishes	*Oxycirrhites* spp.

Mouth-brooders

Cardinalfishes	*Sphaeramia* spp.
Jawfishes	*Opisthognathus* spp.

Pouch-brooders

Seahorses	*Hippocampus* spp.

Health Care and Disease Treatment

Marine fish have survived for many millions of years longer on this planet than humans have and, as we have seen, they are superbly suited to their aquatic existence. In the marine environment a number of lower life forms that live in and on the fish at its expense have also evolved over this time. These organisms come in a number of forms: viruses, bacteria, protozoa, crustaceans, nematodes, fungi etc. Some are pathogens – that is potential disease-causing organisms.

Fish disease can be defined as a deviation from a fish's normal physical condition due to infection, weakness or environmental stress. In most cases it is not in the best interests of these lower life forms actually to cause the death of their fish host, although some do depend on the host's death so that they can spread and proliferate or pass on to the next part of their life cycle. Fish disease pathogens can be categorized under four broad headings: bacteria, fungi, viruses and parasites.

Bacteria

Bacteria are microscopic single-celled living organisms. There are thousands of species, inhabiting every possible environment and they come in a large variety of shapes. There are rods (bacilli), spirals (spirilla), spheres (cocci) appearing either singly or in groups. Some bacteria are capable of locomotion, using flagella or cilia (hairs), which enable them to swim through fluids. Other forms live attached to surfaces or simply drift through the water.

Bacteria are capable of very rapid reproduction by means of, simply, dividing in two. They are reputed to be able to produce billions of new bacteria in a few hours in suitable conditions. Not all bacteria are bad, of course; as we have seen (pages 96-97), nitrifying bacteria are essential in filters if we are to keep marine fish and invertebrates in the aquarium. Some bacteria are, however, disease-causing organisms, and many of these are common in marine aquariums.

Fungi

Fungi constitute another group, which, again, contains tens of thousands of species, ranging from the mushrooms that we eat to single-celled yeast. These organisms are classified as plants, but they cannot produce their own energy (as green plants do) so they need to feed. Some fungi feed off dead matter, while others are parasites, which feed off living organisms. It is this latter group that is potentially pathogenic to fish.

Pathogenic fungi are typically made up of thin threads (hyphae) which form a network (mycelium)

Below left: *This photograph, taken down a microscope, shows stained rod-shaped bacteria, each measuring only a few microns (millionths of a metre).*

Below: *These fungal threads, or hyphae, are about 20 microns in thickness. They intertwine and form a network deep into a fish's tissue.*

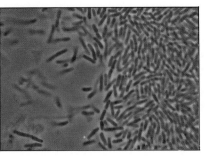

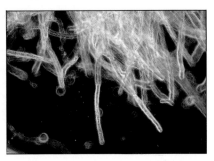

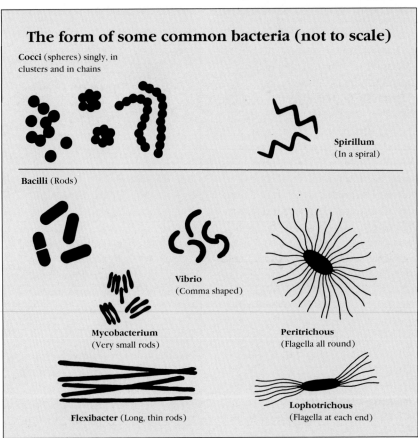

The form of some common bacteria (not to scale)

Cocci (spheres) singly, in clusters and in chains

Spirillum (In a spiral)

Bacilli (Rods)

Vibrio (Comma shaped)

Mycobacterium (Very small rods)

Peritrichous (Flagella all round)

Flexibacter (Long, thin rods)

Lophotrichous (Flagella at each end)

deep into a fish's tissue or through its organs. Fungi reproduce by production of infectious spores, which are distributed into the environment and are very tough, remaining viable for some time, even in adverse conditions. These fungal spores are present in all marine aquariums.

Viruses

These are among the simplest form of organisms. In fact, on their own, they can barely be considered living creatures. They are only millionths of a millimetre in diameter and are obligate parasites, only coming to life and reproducing in the living cells of their host. They do so at their host's expense, causing disease.

Viruses are usually very host-specific and they only affect a few species in an aquarium. Treatment is impossible without destroying the host cell. The fish's immune system is its only defence.

Parasites

Fish parasitism is an association where the parasite is dependent on, and derives benefit from, the host fish, who derives no benefit. Parasites come in a large number of forms: viruses (as seen above), protozoa, crustaceans, roundworms, flukes, tapeworms and trematodes. The association between parasite and fish host can take many forms. Some parasites, for example, spend only part of their life cycle on their host fish.

It is not usually in the best interests of the parasite to kill its own food supply, although there are some exceptions where the death of the fish is necessary for the parasite to complete its life cycle. Parasitic disease is usually caused by an upset in the fish/parasite balance, which, as we will see later, is more likely in an aquarium than in the wild.

The disease defence system

Given the potential onslaught of fish pathogens, you could be forgiven for wondering how marine fish have survived at all. The reason that they do is that they have evolved a very effective defence mechanism, which

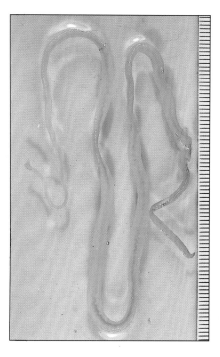

Above: *An adult tapeworm. This parasite has two hosts during its life cycle: fish and a copepod. The adult lives in the fish's gut, robbing it of nutrients.*

is called their immune system. Although not as advanced as mammals, fish do have an immune system that protects their bodies against disease. The first line of defence involves preventing disease organisms physically invading the body. Fish have an effective outer barrier in the form of scales and the layers of the dermis and epidermis of the skin, all of which provide some protection against disease organisms and physical damage. This outer barrier is further improved by a covering of mucus that contains a number of bactericides and fungicides. The mucus membrane is constantly being renewed, which also has the effect of sloughing off debris and dissuading the proliferation of external parasites. The other possible area of infiltration is through the digestive tract. Here, enzyme action and an unsuitable pH produce a hostile environment , which discourages most pathogens.

If one of these barriers breaks down, then pathogens can gain entry, for example through skin wounds and through the gut. In some cases of stress, the gut seizes

up and anaerobic fermentation and enzyme action attack the gut wall, allowing disease organisms to enter the tissue and bloodstream.

There are a number of products in the bloodstream that give some general immunity by immediately attacking any pathogens. These include the anti-viral chemical interferon and C-reactive protein, which acts against bacteria and viruses. The fish's first co-ordinated response to foreign bodies entering the bloodstream through a wound is to seal off the entry sites, thus preventing osmoregulatory problems and hampering the spread of the disease organism throughout the body. This inflammation reaction is achieved by histamines and other products released by damaged cells that cause blood vessels in the area to close up. At the same time, the blood protein fibrinogen and other clotting factors seal the area off with a physical barrier of fibrin. White blood cells are attracted into the area and these ingest the foreign bodies and carry them to the kidney and spleen. Some bacteria have the ability to overcome the inflammation response, either by releasing toxins that destroy the ingesting white cells, or by producing an agent that dissolves the fibrin wall and allows them to spread throughout the body.

In the kidney and spleen, special antibodies are formed that will act specifically against each particular intruder, but the antibody production process may take up to two weeks. Each antibody attaches to its specific intruder and acts against it in a number of ways: it may detoxify the intruder so that it can be ingested by the white blood cells; inactivate its reproductive system so that it cannot proliferate; or attract a blood component (a complement) that helps to destroy the whole antigen cell.

If the specific intruder has been previously encountered, the fish's immune system will react much more quickly; specific antibodies will already exist and will multiply extremely rapidly upon contact with that particular intruder. This is the principle behind vaccination, in which introducing a detoxified disease organism enables the fish to build up specific antibodies and thus increase its chances of survival during a bout of a specific disease.

The effectiveness of a fish's immune system is related to its environment. Poor water quality reduces a fish's immune response.

How disease is caused

Most fish and invertebrates are 'disease carriers'. This means that they contain parasites that their immune system has not completely eradicated, but that have been kept sufficiently in check so that there is no evident sign of disease. The marine aquarium environment also contains a number of pathogenic bacteria and fungi, but again, a healthy fish's immune system ensures that these do not cause disease. Thus, in the aquarium, there exists a delicate balance between the fish and the disease organisms present. Anything that upsets this balance will lead to a disease outbreak.

Stress

The key cause of disease outbreaks in the marine aquarium is fish stress, which is a fish's physiological response to a stressor. These stressors can take the form of poor environmental conditions; bullying by dominant fish; clumsy aquarium maintenance; capture and transportation; lack of suitable hiding places or, even, inappropriate lighting.

The fish's stress response has evolved to deal with 'fight or flight' situations, the response releasing hormones that enable immediate escape from, or attack against, the stress situation. These hormones create short-term problems with osmoregulation and suppress the immune system. However, the stress response is designed to deal with short-lived emergency situations, where the priority is immediate survival, allowing a return to equilibrium later. Full recovery may take up to 14 days from the time of application of the stress. Problems

Above: *These fish are in quarantine. They are carefully examined daily for signs of stress-induced disease before being introduced to a new aquarium.*

occur in an aquarium where the stressor is a continuous chronic condition; for instance where there is no escape from poor water quality or a bullying fish. The continuous stress response, with its negative physiological effects of immunosuppression and increased osmoregulation problems, is quite sufficient to upset the pathogen/fish balance, causing disease outbreak.

One of the keys to a healthy aquarium is the prevention of stress. The prevention of stress can be brought about by careful attention to the following areas: water quality; selection of suitable and compatible inhabitants (see pages 118-119; suitable decor providing sufficient hiding places and space for territorial species (see pages 162-361 for the requirements of individual species); staggered starting of lighting tubes reducing shock to fish at the beginning and end of the day; reduction of the duration and effect of aquarium maintenance and fish capture and handling. Where possible, treat diseased fish in situ, since secondary stress of handling them will reduce the chances of an effective cure.

Fish introductions

Fish present in an aquarium have been challenged over time by all the pathogens present in the aquarium and, if healthy, will have established an immunity to them – a so-called immunobiological balance has been established. Introduction of a new fish or invertebrate into an established aquarium can result in

an upsetting of this immunobiological balance. This is because they may well carry different disease organisms which will challenge the existing population, while they themselves are challenged by new pathogens present in their new environment.

What makes matters worse is that this challenging of the immune system occurs at a time of stress. The introduced fish has been stressed through chasing, netting, bagging, transport and water quality changes. The balance of the aquarium also shifts. For example a new social hierarchy may need to be established and the filtration system has to respond to an increase in waste, which takes some time, so stress from a temporary drop in water quality may occur.

As we have seen above, stress is, itself, a major cause of disease as a result of the suppression of the immune system. The combination of the upset of the immunobiological balance and immunosuppression in the fish as a result of stress, as well as contact with new pathogens, often results in disease outbreaks.

How, then, can we avoid the risk of disease when introducing new fish? Ideally, all new fish should be quarantined. This involves setting up a small, fully filtered aquarium where the new fish is placed for a few weeks to allow it to recover from the stress of capture and movement. As mentioned above, return to physiological equilibrium can take up to 14 days and a fish cannot deal as well with a secondary stressor during this time. The fish can also be monitored for disease outbreak, which may have been caused by the stress that it has experienced. It is recommended (after the first settling-in week) to treat the aquarium with a broad spectrum anti-parasite treatment to try and reduce the chances of introducing new parasites into the established aquarium.

When the fish has been quarantined for the full three-week period, and you are entirely satisfied with its health, introduce it into the main aquarium. Upon introduction, it is sometimes wise to rearrange the tank decor in the established aquarium. This will help confuse the existing fish sufficiently so that the new introduction is on a more even footing in establishing its place in the aquarium hierarchy. After introduction of any fish, careful monitoring of the water quality parameters and social interaction between the fish is necessary to ensure that these have not been upset. Extra vigilance for disease symptoms for a few weeks is also very wise.

Problems of aquarium life

In some cases the enclosed environment of the aquarium is enough to upset the balance between the pathogen and the fish. Finding a host is a hit-and-miss affair for a parasite in the wide ocean expanses. Parasites' reproductive systems compensate by producing a large number of offspring so that the probability of one finding another host is increased. As an example, a single marine White Spot cyst will release up to 256 infective swarmers. Obviously, within the confines of an aquarium, the chances of finding a host are greatly increased and parasitic disease can rapidly escalate to a life-endangering level. The danger of overstocking should be evident from this. Under no circumstances should you stock

more than 2.5 cm of fish for every 9 litres (1 in of fish for every 2 Imp/2.5 US gallons) of aquarium water, for a fish-only system.

A further method for ensuring that the pathogen/fish balance is maintained in favour of the fish is to reduce the viable pathogen numbers. There are a couple of ways of doing this. For example, pathogen numbers can be controlled by using ozone or UV sterilization techniques. These are linked into the filtration system and control the number of disease pathogens in the water passing through them. If they are set up properly (see pages 100-101) they will help to destroy water-borne bacteria, fungal spores and free-swimming infective parasite life cycle stages. Alternatively, some fish-only marine aquarists maintain a preventative level of medication in the water (usually a copper-based

medicine) to control parasite numbers. Keeping the true Cleaner Wrasse (*Labroides dimidiatus*) with large fish can sometimes keep down the external parasite numbers, since these fish feed on the parasites living on the gills and skin of their aquarium mates.

Disease diagnosis

The early recognition of disease and accurate diagnosis are both critical if fish losses are to be kept to a minimum. Unfortunately, many marine fishkeepers do not notice the onset of disease until it is so far advanced that successful treatment is unlikely. It is therefore critical to know your own fish; careful daily observation will enable you to recognize your fish's normal behaviour and appearance so that any deviation from normal will act as an alarm and allow you to investigate the cause.

Disease recognition and diagnosis

Symptom		White Spot	Marine Velvet	Black Spot	Gill/skin parasites	Intestinal worms	Microsporidian infection	Bacterial Fin Rot	TB/Wasting disease	Vibriosis	Lymphocystis	Ichthyophonus	Head and Lateral Line Erosion (HLLE)	Poisoning
Gasping/coughing		■			■									■
Flicking/rubbing		□	□		□									■
Overall change of colour							■		■					■
White patches														
Cloudy eyes		■												■
Erratic swimming							■			■				■
Fin erosion								■						
Ulcers/holes										■		□		
Not feeding							■		■	■		■		
'Pop eyes'									■					
Distended stomach					■									
Emaciation									■					
Visible spots	White (1mm)	□												
	Black			□										
	Gold/brown peppering		□											
Warty growths/lumps											□			
Reddened areas								■		■				
Listless		■	■		■					■				■
Sand paper skin effect		■	■											
Visible wormlike attachments				□		□								

■ Possible symptoms   □ Positive symptoms

Marine Fish Diseases

PARASITIC INFESTATIONS

Marine Velvet

Cause Velvet is caused by a dinoflagellate protozoan called *Amyloodinium ocellateum*. These protozoa are microscopic, reaching a maximum length of 150 micrometers.

Life cycle The adult protozoan attaches by a root-like structure to a fish's fins, gills, and skin, producing a dusting of just-visible brown to gold spots on the fish. When mature, the protozoa fall off the fish, swell and undergo division, producing up to 250 free-swimming parasitic dinospores. These dinospores can survive for a few days while searching for a fish host to attach to. Rapid multiplication of Marine Velvet protozoa in the confines of the aquarium, where finding a host is easier, can result in severely infected fish. Mortality is mainly caused by suffocation due to heavy gill infestation.

Treatment The treatment-resistant cyst stage makes eradication difficult and necessitates prolonged treatment. (See parasite treatment table page 156).

Marine White Spot

Cause Marine 'Ich' is caused by the parasite *Cryptocaryon irritans*. This is a ciliate protozoan parasite.

Life cycle The adult parasite (trophont) invades the skin and gills of its host, living off the fish's tissue. The fish forms a white cyst around the individual protozoan to limit its effect. This gives rise to the easily recognized symptoms of white spots which form on the fins, gills and skin. The adult protozoan breaks out and sinks to the bottom of the aquarium and forms a protective capsule. Its reproduces over four to 22 days by division, forming 100 to 400 infective swarmers. These swarmers are released to swim and

The life cycle of *Amyloodinium*, the Marine Velvet parasite

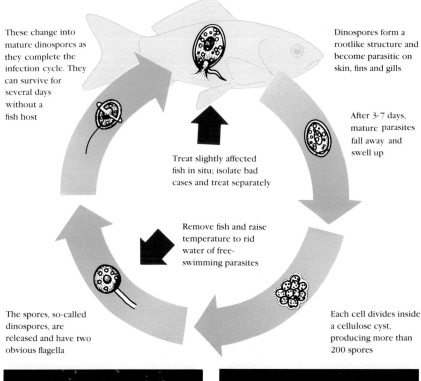

These change into mature dinospores as they complete the infection cycle. They can survive for several days without a fish host

Dinospores form a rootlike structure and become parasitic on skin, fins and gills

After 3-7 days, mature parasites fall away and swell up

Treat slightly affected fish in situ; isolate bad cases and treat separately

Remove fish and raise temperature to rid water of free-swimming parasites

The spores, so-called dinospores, are released and have two obvious flagella

Each cell divides inside a cellulose cyst, producing more than 200 spores

Above: *Marine Velvet is a common disease. A light dusting of the* Amyloodinium *parasite can be seen on this fish's skin and fins.*

Above: *The whitish nodules covering the head and fins of this fish are caused by* Cryptocaryon irritans *and give this disease its apt name Marine White Spot.*

find a host, which they must do within 24 hours or else die. A heavy infestation will result in gill damage, preventing the fishes' respiration and upsetting their osmoregulation system, and can thus lead to death.

Treatment Like Marine Velvet, White Spot can be very difficult to eradicate because most treatments are only effective against the free-swimming infective and adult stages. The encysted stages on the fish and in the gravel provide a reservoir for future outbreaks. It is essential, therefore, to retain a level of

medication in the water for up to a month, which will ensure that all the parasites present will have passed through the susceptible free-swimming stage. (See parasite treatment table, page 156).

Gill and Skin Parasites

Cause There are a number of other parasites of marine fish that are less often encountered in the aquarium. These include copepods and monogenetic trematodes, i.e. Dactylogyrid species, such as *Benedina*, which is a 5mm (0.2in)

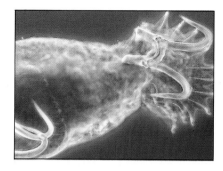

Above: *These ferocious looking hooks are used by this Dactylogyrid parasite to attach itself to the fish's skin.*

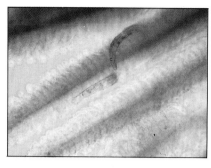

Above: *Another Dactylogyrid parasite attached to the gills of its fish host. Heavy infestation causes breathing problems.*

long worm-like parasite; Tubellarian worms, such as in Black Spot Disease and microscopic ciliated protozoa, i.e. *Brooklynella hostilis* and trichodinid species.

Life cycle These parasites spend most of their life cycle on the fish's gills, fins and body, feeding off tissue and body fluids. An explosion in numbers of these parasites leads to extreme irritation. Common symptoms are related to the irritation caused by these parasites. There is usually reddening of the skin and a heavy production of body slime, which the fish secretes in an effort to sweep away the pests. Fish

also flick and scratch against objects to get rid of the irritation. Where infestation of the gill occurs, visible damage and slime production may be noted, and the fish have difficulty breathing. Mortality is usually caused by suffocation or osmotic imbalance.

Treatment See parasite treatment table (page 156).

Microsporidian Infection

Cause Protozoa such as *Pleistophora* species.

Life cycle Microsporidian spores are ingested by the fish. This usually

happens when they are cannibalizing a previous fish host. In the gut, the spores release infective stages which penetrate and enter the bloodstream. They then enter a host cell and multiply rapidly, swelling the cell until it ruptures. This cell rupture releases more spores, causing further infection or, if the cells are close to the surface of the fish skin, may release spores into the environment. The host will eventually die. The cause of the infection gives rise to the production of white/grey tumourous masses that are visible when they occur close to the skin surface.

Treatment There is no treatment for this disease, therefore early diagnosis and rapid removal of suspected fishes to quarantine tanks is essential. If initial diagnosis is confirmed, destroy the fish and disinfect the quarantine tank with a disinfectant such as Chloramin T. Fish can be destroyed painlessly by overdosing with a product that has an anaesthetic action such as Phenoxethol at 1ml per litre.

Internal Worm Parasites

Cause There are a number of different species of internal worm parasites. These include roundworms, i.e. parasitic nematodes (these are roundworms that have a cotton-like appearance); tape worms, i.e. cestodes (these are white tape-like worms); thorny-headed worms, i.e. acanthocephalans.

Life cycle These parasites of the gut normally have relatively complex life cycles involving more than one host, so heavy infestations are rare in the aquarium. It has, however, been estimated that over 80 percent of all wild fish contain some gut parasites. Heavy infestations prevent effective digestion and actively rob the fish of nutrients so feeding fish may still be emaciated. Alternatively, massive proliferation of gut parasites may cause distending of the gut.

Treatment See parasite treatment table (page 156).

Below: Hippocampus erectus *showing white/grey tumourous masses – a clear sign of a Microsporidian infection.*

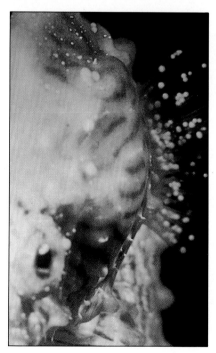

Below: *The 'head' of a cestode worm parasite with its tiny 'jaws' that severely damage the host's gut wall.*

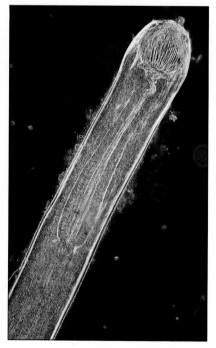

BACTERIAL DISEASES

Bacterial Finrot

Cause *Aeromonas*, *Pseudomonas* and *Mycobacter* such as *Flexibacter maritimus*.

Bacterial Finrot is typically a mild external bacterial infection. It normally causes a reddening of the base of the fins and anus. The symptom that gives rise to the name Finrot is clear erosion or rotting of the fins and the fin rays. In severe outbreaks, bacterial infection and breakdown of the gills may occur, and the fish's mouth be eaten away.

Treatment

Bacterial Finrot and gill disease are usually very clearly associated with stress, often brought on by poor water quality. Identify the stressor and remove it. See bacteria treatment table (page 157).

Vibriosis

Cause Bacteria of the *Vibrio* genus, e.g. *Vibrio anguillarum*. Similar symptoms are also caused by *Pseudomonas* species, such as *Pseudomonas fluorescense*. *Vibrio* bacteria are common in the gut flora of healthy fish and in the marine environment. *Vibrio* is only a

Above: *This is a very sick fish. A parasite infection has caused stress and this has triggered a secondary* Vibrio *bacterial infection.*

pathogen when stress allows infection from the gut or via wounds. Incubation of *Vibrio* cysts can be as little as three days, depending on temperature, virulence of the *Vibrio* strain and the level of stress the fish is under. *Vibrio* and *Pseudomonas* species produce general bacterial septicaemia or organ and tissue breakdown. Symptoms vary depending on the course of the

disease. They can be as little as lethargy, darkening of colour, anorexia and sudden unexpected death, or as severe as ulceration, abdominal swelling, due to build up of fluids, swollen eyes (popeyes) and darkening of the fish.

Treatment Treatment of *Pseudomonas* infection and Vibriosis is difficult. It must be caught early and treated with

Below: *This Squirrelfish is suffering from skin ulcers and 'dropsy', both of which are clear symptoms of a severe bacterial infection.*

effective antibiotics to avoid the typical mortality level of over 50 percent of the aquarium inhabitants. Apart from treating the disease, identifying and reducing the stress which caused the outbreak is essential. See bacteria treatment table (page 157).

Marine Fish T.B/Wasting Disease

Cause Wasting disease is caused by a bacterium called *Mycobacterium marinum*. This is the marine fishes' equivalent of tuberculosis. Infections in aquariums are introduced by a carrier fish, which contains the bacteria but shows no obvious symptoms of the disease. An upset of the fish's immune system then leads to the outbreak of Wasting Disease. The bacteria are transferred from fish to fish, either by ingestion of infected material, e.g. when picking at dead fish, or possibly through water-borne bacteria entering via skin wounds.

The incubation period for marine fish tuberculosis may be up to six weeks, so a two-month quarantine

Below: *The 'cauliflower' growths on the tail of this Angelfish are caused by the virus* Lymphocystis. *Fortunately, the infection is rarely fatal.*

period would be necessary to reduce the chances of introducing a diseased fish. T.B. is an internal systemic bacterial infection, with the bacteria spreading throughout the fish's tissue and organs, forming tubercles and lesions.

Treatment Again, this disease must be caught early and treated with the appropriate antibiotic to prevent fish mortality, which can vary from 10 to 100 percent, depending on the susceptibility of the fish in the aquarium. See bacteria treatment table (page 157).

FUNGAL DISEASES

Ichthyophonus (Ichthyosporidium)/CNS Disease.

Cause Ichthyophonus is an internal infection caused by a fungus called *Ichthyophonus hoferi*. The causative organism is an obligate parasite with a complex life cycle. The fungal cysts are taken into the fish orally, either by picking up fish faeces or cannibalizing dead fish. In the digestive system, the cysts will burst and infective amoebae pass through the gut into the bloodstream, or pass with faeces into the environment. In the bloodstream, the amoebae are

transported to the fish's organs, where they infect tissue, feed and form spherical resting cysts. The fungi can affect the central nervous system (hence its common name of CNS Disease), causing the fish to behave erratically.

Ichthyophonus carriers are the main method of introduction of this disease. This is usually an infected fish, with a slight to moderate infection, which exhibits no symptoms. Poor conditions and stress can then lead to an outbreak of the disease being carried.

Treatment Typically with Ichthophonus, a few fish die over a long period of time. Treatment of this disease is extremely difficult, and effecting a cure is unusual. Food soaked in Phonoxethol is reputed to be effective.

VIRAL DISEASES

Lymphocystis (Cauliflower Disease)

Cause Cauliflower Disease is caused by a virus called *Lymphocystis*. The *Lymphocystis* virus enters through lesions in the fish's skin, or, possibly, via the oral route. The virus infects connective tissue cells and is very slow to develop. Symptoms may take up to 10 days to become evident. The infected cells swell and enlarge, forming warty lumps like cauliflowers. These lumps can take three to four weeks to reach their full size.

The infection is not malignant and rarely fatal. Unfortunately, the slow development of full symptoms may mean that all the fish in an aquarium may be infected before the problem is noted. Infection is transmitted by viruses passed on from disintegrating old tissue from the growths.

Treatment Remove the fish suspected of carrying the *Lymphocystis* virus as soon as symptoms are noted. Place the infected fish in a quarantine aquarium and leave it there while its immune system gets rid of the virus; this may take several months.

OTHER FISH/HEALTH PROBLEMS

Head and Lateral Line Erosion – HLLE

Cause Head and Lateral Line Erosion is thought to be caused by poor environmental conditions, nutrient deficiency and, possibly, a parasite (this may be the marine equivalent of *Hexamita*, which causes Hole-in-the-Head disease in freshwater fish). However, no specific parasite species has ever been isolated from an HLLE attack. Holes develop and enlarge in the sensory pits around the head and down the lateral line of the fish. The progress of the disease is generally relatively slow, and fish do not appear unduly disturbed. Advanced cases, however, will cause severe osmoregulation problems and, possibly, death of the fish.

Treatment It is important to investigate the water quality and make any necessary improvements. Increase the greenstuff in the fish's diet and put vitamin supplements in the food. It has been suggested that you should treat with Flagyl. This is probably because this is the treatment of choice for freshwater Hole-in-the-Head Disease.

Poisoning

Cause This can be caused by a number of factors: metabolite build-up (e.g. ammonia and nitrite levels); overdosing or incompatible medications; heavy metals, chlorine or chloramine from untreated tap water; household chemicals polluting the aquarium (e.g. furniture polishes, paint/varnish fumes, smoke, etc).

Symptoms of fish poisoning depend of the level of the toxic pollutant present. Low levels may just stress the fish, reducing the effectiveness of their immune system, and irritate the skin and gills sufficiently to open the way for a secondary infection. Higher levels will directly poison the fish, causing erratic behaviour, darting around

Above: *A Blue-striped Grunt with Head and Lateral Line Erosion (HLLE).*

the aquarium and jumping, seeking an escape. High gill movement rates and gasping at the surface are also caused as delicate gill membranes are damaged.

Treatment Rapid action is essential to save fish from poisoning. They must either be removed to unpolluted water, or the pollutant must be removed from the aquarium using water changes and adsorpative filter media, such as activated carbon or artificial polymer pads.

Below: *A clownfish with a tumour on its lower lip. The condition is untreatable but, fortunately, rarely fatal.*

Disease Treatment

There are a number of methods for treating marine fish diseases. The method you adopt will depend on the specific disease, the medication of choice and your aquarium set-up. In order of preference, treatment methods are as follows:

Aquarium long bath With this method, the water-soluble treatment is added directly to the water of your established aquarium. The dose is carefully formulated so that it is non-toxic to the fish for the full duration of the treatment, but effective against the pathogen. Long-bath treatments have the advantage of being simple to administer, minimizing further stress on the sick fish, and eradicating disease pathogens both on and off the fish. Most proprietary treatments are administered in this way.

In some circumstances, this method of treatment is impossible because of one of the following reasons: active ingredients damage a mature filtration system; medication is absorbed by decorative material, organic or filter media; medication is toxic to some of the aquarium inhabitants (see *Treatment in the mixed aquarium*, page 156); medication is too expensive to use

with this method; medication needs to be administered by some other route to effect a cure, either because it is insoluble, or because it needs to be taken internally.

Many of these drawbacks can be overcome by removing the fish to a separate specially set-up hospital aquarium, where they can be 'long-bath' treated.

Dip method This method is more aggressive and more stressful than the long-bath method. Medications are used at much higher levels i.e. above the long-term toxicity level for fish. The length of exposure is critical so that the pathogens are killed, but the fish is removed before it is damaged. The other disadvantage of this method is that it only treats the disease on the fish, not the infective agents which may still be in the aquarium, and so it is not suitable for treating all diseases. It is the only way to administer a freshwater bath, which we shall look at in more detail later.

Internal medication Some medications can be taken up from the water, through the gills and the water drunk by the fish, sufficiently effectively to result in a cure.

Other medications, however, need to be administered differently. Antibiotics and internal parasite products can be injected, but this is usually beyond the capabilities of the untrained marine fishkeeper. The only alternative route for getting medications inside the fish is by mouth. Feeding fish medicated food has the advantage of being a very low-stress method, and, in treating some diseases, is the only chance of effecting a cure. However, there are a number of difficulties: sick fish often have a depressed appetite; it is hard to ensure that each fish gets a fair share of the food and, consequently, an effective dose; it is difficult to calculate correct dosages and prepare the medicated food. Often, oral dosage is quoted as so many grams of medication per kilograms of fish to be treated per day, but how many aquarists know the weight of their fish? Where possible, use a long-bath antibiotic treatment. Take advice from a vet in preparing your own medicated food if there is no alternative.

Below: *This fish is being injected with an antibiotic. This method of treatment should not be attempted except by vets or very experienced fishkeepers.*

Freshwater dips

Freshwater dips work by killing the pathogens as a result of a massive osmotic shock. When placed in the freshwater, they take water up so fast, that they literally pop.

Prepare a 5-litre (1.1 Imp./1.4 US gallon) container full of fresh water from the tap, matching the temperature of the water with that of your aquarium. Add sodium bicarbonate to increase the pH to match that of your aquarium (1 tsp per 4.5 litres/1 Imp./1.25 US gallons as a guideline). Then add 1 litre of sea water. These measures are taken to reduce the stress on the fish as much as possible.

To treat a fish, capture it in a net and place it into the fresh water bath. Fish respond differently to the immediate osmotic shock. Typically, angelfishes initially lose orientation, lying on their sides, but they will then recover.

It is essential to keep a very close eye on the fish for the duration of the bath. After the initial shock, any signs of loss of ability to maintain balance should be the trigger for immediate removal of the fish back to sea water. (See parasite treatment table on page 156 for dip times).

Copper treatment

Copper, like most medications, is toxic to both the fish and the pathogen. However, the margin of safety is much lower in the case of copper, so although it is a very effective anti-parasite cure, it does require careful use. Matters are complicated by the fact that copper is not very stable in marine systems.

Copper combines easily with calcium carbonates, which prevent its medication action. Calcium carbonates are present at high levels in marine water and are the building blocks of tufa rock and coral sand. So, in an established display aquarium, the active copper medication level drops rapidly as copper is bound up by the calcium carbonates. It is therefore better to use the copper treatment in a separate, bare, hospital aquarium.

When treating White Spot or Velvet, where parts of the life cycle will still be in the display aquarium,

Below: *The osmotic shock caused by entering freshwater kills pathogens (they take up water so fast that they pop), but freshwater dips are also stressful for the fish and should be carried out with great care.*

raise the display aquarium temperature to 80°F (c 27°C) to reduce the life cycle duration and leave the display tank bare of fish for at least a week. The infective stages will have hatched out during this week, failed to find a host, and died. Whether the copper is used in the display aquarium or in a hospital aquarium, a copper test kit should be used to ensure the necessary copper level is maintained in the aquarium for the required time.

It is wise to remove any remaining copper medication after the treatment is finished through water changes and the use of carbon or a polymer adsorption filter medium. Coral sand can act as a store of copper, which is dumped back into the aquarium at a later date, so precautionary use of carbon in the filter, or replacement of a portion of the coral sand after treatment with copper, is wise.

Copper cures come in two forms – free ionic copper and chelated copper, where the element is chemically bound up with another substance. Chelated copper is more stable than free copper and less toxic, so it can be used at higher levels. However, chelated copper

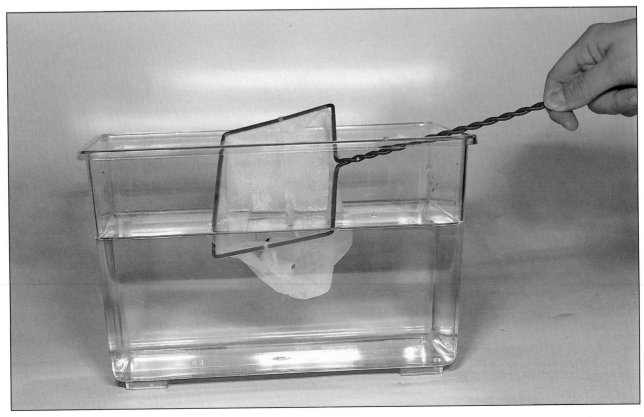

cures should not be used with ozone or UV. (Remember, too, that copper is toxic to invertebrates.)

Antibiotics

Antibiotics are only available with a veterinary prescription in the UK, while in the US they are available on general sale. Regardless of the legal situation, the use of antibiotics should be very carefully considered.

Antibiotics are valuable allies in the war against disease, but misuse can have a significant impact on both fish and human health. The main danger associated with antibiotic treatments is that inappropriate use and incorrect dosage can promote the development of resistant strains of disease organisms. The resistance is generated as rapidly multiplying pathogens undergo mutation and throw up strains that can withstand the medication. Sulphonamide antibiotics are particularly prone to the development of resistant strains.

Antibiotics can be administered by mouth, injection or long-bath treatment. The last method is to be preferred because of the simplicity of administration. However, long-bath treatment is not always

possible, since some antibiotics are not soluble in water. Another key issue is to ensure the antibiotics being used are stable in marine systems. For instance, high pH and hard water conditions rapidly cause the inactivation of Tetracycline and Sulphonamides.

Below: *A good external power filter and extra aeration are recommended for the treatment or quarantine tank.*

General guidelines for the use of antibiotics:

- Never use antibiotics as a preventative treatment.
- Treat affected fish in a separate 23-46 litre (5-10 Imp/6.3-12.5 US gallon) hospital aquarium, as antibiotics usually damage filtration systems.
- Do not use carbon, ozone, UV filtration or ion-exchange resins in the filter as these will inactivate antibiotics.
- Do not use other treatments at the same time as using antibiotics. Both copper and organic dyes, for example, inactivate some antibiotics.
- Where possible, accurately diagnose the disease and use an antibiotic with a targeted spectrum, rather than broad-spectrum products such as Tetracycline and Nitrofurazone.
- Discard all antibiotic-loaded water after treatment, or use carbon to remove the antibiotic.
- Always complete the full course of antibiotic treatment.
- Chloramphenicol is probably the most effective general use antibiotic. However, great care should be taken when handling it, since allergies to this product can be fatal.

Medicated Food

Some medicated foods are available in commercial preparations, but most have to be home-made. To make medicated foods, moisten a commercial flake food or chopped fresh food. Thoroughly mix a known amount of antibiotic into the food and add some gelatine. Spread the resultant mixture onto a tray in thin sheets and freeze, then divide each sheet up into dosage-sized cubes.

A basic treatment tank

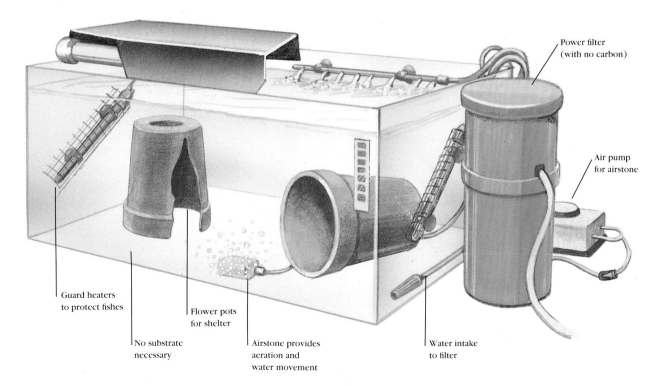

Power filter (with no carbon)

Air pump for airstone

Guard heaters to protect fishes

Flower pots for shelter

No substrate necessary

Airstone provides aeration and water movement

Water intake to filter

Treatment in a mixed aquarium

Fish disease treatment is greatly complicated in the mixed aquarium, so it is very important that you appreciate the risks and problems involved if you are contemplating a mixed aquarium.

Many of the active ingredients used in fish disease treatments are toxic to invertebrates (see page 159), particularly copper, which is universally used in treating common parasite problems. If you have a mixed aquarium, you should adhere to the following points:

• Quarantine all fish prior to adding them to a mixed tank for at least two weeks at 80°F (c 27°C) and dose them with a preventative treatment of a proprietary White Spot and Velvet cure.

• Stock the aquarium at a much reduced rate. This makes it harder for parasites to find a host. An absolute maximum stocking level is 2.5cm per 27.3 litres (1in per 6 Imp/7.5 US gallons).

• Avoid fish species that are susceptible to White Spot or Velvet, such as tangs.

• If a disease outbreak does occur, remove all the fish from the display aquarium and treat them in a separate, well-filtered hospital tank (you can use the same set-up as the quarantine aquarium). Do not return fish to the display aquarium for at least three weeks, by which time all infective stages of parasites should have died.

It must be noted that, if you do

Above: *An attractive marine reef aquarium, with fish swimming among corals. Although this is the dream of many, such a set-up does have inherent risks, which the prospective marine aquarist should be aware of.*

not take these precautions, you will be forced to leave your fish's own immune system to attempt to conquer the parasite infection alone.

Parasite treatments

CHEMICAL	DISEASE	METHOD	LEVEL	NOTES
Mebendazole	Tapeworm	Food	25-50 mg/Kg	5 Days
Yomesan (Niclosamide)	Tapeworm	Food	50 mg/10gm Fish	
Flagyl Metronidazole	HLLE	Long Bath	6.6 mg/l	
Formalin	Crustacean & Skin/Gill Parasites	Dip	200-250 mg/l	30-60 min Check every 15 min
Fresh Water	Velvet and Protozoan Parasites	Dip	See page 154	3-10 mins
Methylene Blue	Gill/Skin Parasites	Long Bath	2.9 mg/l	Affects filters
Trichlorofon	Crustaceans Black Spot	Long Bath	0.25 mg/l 1.0 mg/l	Use with care
Chloramin T	White Spot	Long Bath	10 mg/l	Use with care

Bacteria Treatments

CHEMICAL	DISEASE	METHOD	LEVEL	NOTES
Furanace	Finrot	Food	0.05-1gm/Kg	10 Days
	Vibrio	Food	0.05-1gm/kg	10 Days
Ciprofloxin	Finrot	Long Bath	6.6mg/l	*
Ciproxin Tablets	T.B.	Long Bath	6.6mg/l	*
Kanamycin	T.B.	Food	100mg/Kg	5-10 Days
		Long Bath	12-13mg/l	5 Daily doses 25% water change per day
Chloramphenicol	Vibrio	Long Bath	6.6mg/l	*
Erythromycin	Vibrio	Long Bath	13.2mg/l	*
Augmentin	Finrot	Long Bath	6.6mg/l	*
(Amoxycillin Trihydrate)	Vibrio	Long Bath	6.6mg/l	*
Nitrofurazone	Vibrio	Food	56mg/Kg	10 Days

* Treat on days 1, 3 & 5, carry out 25% water changes on days 3 & 5 before treating.

Parasite treatments Most proprietary treatments will deal with White Spot, Velvet, protozoan parasites and monogenetic trematodes.

Again, note that, in the US, organophosphates such as Trichlorofon, which treat crustacean and Black Spot parasites, are available in proprietary form. Look for products containing the following active ingredients: Quinine, Copper, Acriflavine, Formaldehyde, Methylene Blue and Malachite Green.

Bacterial treatments In general, proprietary treatments will only treat minor bacterial infections such as Finrot – very few will actually treat Vibriosis.

In the US, a number of the antibacterial/antibiotic products mentioned below are available in proprietary form. Look for products containing Phenoxetol, Malachite Green, Acriflavine, Formaldehyde and Benzalkonium chloride.

Fungus treatments *Ichthyophonus* infection is the only fungal disease generally found in marine aquariums and, as has been noted, the only reputed treatment is to soak the fish's standard food in a solution of Phenoxetol, which is available in a number of proprietary treatments.

Above: *A Lionfish is suffering from Marine Velvet disease. This parasitic infestation should be successfully treated with a proprietary copper-based anti-velvet product.*

Invertebrate Health

Keeping marine invertebrates is a relatively recent development of the aquatic hobby. As a result, very little is currently known about disease pathogens that affect tropical reef invertebrates, and still less about effective disease treatment methods. What little is known has been revealed by studies into the health of commercially important species, which include crustaceans, such as crabs, lobsters and shrimps, and molluscs, such as oysters, abalones and mussels. Investigation into the diseases that affect these farmed species have revealed that they are affected, just like fish, by a wide range of pathogens. These include:

● Parasites – *Pleistophora*, which causes Cotton Disease in shrimps; the protozoan *Heximita nelsoni* which affects shellfish; and even a snail, *Stilifer linckiae*, which preys on the Blue Linckia Starfish.

● Viruses

● Bacteria, such as *Aeromonas viridens*, which causes a disease of lobsters called Gaffkemia; *Aeromonas*, *Pseudomonas* and *Vibrio* bacteria, well-known as fish pathogens, that have been isolated from bacterial infections in invertebrates.

● Fungi such as *Aphanomyces astaci*, the 'Crayfish Plague'; and *Fusarium*, which affects the gills of crustaceans.

Treatment of these diseases has been investigated, particularly for crustaceans like shrimps. Not surprisingly, similar treatments to those used for fish have been found effective, such as Malachite Green (1mg/l) plus Formalin (25mg/l) for parasites, and Erythromycin (1.5mg/l) and Furazolidone (2.0mg/l) for bacterial problems.

The one clear lesson that has arisen from investigation into invertebrate health is that, in common with fish disease, onset is directly linked to stress caused by poor environmental conditions (overstocking and deteriorating water quality), nutritional deficiencies and incompatible combinations of species.

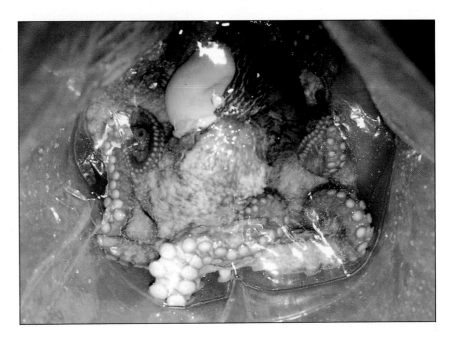

With the lack of information on diagnosis and treatment of invertebrates in an aquarium, invertebrate keepers must concentrate their efforts on prevention of disease. Pay particular attention to the following.

Select healthy specimens Look for, and avoid purchasing: anaemic, abnormally coloured corals (these have probably already lost their algae); withered and flaccid appearance; excess mucus secretions, especially around the base; localized white circular areas (these are already infected areas); eroded areas where you can see the coral skeleton; obvious signs of damage or appendages missing. (See also *Choosing and Maintaining Invertebrates*, pages 120-125.)

Above: *The ulcers on the mantle of this octopus show that invertebrates are as susceptible as fish to bacterial infection.*

Perfect water quality The following levels should be adhered to for invertebrates.

Temperature	24-27°C (75-80°F)
pH	8.1-8.3 (stable)
Specific gravity	1.021-1.024 (stable)
Ammonia	0
Nitrite	0
Nitrate	<5 mg/l
Phosphate	<0.05 mg/l
Heavy metals	0
Redox potential	350-400 mV
Alkalinity	400 mg/l (KH 14)

Good lighting Since many corals contain symbiotic algae, which require high-light intensity of the right spectrum to generate life-giving energy, good lighting is essential. Inadequate light causes these algae to die and their coral host soon follows.

Adequate water movement This is essential to supply food to filter feeders and to help clear excreted toxic waste away from the sessile invertebrates.

Left: *An invertebrate that is clearly damaged, like this starfish, should not be selected for the aquarium.*

Good nutrition A correct diet is essential if the invertebrates are not to weaken and die. It is essential to feed a correctly balanced diet (see pages 131-133) directly to corals. Avoid over-feeding, which will affect water quality. Well-lit corals require remarkably little food.

Corals also get essential micro-nutrients and trace elements by absorption from the water around them. A quality micro-nutrient mix should be fed regularly to prevent deficiencies. This should contain vitamins and trace elements, including strontium and molybdenum, which are essential for the health of hard coral. Hard corals have a calcium carbonate exoskeleton. To grow and flourish, they need a ready supply of calcium

carbonate from the water. To ensure sufficient calcium is available, some experts recommend that you add a little limewater or 'Kalkwasser' (calcium hydroxide solution) to the aquarium every day.

Many disease problems in invertebrates are caused by secondary infection of areas that have been physically damaged. This

highlights the importance of careful handling of invertebrates. It is particularly important to check for damage of the basal discs of anemones, since such damage will almost inevitably lead to death.

Below: *Such a display of healthy invertebrates is the result of careful attention to aquarium conditions.*

The toxicity of disease treatments to invertebrates	
Toxic Products	**Non-Toxic Products**
Acriflavine at 1mg/l	Quinine
Organophosphates at 0.25mg/l	Nifurpirinol
Copper at 0.05mg/l	Nitrofurazone
	Formalin
	Malachite Green

PART FOUR

TROPICAL MARINE FISHES

Although nearly three-quarters of the earth's surface is kept covered by salt water, relatively few marine fishes are kept in the aquarium, and of these, the vast majority are tropical species. Nevertheless, such is the appeal of this small group of fishes that the marine fishkeeping hobby has flourished and continues to attract an ever-growing number of enthusiasts around the world.

The most striking tropical marine fishes are native of the coral reefs and coastal waters, where collection is quite easy. Fishes from the deepest waters usually grow too large for the aquarium, and also present too many collection and transportation problems. The majority of suitable fishes come from the Indo-Pacific Oceans, the Caribbean area of the Northern Atlantic Ocean, and the Red Sea.

The fish featured in this section should all be readily available to the home aquarist. Although not an exhaustive survey, it does provide a typical illustration of each of the more commonly kept families of tropical marine fishes. If you cannot find the specific fish you are looking for, similar species will usually provide enough useful data.

Many species have been omitted on the grounds that they have proved extremely difficult or impossible to keep in the home aquarium. Others are, or are likely to become, the subject of various import bans in some countries. Appendix Two on pages 380-381 lists some of those species still currently available but not recommended for the aquarist, and you may have to make a decision, based on conscience and practicality, as to whether or not to keep these species.

Left: *The exotic beauty of the Queen Angelfish is undeniable and typifies what attracts many people to the marine fishkeeping hobby. Chosen wisely, fish such as this can thrive in captivity and give pleasure for many years.*

Family: ACANTHURIDAE

Surgeons and Tangs

Family characteristics

Members of this family are characterized by their high profile and laterally compressed, oval bodies. In addition, they have very sharp 'scalpel-like' erectile spines on the caudal peduncle (hence the name 'Surgeons'), which are used during inter-territorial disputes and in defence. The dorsal and anal fins are long-based and the eyes are set high on the head. The scales often end in a small protruberance, giving a rough feel to the skin. In their natural habitat these fishes may grow up to 400mm (16in), but aquarium specimens usually attain only half the size, if that, of their wild counterparts.

Although there are no drastic colour changes between juveniles and adults in most species, the Caribbean Blue Tang (*Acanthurus coeruleus*) has a yellow juvenile form. Since the adult colour occurs at no predetermined age or size, small fishes can show adult coloration while larger specimens retain their immature colours. When the change occurs, the body is the first area to show the blue adult colour, followed by the caudal fin. Thus, for a period there is an intermediate stage; which has a blue body with a yellow caudal fin.

Although external differences between the sexes are normally rare, some darkening of the male's colours during breeding is quite usual. Size is not a reliable indication of the sex of the fishes; sometimes the male is larger, sometimes the female. The pelagic (free-floating) eggs that result from the typical ascending spawning actions of two fishes (or a group of fishes) take a long time – possibly months – to pass through the planktonic stage. This means that, although spawning in captivity may occur, rearing the fry may prove to be much more difficult.

Diet and feeding

Surgeons and tangs need to be fed several times each day, especially if there is insufficient algal growth for them to browse upon. In fact, algae are such an important element of their diet that you should not introduce them into an algae-free aquarium.

Young fishes grow very quickly and will starve if denied ready nourishment. Although many species are herbivorous, others will eat small animals too, which means that once they have become accustomed to feeding in captivity they will take many of the established dried, frozen and live foods.

Aquarium behaviour

Surgeons and tangs live in shoals around the coral reefs of the world. In the aquarium, however, they forsake this gregariousness and will quarrel among themselves, unless you provide a suitably spacious tank. Established species often resent new fishes introduced into the aquarium; smaller fishes may well get off with a warning but similarly sized fishes may suffer attacks. Young surgeon and tang specimens, whose scalpels are, fortunately, not as dangerous as those of adults, mount threatening motions against newcomers, but, thankfully, these displays are generally shortlived.

All members of this family are somewhat prone to Oodinium and White spot infection, especially in less than perfect water conditions. Unfortunately, the copper-based medications that are used to treat such infections generally preclude the suitability of these fishes in mixed fish and invertebrate aquariums (see *Health Care and Disease Treatment*, pages 144-159).

SCIENTIFIC CLASSIFICATION AND THE HOBBYIST

It is most unfortunate that scientific classification does not necessarily correspond to that used by the hobbyist. The academic, scientific system has to deal with all known organisms in a universal way. There has to be as little uncertainty as possible so that scientists of different nations, speaking different languages, all know which organism is under discussion or being talked about. The hobbyist, by contrast, deals with a small and eclectic sample of the world's fauna and flora and this is reflected in his or her terminology, which suits that localized interest. There is nothing wrong with this attitude because it has evolved to satisfy and cope with their needs and enables enthusiasts to communicate with each other. But only within limits. For an English-speaking hobbyist to communicate with precision to another whose native tongue is not English there has to be resort to scientific names. That species you may know as the Harlequin Tuskfish is not going to evoke any response from a Russian speaker, nor even, necessarily, from an Australian. However, *Lienardella fasciata* (this species' scientific name) has a universality that transcends native languages. Hence the important need for the scientific classification.

Unfortunately, although scientific classification has the advantages of being universal and having reference points, this does not mean to say that there is universal agreement. Indeed, there are some areas that are the subject of particular scientific dispute. For example, a series of 'very similar' fish known by only a handful of specimens from just a few sites over a wide area may belong to a single widespread but variable species, or be specimens of many, closely related species; only further study could decide and later information may change the early conclusions.

Acanthurus achilles
Achilles Tang; Red-tailed Surgeon

☐ **Distribution:** Pacific.

☐ **Length:** 250mm/10in (wild), 180-200mm/7-8in (aquarium).

☐ **Diet and feeding:** Will accept the usual protein foods, such as gamma-irradiated frozen foods (*Mysis* shrimp, plankton, krill, etc.) and live brineshrimp, plus algae and other greenstuff, such as blanched lettuce and spinach. Shy grazer.

☐ **Aquarium behaviour:** Normally peaceful, but very delicate. Compatible with most fish, but may fight at first with other members of its own family. Do not add to tank until the first fish are established.

☐ **Invertebrate compatibility:** Not recommended.

The brown body is offset by yellow-red baselines to the dorsal and anal fins. The white marking on the gill cover behind the eye and the dull white patch on the chest are shared by other members of the family, but the feature that positively identifies this fish is the teardrop-shaped orange-red area on the caudal peduncle, in which the scalpels are set. Like most members of the family, this is a beautiful fish, but young specimens do not have nearly as many red markings as the adults.

Below: Acanthurus achilles
Members of the Acanthuridae family are easily distinguished by their oval shape. Apart from one or two species – many like this Achilles Tang – are brilliantly coloured, with beautiful body patterns.

Acanthurus coeruleus
Blue Tang

☐ **Distribution:** Western Atlantic.

☐ **Length:** 300mm/12in (wild), 150mm/6in (aquarium).

☐ **Diet and feeding:** Mainly algae. Bold grazer.

☐ **Aquarium behaviour:** Small specimens may become bullies if established in the aquarium ahead of other fishes, but this tendency generally decreases with time.

☐ **Invertebrate compatibility:** Not recommended.

Young fishes are yellow with blue markings around the eye. As they age, the fish develop narrow blue lines, the adult fish being darker blue than the 'almost adult' fish. The scalpels on the caudal peduncle are ringed with yellow or white in mature fishes.

Above: Acanthurus coeruleus (adult)
In adulthood, the Blue Tang may lose its territorial nature and become slightly more sociable. Its dark blue coloration defies the bright yellow of the juvenile.

Below: Acanthurus coeruleus (juvenile)
The young fish is bright yellow with blue-rimmed eyes – not to be confused with Zebrasoma flavescens (see page 169). Small specimens can be quite aggressive.

Acanthurus glaucopareius
Goldrim Tang; Powder Brown

☐ **Distribution:** Mainly the Pacific Ocean, but is sometimes found in the eastern Indian Ocean.

☐ **Length:** 200mm/8in (wild).

☐ **Diet and feeding:** Algae. Bold grazer.

☐ **Aquarium behaviour:** Normally peaceful.

☐ **Invertebrate compatibility:** Not recommended.

It is fairly easy to identify this fish by the white area on the cheeks. Yellow zones along the base of blue-edged dorsal and anal fins may extend into the base of the caudal fin. A yellow vertical bar crosses the caudal fin.

Below: Acanthurus glaucopareius
The markings on this fish are extremely fine. There is some justification in defining it as a Powder Brown as the colour patterning is similar to that of the Powder Blue Surgeon, Acanthurus leucosternon. Acanthurus glaucopareius is easily distinguished by the white cheek patches, however. It is a fairly easy species to keep in the aquarium, being generally quite peaceful, but should not be kept with invertebrates.

Acanthurus leucosternon
Powder Blue Surgeon

☐ **Distribution:** Indo-Pacific.

☐ **Length:** 250mm/10in (wild), 180-200mm/7-8in (aquarium).

☐ **Diet and feeding:** Protein foods and vegetable matter. Bold grazer.

☐ **Aquarium behaviour:** Keep only one in the aquarium. Dealers usually segregate juveniles to prevent quarrels developing.

☐ **Invertebrate compatibility:** Not recommended.

Above: Acanthurus leucosternon
One of the most familiar of all surgeonfishes, the Powder Blue Surgeon is a firm favourite with hobbyists. It is best kept on its own (as any two will fight) and needs plenty of room, excellent water conditions and sufficient vegetable matter in its diet.

This is a favourite surgeon among aquarists. The oval-shaped body is a delicate blue; the black of the head is separated from the body by a white area beneath the jawline. The dorsal fin is bright yellow, as is the caudal peduncle. The white-edged black caudal fin carries a vertical white crescent. The female is larger than the male. In common with all surgeons, it requires plenty of space and optimum water conditions.

Acanthurus lineatus
Clown Surgeonfish; Blue-lined Surgeonfish; Pyjama Tang

☐ **Distribution:** Indo-Pacific.

☐ **Length:** 280mm/11in (wild), rarely above 150mm/6in in the aquarium.

☐ **Diet and feeding:** Algae. Bold grazer.

☐ **Aquarium behaviour:** Small specimens can be quarrelsome. Keep only one per tank or, alternatively, try keeping several together, rather than a pair, if the aquarium is large enough, on the assumption that there is safety in numbers.

Acanthurus sohal
Zebra Surgeon; Majestic Surgeon

☐ **Distribution:** Red Sea.

☐ **Length:** 250mm/10in (wild), 180mm/7in (aquarium).

☐ **Diet and feeding:** Algae. Bold grazer.

☐ **Aquarium behaviour:** Small specimens can be quarrelsome; keep only one per tank.

☐ **Invertebrate compatibility:** Not recommended.

Above: Acanthurus sohal
This striking fish is not a common sight in aquatic dealers' tanks, but its beautiful body lines make it very noticeable when it does appear. As with other surgeonfishes, young specimens can be quarrelsome.

This smart fish is similar in body shape to *A. lineatus*. Its blue-edged, blue-black fins add an outline to the pale body, and the scalpels are marked with a vivid orange stripe. The upper part of the body and head are covered with a series of parallel dark lines. A rare but beautiful fish.

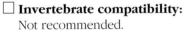

☐ **Invertebrate compatibility:** Not recommended.

The yellow ground colour of the body is covered with longitudinal dark-edged, light blue lines. The pelvic fins are yellow.

Right: Acanthurus lineatus
This is one of the species of the Acanthuridae family that has a split level of coloration; there is a lighter area to the lower body with decorative parallel longitudinal lines above. Like other surgeonfishes, it appreciates some coral or rockwork to provide welcome sheltering places. Ideally, keep only one of these fish in an aquarium (unless you have a very large tank for several), as small specimens, in particular, can be very quarrelsome.

Naso lituratus
Lipstick Tang; Lipstick Surgeon

☐ **Distribution:** Indo-Pacific.

☐ **Length:** 500mm/20in (wild), 200mm/8in (aquarium).

☐ **Diet and feeding:** Protein foods and greenstuff. Bold grazer.

☐ **Aquarium behaviour:** Normally peaceful.

☐ **Invertebrate compatibility:** Not recommended.

The facial 'make-up' of this fish is quite remarkable; the lips are red or orange and a yellow-edged, dark brown-grey mask covers the snout and eyes. The front of the narrow dorsal fin is also bright yellow. The basic colour of the dorsal fin varies according to geographical origin of the fishes; Hawaiian specimens have a black dorsal, while in those from the Indian Ocean the dorsal is orange. The two immovable, forward-pointing 'scalpels' on each side of the caudal peduncle are set in yellow patches.

Paracanthurus hepatus
Regal Tang

☐ **Distribution:** Indo-Pacific.

☐ **Length:** 250mm/10in (wild), 100-150mm/4-6in (aquarium).

☐ **Diet and feeding:** Algae. Bold grazer.

☐ **Aquarium behaviour:** May occasionally be aggressive towards members of the same species.

☐ **Invertebrate compatibility:** Not recommended.

The brilliant blue body has a black 'painter's palette' shape marking, but the most striking feature of this species is the bright yellow wedge section in the caudal fin. The dorsal and anal fins are black-edged, and the pectoral fin is yellow-tipped.

Right: Paracanthurus hepatus
The striking black markings on the blue body and its yellow caudal fin make positive identification of the Regal Tang very easy. This is one of the few members of the Acanthuridae family that can generally be kept safely in the company of its own species.

Below: Naso lituratus
The extremely well-defined facial markings of the Lipstick Tang are quite remarkable – worthy of any beautician's salon! The remainder of the streamlined fish is no less attractive, with a lyre-shaped caudal fin and double scalpels set in vivid patches.

☐ Invertebrate compatibility: Not recommended.

It is unusual to find a marine fish of a single colour, but the vividness of the bright yellow makes up for any lack of pattern. This species can be distinguished from juvenile forms of *A. coeruleus* by the absence of blue around the eyes, although a more obvious guide is the difference in shape of the head and mouth.

Zebrasoma flavescens
Yellow Tang

☐ Distribution: Pacific Ocean.

☐ Length: 200mm/8in (wild), 100-150mm/4-6in (aquarium).

☐ Diet and feeding: Algae. Bold grazer.

☐ Aquarium behaviour: Highly territorial. Keep either a single fish or a group of six or more per large tank.

Below: Zebrasoma flavescens
The long snout of this species enables it to graze effortlessly on luxuriant growth of algae. In common with Paracanthurus hepatus, *the Yellow Tang can be kept in a shoal, provided that enough space is available in the aquarium.*

Zebrasoma veliferum
Striped Sailfin Tang

☐ **Distribution:** Indo-Pacific, Red Sea.

☐ **Length:** 380mm/15in (wild), 180-200mm/7-8in (aquarium).

☐ **Diet and feeding:** Protein foods and greenstuff. Bold grazer.

☐ **Aquarium behaviour:** Normally peaceful, but may be aggressive with large fish. Young fish are more adaptable and so do better in captivity than adults.

☐ **Invertebrate compatibility:** Not recommended.

The main feature of this species is the large sail-like dorsal fin and almost matching anal fin; both are patterned. Coloration of both fins and body may be variable in shades

of brown overlaid with several vertical bands.

Below: Zebrasoma veliferum
This juvenile lacks the facial markings of the adult fish. The typical tang body shape is clearly visible. Juveniles do better than adults in captivity.

Zebrasoma xanthurum
Purple Sailfin Tang; Emperor Tang

☐ **Distribution:** Indo-Pacific, Red Sea.

☐ **Length:** 200mm/8in (wild).

☐ **Diet and feeding:** Protein foods and greenstuff. Bold grazer.

☐ **Aquarium behaviour:** Normally peaceful.

☐ **Invertebrate compatibility:** Not recommended.

Above: Zebrasoma xanthurum
This incredibly beautiful fish is generally imported close to full adult size and you will need an appropriately large tank to accommodate it.

The body colour may vary from purplish blue to brown, depending on the fish's natural habitat. A number of dark spots cover the head and front part of the body. The caudal fin is bright yellow.

Family: APOGONIDAE

Cardinalfishes

Family characteristics

Cardinalfishes are generally slow-moving, often nocturnal fishes that hide among coral heads during the day. However, at the approach of a net, they can move very fast! They are usually found on coral reefs, but some frequent tidal pools and one species enters fresh water.

Unusually for a marine fish, the two separate dorsal fins are carried erect. This feature, together with the large head, mouth and eyes, is a characteristic of the family.

Reproduction is by mouthbrooding. The male generally incubates the eggs, although in some species this task is undertaken by the female. In other species within the family, both sexes share the responsibility.

Diet and feeding

Once acclimatized to aquarium conditions, cardinalfishes will eat most live and other foods (but never flake food). Do not keep them with fast-swimming boisterous fishes or they will lose out in the competition for food. It is a good idea to feed cardinalfishes late in the evening, since this will suit their nocturnal lifestyle and may result in a greater willingness to accept new foods.

Aquarium behaviour

Hardy fishes that should be acclimatized gradually to the bright lights of the main aquarium. Cardinalfishes are an ideal choice for the beginner.

Apogon maculatus
Flamefish

☐ **Distribution:** Western Atlantic.

☐ **Length:** 150mm/6in (wild), 750mm/3in (aquarium).

☐ **Diet and feeding:** All foods. Shy feeder.

☐ **Aquarium behaviour:** Prefers a quiet aquarium with fishes of a similar disposition.

☐ **Invertebrate compatibility:** Ideal for the invertebrate aquarium.

Left: Apogon maculatus
The strikingly coloured Flamefish is much slimmer than the more common Pyjama Cardinalfish, Sphaeramia nematopterus. *Like the latter, it is a nocturnal species by nature, and prefers to share its aquarium with less boisterous fishes. It will usually settle well to aquarium life once established.*

This bright red fish with two white horizontal lines through the eye is very easy to identify. It has two dark spots on the body, one below the second dorsal fin and the other on the caudal peduncle (although faint at times). It is a nocturnal feeder and, although shy, usually settles down well in captivity.

Sphaeramia nematopterus
Pyjama Cardinalfish; Spotted Cardinalfish

☐ **Distribution:** Indo-Pacific.

☐ **Length:** 100mm/4in (wild), rarely seen above 75mm/3in in the aquarium.

☐ **Diet and feeding:** Most foods, but not flake.

☐ **Aquarium behaviour:** Do not keep with larger boisterous species, which would upset the tranquil lifestyle of these fishes.

Above: Sphaeramia nematopterus
The body shape of this fish might be reminiscent of the freshwater tetras, although it boasts an extra dorsal fin and much larger eyes.

☐ **Invertebrate compatibility:** Ideal for the invertebrate aquarium.

This fish has three distinct colour sections to its striking body, each dissimilar to the next, almost as if it had been assembled like an 'identikit'. The large head section, back to the first of the two dorsal fins, is yellow-brown in colour. A dark brown vertical band joins the first dorsal fin to the pelvic fins. A spotted paler area covers the rear of the fish. The large eyes indicate a naturally nocturnal behaviour. It may be necessary to acclimatize this species with live foods, but, once settled in the aquarium, it will eat well. However, do not offer flake food. This fish was formerly known as *Apogon orbicularis*.

Family: BALISTIDAE

Triggerfishes

Family characteristics

Members of this family have acquired their common name from the characteristic locking and unlocking mechanism of the first dorsal fin. This fin is normally carried flat in a groove, but it can be locked into position by a third ray, thus preventing the fish from being eaten or withdrawn from a crevice in which it has taken refuge.

Triggerfishes are relatively poor swimmers. They achieve propulsion by undulating wave motions of the dorsal and anal fins, the caudal fin being saved for emergency accelerations when required. The pelvic, or ventral, fins are absent in most species, or are restricted to a single spine or knoblike protruberances.

Body coloration can range from the dull to the psychedelic. The patterning around the mouth is typically exaggerated, probably to deter rivals or predators. The teeth are very strong and often protrusive – ideal for eating shelled invertebrates and sea urchins. You should not keep triggerfishes with invertebrates in the home aquarium. Also take care that they do not nip your fingers!

Reproduction takes place in pits dug in the sand within the territory of one of the female fish. These territories, in turn, are all enclosed within the dominant male's greater territory. The eggs, presumed to be demersal (i.e. heavier than water), are released either in an ascending swimming action or over a preselected site.

In the sea, triggerfishes live alone and are intolerant of similar species in the aquarium. They may adopt peculiar resting positions, headstanding or even lying on their sides.

Diet and feeding

Triggerfishes are greedy feeders, accepting anything that is offered. Natural foods taken by the bottom-feeding species of the family include echinoderms such as starfishes and sea urchins, which they devour complete with the spines. Triggerfishes consume the Crown-of-Thorns Starfish in a specific manner – they first blow the starfish over so that its spines are out of the way and then eat the soft unprotected underbelly. Species that occupy the middle and upper waters of the tank take plankton and green foods. Suitable aquarium foods include frozen foods, chopped earthworms and live river shrimp.

Aquarium behaviour

The behaviour of triggerfishes in the aquarium varies from peaceful to unaccommodatingly aggressive, depending on the species. Fishes rest in crevices or caves at night, and so it is advisable to aquascape the aquarium to allow for this. Do not be surprised, however, if the fish take advantage of your thoughtfully provided refuges when you try to net them! In nature, they favour underwater cliff faces; this is especially so in the Caribbean species.

Balistapus undulatus
Undulate Triggerfish; Orange-green Trigger; Red-lined Triggerfish

☐ **Distribution:** Indo-Pacific.

☐ **Length:** 300mm/12in (wild), 200mm/8in (aquarium).

☐ **Diet and feeding:** Corals, crustaceans, molluscs, sea urchins. Bold grazer.

☐ **Aquarium behaviour:** This is the most aggressive triggerfish of all; keep it out of aquariums that contain invertebrates and most fish. Using its powerful jaws, *B. undulatus* is in the habit of picking up lumps of coral and distributing them elsewhere in the aquarium. However, despite its aggressive behaviour and potential size – in a large tank it will grow to about 300mm/12in –

it is a rewarding fish to keep, becoming quite tame and enjoying a lot of fuss from its owner.

☐ **Invertebrate compatibility:** This species must not be kept with invertebrates, unless they are destined to be its food!

This species was first discovered in 1797 by the Scottish explorer Mungo Park. In the wild, it is found over a wide area of the Indo-Pacific, although not around Hawaii. The body coloration of this fish – always striking – can vary quite markedly, as its common names suggest. Indian Ocean variants have orange tails, while Pacific specimens have orange-rayed green caudal fins. Males are larger, with no orange banding on the head. Several large spines are arranged in two rows on the caudal peduncle.

Balistes bursa
White-lined Triggerfish; Bursa Trigger

☐ **Distribution:** Indo-Pacific.

☐ **Length:** 250mm/10in (wild), 150mm/6in (aquarium).

☐ **Diet and feeding:** All foods. Bold.

☐ **Aquarium behaviour:** Unsociable towards other Triggerfishes. Aggressive in general to other fishes.

☐ **Invertebrate compatibility:** No.

The red and yellow lines on the head joining the eye to the pectoral fin and the snout to the pectoral fin are the principal clues to the identification of this fish. An area of

Above: Balistapus undulatus

It is easy to understand why this fish is popular, but its striking coloration, and potential tameness should be considered along with another of its traits – it is very aggressive.

Right: Balistes bursa

The coloration around the mouth, together with the lighter body colours, accentuates and apparently enlarges the actual size of the mouth – a good deterrent against would-be predators.

light blue runs below the horizontal line from snout to vent. The fins are virtually colourless. Males are larger and more colourful than females.

Balistes vetula
Queen Triggerfish; Conchino; Peja Puerco

☐ **Distribution:** Tropical western Atlantic.

☐ **Length:** 500mm/20in (wild), 250mm/10in (aquarium).

☐ **Diet and feeding:** Crustaceans, molluscs, small fishes, usual frozen foods, etc. Bold; will take good-sized pieces.

☐ **Aquarium behaviour:** Do not keep with small fishes. Although peaceful with other species, it will quarrel with its own kind.

☐ **Invertebrate compatibility:** No.

Dark lines radiate from around the eyes and there are striking blue facial markings. The tips of the dorsal fin and caudal fins become filamentous with age, especially in the male, which is larger and more colourful than the female. This beautifully marked species may become hand-tame in captivity.

Right: Balistoides conspicillum
With its spectacular spotted markings, the Clown Trigger is an unmistakable fish. Balistoides conspicillum *is a bold, but also aggressive species, so do not keep it with small fishes.*

Below: Balistes vetula
A characteristic of triggerfishes is that many will become hand tame in time. Exercise care when hand feeding, however, for their slightly protrusive teeth are very sharp. Try impaling pieces of food on a cocktail stick before offering them to the fish – it will be safer than hand feeding.

Balistoides conspicillum
Clown Trigger

☐ **Distribution:** Indo-Pacific.

☐ **Length:** 500mm/20in (wild); 250mm/10in (aquarium).

☐ **Diet and feeding:** Crustaceans, molluscs. Bold.

☐ **Aquarium behaviour:** Aggressive. Do not keep with small fishes.

☐ **Invertebrate compatibility:** No.

This is a stunningly beautiful and easily recognizable species, with its large white-spotted lower body. The 'brightly painted' yellow mouth may serve to deter potential predators, while the disruptive body camouflage may assist species recognition. Both the dorsal and anal fin are basically pale yellow and the caudal fin is dark edged. The pectoral fin is clear.

Melichthys ringens
Black-finned Triggerfish

☐ **Distribution:** Indo-Pacific.

☐ **Length:** 500mm/20in (wild), 250mm/10in (aquarium).

☐ **Diet and feeding:** All foods. Bold grazer.

☐ **Aquarium behaviour:** Peaceful. A very gentle triggerfish.

☐ **Invertebrate compatibility:** No.

The body is brownish and the fins are black, but it is the white lines at the base of the dorsal and anal fins and the white-edged caudal fin that distinguish this species.

Below: Melichthys ringens
Although not usually imported in great quantities, the Black-finned Triggerfish is certainly worth waiting for.

Above: Rhinecanthus aculeatus
Whether used for camouflage, species recognition or as a deterrent, the exaggerated patterns of the Picasso Trigger make it easy to distinguish from other members of the family.

Left: Odonus niger
This species is very colour variable, depending on the collection area; accordingly, it has also occasionally been called the Blue Triggerfish or the Green Triggerfish.

Odonus niger
Black Triggerfish

☐ **Distribution:** Indo-Pacific and Red Sea.

☐ **Length:** 500mm/20in (wild), 250mm/10in (aquarium).

☐ **Diet and feeding:** All foods. Bold.

☐ **Aquarium compatibility:** Fairly sociable and relatively peaceful.

☐ **Invertebrate compatibility:** No.

The body coloration of *O. niger* can vary from blue to green from day to day. The red teeth are often quite conspicuous. Propulsion is achieved by undulations of the dorsal, anal and caudal fins rather than by body movements.

Rhinecanthus aculeatus
Picasso Trigger

☐ **Distribution:** Indo-Pacific.

☐ **Length:** 300mm/12in (wild), 230mm/9in (aquarium).

☐ **Diet and feeding:** Crustaceans, molluscs, 'sea meat' foods. Bold.

☐ **Aquarium behaviour:** Aggressive towards members of the same species and towards other fish of the same size.

☐ **Invertebrate compatibility:** No.

The 'avant garde' colours of this fish make it a popular species. A number of diagonal white bars slant upwards and forwards from the anal fin. The mouth and jawline are accentuated with colour and a blue and yellow-brown stripe across the head connects the eye with the pectoral fin base. This fish may emit a distinctive whirring sound when it is startled.

Family: BLENNIIDAE

Blennies

Family characteristics

Bluntheaded, elongate and constantly active, Blennies make a cheerful addition to the aquarium, although they should be kept in a species tank rather than in a community collection. They naturally frequent inshore waters, hiding in any handy cave or crevice, not always bothering to follow the tide out to sea. Provide suitable living quarters in the aquarium by arranging rockwork to form plenty of hiding crevices and 'observation posts'.

The dorsal fin is long and there are cirri (hairy, bristle-like growths) above the high-set eyes. The skin is slimy, hence the alternative name of slimefishes.

Male blennies tend to be larger and more colourful than females. During breeding, the male may undergo changes in colour during both the pre- and post-spawning periods. One member of the family, the False Cleanerfish, lays its eggs in any handy shelter – empty shells are particularly acceptable – and the eggs are guarded by the male.

Diet and feeding

Blennies are completely omnivorous, eating everything from algae to small fishes and bits of large ones! They will even take dried foods with apparent relish.

Aquarium behaviour

Some blennies are very territorial and aggressive to any other fishes; *Ophioblennius* is a typical case that should not be kept with any fish less than twice its size.

Aspidontus taeniatus
False Cleanerfish; Sabre-toothed Blenny; Cleaner Mimic

☐ **Distribution:** Indo-Pacific.

☐ **Length:** 100mm/4in (wild).

☐ **Diet and feeding:** Skin, scales and flesh – preferably from living, unsuspecting victims! Sly and devious.

☐ **Aquarium behaviour:** Do not keep with other fishes.

☐ **Invertebrate compatibility:** Ideal.

Using its similarity in size, shape and colour to the true Cleanerfish, *Labroides dimidiatus*, this fish approaches its victims, which expect the usual 'cleaning services'; instead they end up with a very nasty wound and a little bit wiser. Easily distinguished by its underslung mouth, which gives it a shark-like appearance. Because of its predatory nature, this species is not recommended for inclusion in the aquarium under any circumstances. It is featured here primarily as a warning, so that you can avoid it.

Below: Aspidontus taeniatus
This fish should be kept on its own in an aquarium; it uses its coloration to mimic the true Cleaner Wrasse, Labroides dimidiatus, *and denude its unsuspecting victims of lumps of flesh! It can be distinguished by its underslung, sharklike mouth.*

Ecsenius bicolor
Bicolor Blenny

☐ **Distribution:** Indo-Pacific.

☐ **Length:** 100mm/4in (wild).

☐ **Diet and feeding:** Frozen marine foods and algae. Bottom-feeding grazer.

☐ **Aquarium behaviour:** Shy with larger fishes; needs hiding places.

☐ **Invertebrate compatibility:** Ideal.

The front half of this fish is brown, the rear orange-red. During spawning, the male turns red with white bars. The female's breeding colours are light brown and yellow-orange. After spawning, the male becomes dark blue with light patches on each side of the body.

Ecsenius midas
Midas Blenny

☐ **Distribution:** Red Sea and Indian Ocean.

☐ **Length:** 100mm/4in (wild).

☐ **Diet and feeding:** Will accept most marine frozen, live and flake foods quite readily.

☐ **Aquarium behaviour:** Can be territorial, but is usually peaceful.

☐ **Invertebrate compatibility:** Ideal.

Above: Ecsenius midas
When swimming, the Midas Blenny is rather reminiscent of a tiny Moray Eel, although the two are totally unrelated. E. midas *is a fish full of character that seems almost completely fearless once settled into a new home, even in the face of much larger fish. Food is definitely top of its list of priorities.*

This endearing fish requires plenty of holes into which to retreat and use as vantage points to observe the rest of the aquarium scene. *E. midas* is an ideal beginner's species, which does well in fish-only and invertebrate aquariums.

Below: Ecsenius bicolor
The rearmost body colour of this shy fish is often hidden, since it spends much of the time in the many hiding places you will need to provide in the aquarium. Given time, however, this endearing fish becomes more confident.

Ophioblennius atlanticus
Redlip Blenny

☐ **Distribution:** Tropical western Atlantic.

☐ **Length:** 120mm/4.7in (wild).

☐ **Diet and feeding:** Frozen marine foods and algae. Bottom-feeding grazer.

☐ **Aquarium behaviour:** Territorial, and it chases everything.

☐ **Invertebrate compatibility:** Ideal.

Keep the Redlip Blenny in a community of hardy fishes and provide plenty of hiding places for it in the aquarium.

Left: Ophioblennius atlanticus
Very common off the West Indies, the Redlip Blenny varies in colour from very dark, as here, to almost white.

Petroscirtes temmincki
Striped Slimefish; Scooter Blenny

☐ **Distribution:** Indo-Pacific.

☐ **Length:** 100mm/4in (wild).

☐ **Diet and feeding:** Algae, small animals. Bottom grazer.

☐ **Aquarium behaviour:** Can be kept in small groups of the same species.

☐ **Invertebrate compatibility:** Ideal.

The body shape is that of a typical blenny, with the eyes set up high on the head. There are no cirri. The coloration is black with white blotches, plus bright blue spots on the head region. Males have an elongated 'flag-shaped' dorsal fin with which they signal to attract females and deter rival males.

Left: Petroscirtes temmincki
The unattractively named Striped Slimefish is nevertheless endearing.

Above: Salaria fasciatus
It is a pity that blennies are so keen on hiding away in caves and under rocks, for it is only when they emerge into more open areas that you can see their body colour patterns, long finnage and cirri – crestlike growths above the eyes.

Salarius fasciatus
Banded Blenny

☐ **Distribution:** Indo-Pacific.

☐ **Length:** 100mm/4in (wild).

☐ **Diet and feeding:** Small animal foods and algae. Bottom grazer.

☐ **Aquarium behaviour:** Requires plenty of hiding places.

☐ **Invertebrate compatibility:** Ideal.

The elongate body is covered with mottled light and dark brown vertical bands, extending into the long-based dorsal fin. The eye patterning – radiating stripes around the rim – is a particular feature.

Family: CALLIONYMIDAE

Mandarinfishes and Dragonets

Family characteristics

Mandarinfishes and the related dragonets are small, mainly bottom-dwelling species that often bury themselves in the sand during the day. Sometimes they will perch on a firm surface not too far away from the aquarium floor.

Sexing these fish is fairly straightforward, males having longer dorsal and anal fin extensions and brighter colours than females. Some reports indicate that fertilization – at least in some species – may be internal. The eggs are scattered in open water, and are normally described as pelagic (in contrast to so-called demersal eggs, which are deposited on a surface). Although this behaviour has been observed in aquariums, no fry have yet been raised.

Avoid buying badly emaciated specimens: they usually fare very poorly.

Diet and feeding

Members of this family feed predominantly on small marine animals, such as crustaceans, that live among the debris on the seabed. An established aquarium with a good population of micro-organisms, and the occasional supplement of rotifers or brineshrimp nauplii, is therefore essential for these fishes.

Aquarium behaviour

Mandarinfishes should be kept singly or in matched pairs. They are ideally suited to a quiet aquarium containing fishes of a similar disposition. Seahorses make suitable companions. Avoid bare tanks housing large fish, as these very rarely provide the best conditions for the comfort and welfare of these species.

Synchiropus picturatus
Psychedelic Fish

☐ **Distribution:** Pacific.

☐ **Length:** 100mm/4in (wild).

☐ **Diet and feeding:** Small crustaceans and algae. Shy bottom-feeding grazer.

☐ **Aquarium behaviour:** Likely to be intolerant of their own kind. May be better able to cope with livelier tankmates than *Synchiropus splendidus*.

Left: Synchiropus picturatus
The bold-patterned Psychedelic Fish is an attractive subject for a quiet tank.

☐ **Invertebrate compatibility:** Ideal.

The basically green body is adorned with lemon-edged, darker green red-ringed patches. This species is found in the Philippines and Melanesia. It is less common than *S. splendidus*.

Synchiropus splendidus
Mandarinfish

☐ **Distribution:** Pacific.

☐ **Length:** 100mm/4in (wild).

☐ **Diet and feeding:** Small crustaceans and algae. Shy bottom-feeding grazer.

Above: Synchiropus splendidus
The Mandarinfish is more common and more gaudily coloured than its relative. Like S. picturatus, *it prefers a quiet tank.*

☐ **Aquarium behaviour:** It is best kept in a quiet aquarium away from larger, more lively fishes.

☐ **Invertebrate compatibility:** Ideal.

This fish has much more red in its coloration – in random streaks around the body and fins – than *S. picturatus*. Males usually develop a longer dorsal spine. It is said that the skin mucus of *Synchiropus* species is poisonous, a fact often signalled in gaudily patterned fishes.

Family: CHAETODONTIDAE

Butterflyfishes

Family characteristics

Due to their close similarity, butterflyfishes and angelfishes (Pomacanthidae, see pages 242-261) were once classified as one group. Although there are now sufficient differences for them to be considered separately, both families continue to share a great deal of common ground.

Butterflyfishes possess an oval body that is extremely laterally compressed. These features, together with the terminal mouth, provide a strong clue as to their natural habitat of coral heads, where their thin-sectioned bodies can easily pass between the branches. Their amazing colour patterns camouflage vulnerable parts of their bodies and assist in species identification.

There appear to be no external differences between the sexes, although at breeding times the females may become noticeably swollen with eggs. Recorded observations of spawning activity in the wild are fairly scarce, but most reports indicate that the majority of species ascend the water column in pairs or small groups and release eggs and sperm simultaneously. Some species have been seen to engage in a 'chasing' ritual, whereby the female is pursued by a male around the lower water layers. In response to being nudged by the male, the female releases eggs, which the male fertilizes as they pass by him. In all cases, the eggs then float briefly until they hatch. The larvae then feed and develop in the planktonic layers for several months before migrating back down to the reef floor. Reports of aquarium spawnings are extemely rare and, as yet, no larvae of any species have been reared successfully.

Diet and feeding

Most members of the Chaetodontidae family are grazing fishes that feed on algae, sponges and corals; some are omnivorous, however, and include small and planktonic animals in their diet. You may need to feed young fishes several times a day with live brineshrimp. Larger fishes should be offered a variety of marine fare, including live brineshrimp and *Mysis* and frozen foods of a suitable size. Sponge-based frozen foods may prove very popular with some species. It is a good idea to witness all butterflyfishes feeding confidently before purchasing.

One or two species have evolved long snouts for reaching even further into crevices for food.

Aquarium behaviour

Although very attractive, these fishes are not suitable for inexperienced fishkeepers. They can be difficult to maintain in captivity, particularly the algae- and sponge-eating species, and some have proved impossible to sustain (see pages 380-381). These species are easily upset by small changes in water quality, usually showing any dissatisfaction with aquarium life by going on hunger strike; one day they are quite content with the diet you provide and the next day they simply will not touch it.

Juvenile specimens may be far easier to acclimatize to aquarium life and subsequently do much better in the long term. Ensure that the aquarium has sufficient retreats and hideaways to give the fishes some form of security.

Butterflyfishes vary tremendously in their tolerance (or intolerance) of members of their own, or similar, species. Most are fairly peaceful, but it is always wise to make sure of this before buying them. Living corals and sea anemones will not last long in the same aquarium with butterflyfishes.

Butterflyfishes may undergo colour changes at night; the usual transformation is the appearance of darker splodges over parts of the body.

Chaetodon auriga
Threadfin Butterflyfish

☐ **Distribution:** Indo-Pacific, Red Sea.

☐ **Length:** 200mm/8in (wild), 150mm/6in (aquarium).

Left: Chaetodon auriga
Juvenile Threadfin Buttterflyfishes do not have the long filament from the rear of the dorsal. Adults from the subspecies C. a. auriga *from the Red Sea may lose the eye-spot on the dorsal fin (though it remains in Indo-Pacific specimens). Like other butterflyfishes, it is not safe with invertebrates.*

☐ **Diet and feeding:** Crustaceans, coral polyps and algae in the wild. Offer suitable live and frozen foods in the aquarium. Grazer.

☐ **Aquarium behaviour:** Peaceful, but shy.

☐ **Invertebrate compatibility:** No, not to be trusted.

A black bar crosses the eye, and the mainly white flanks are decorated with a 'herring-bone' pattern of grey lines. The anal, dorsal and front part of the caudal fin are yellow. The common name of this species refers to a threadlike extension to the dorsal fin. *C. auriga* can be weaned on to food of your choice by gradual substitution.

Chaetodon chrysurus
Pearlscale Butterflyfish

☐ **Distribution:** Indo-Pacific.

☐ **Length:** 150mm/6in (wild).

☐ **Diet and feeding:** Crustaceans, vegetable matter. Grazer.

☐ **Aquarium behaviour:** Peaceful.

☐ **Invertebrate compatibility:** No, not to be trusted.

The scales on this species are large and dark-edged, giving the fish a lattice-covered, or checkered, pattern. The main feature is the bright orange arc connecting the rear of the dorsal and anal fins; a repeated orange band appears in the caudal fin. The fish's habitat is thought to be nearer to Africa, Mauritius and the Seychelles rather than spread widely over the Indo-Pacific area. The Red Sea *C. paucifasciatus* has a faint spot in the dorsal fin and a slightly different shape to the orange area (which is, in fact, almost red). *C. mertensii* and *C. xanthurus* are also very similar in appearance to *C. chrysurus*.

Below: Chaetodon chrysurus
If any fish is truly aptly named, this beautiful butterflyfish must be it, for the scales, seen at the right angle, are of a wonderfully pearly hue.

Chaetodon collare
Pakistani Butterflyfish

☐ **Distribution:** Indian Ocean.

☐ **Length:** 150mm/6in (wild), 100mm/4in (aquarium).

☐ **Diet and feeding:** Most frozen marine foods and greenstuff.

☐ **Aquarium behaviour:** May be intolerant of other members of its own, or other, species.

☐ **Invertebrate compatibility:** No, not to be trusted.

The brown coloration of *C. collare* is unusual for a butterflyfish. It is reputed to be difficult to keep, although not all authorities agree on this. Species from different locations may have different feeding requirements, the species from rocky outcrops being easier to satisfy in captivity than those from coral reefs. Not a suitable fish for the beginner.

Below: Chaetodon collare
From the coast of East Africa, right across the Indian Ocean to Melanesia, this fish is a common sight around the coral reefs. Beginners to marine fishkeeping should avoid this species.

Chaetodon ephipippium
Saddleback Butterflyfish

☐ **Distribution:** Indo-Pacific.

☐ **Length:** 230mm/9in (wild).

☐ **Diet and feeding:** Crustaceans. Grazer.

☐ **Aquarium behaviour:** May be intolerant of other members of its own, or other, species.

☐ **Invertebrate compatibility:** No, not to be trusted.

Easily recognizable by the white-edged dark 'saddle' that covers the rear upper portion of the body and dorsal fin. The females become plumper at breeding time. It may be difficult to accustom these fish to a successful aquarium feeding pattern.

Chaetodon falcula
Double-saddled Butterflyfish; Pig-faced Butterflyfish

☐ **Distribution:** Indian Ocean.

☐ **Length:** 150mm/6in (wild), 100-125mm/4-5in (aquarium).

☐ **Diet and feeding:** Crustaceans, coral polyps, algae. Grazer.

☐ **Aquarium behaviour:** Aggressive towards similar species.

☐ **Invertebrate compatibility:** No, not to be trusted.

Two dark saddle markings cross the top of the body. The dorsal, anal and caudal fins are yellow, and there is a black spot or bar on the caudal peduncle. The body and head are white. A vertical black bar runs down the side of the head and there are many vertical thin lines on the body. This species is often confused with *C. ulietensis*, which is found across a wider area of the Pacific Ocean, and has slightly lower reaching 'saddles' and less yellow on top of the body and dorsal fin. Both species can be found available.

Above: Chaetodon ephipippium
This is one of the more sensitive butterflyfishes; make sure a specimen is feeding properly before you purchase it.

Below: Chaetodon falcula
In the wild, this fish is confined to the Indian Ocean. Like most species in this family, it is not for the beginner.

Chaetodon frembli
Blue-striped Butterflyfish

☐ **Distribution:** Indo-Pacific, Red Sea.

☐ **Length:** 200mm/8in (wild).

☐ **Diet and feeding:** Crustaceans, coral polyps, algae. Grazer.

☐ **Aquarium behaviour:** Calm community fish.

☐ **Invertebrate compatibility:** No, not to be trusted.

The yellow body is marked with upward slanting diagonal blue lines. A black mark appears immediately in front of the dorsal fin, and the black of the caudal peduncle extends into the rear of the dorsal and anal fins. The caudal fin has white, black and yellow vertical bars. This butterflyfish lacks the usual black bar through the eye. Not an easy species to keep.

Above: Chaetodon frembli
This species does not share the characteristic butterflyfish pattern, in which the real eye is hidden in a vertical dark band.

Chaetodon kleini
Klein's Butterflyfish; Sunburst Butterflyfish

☐ **Distribution:** Indo-Pacific.

☐ **Length:** 125mm/5in (wild), 100mm/4in (aquarium).

☐ **Diet and feeding:** Crustaceans, coral polyps, algae. Grazer.

☐ **Aquarium behaviour:** Peaceful.

☐ **Invertebrate compatibility:** No, not to be trusted.

Left: Chaetodon kleini
The obvious wide bands of the juvenile have faded by the time the fish reaches adulthood, as here. Only the black bar remains as a contrast to the sunburst colour of the body. Klein's Butterflyfish is thought to be the easiest member of the family to keep.

This is a more subtly coloured fish; its black eye bar is followed by a grey bar and the white forebody changes to a golden yellow. The dorsal and anal fins, plus the front part of the caudal fin, are a matching gold-yellow. The mouth is black. *C. kleini* is considered to be the easiest of all butterflyfishes to keep, being hardy once it has settled into the aquarium.

Chaetodon lunula
Racoon Butterflyfish

☐ **Distribution:** Indo-Pacific.

☐ **Length:** 200mm/8in (wild)

☐ **Diet and feeding:** Crustaceans, coral polyps, algae. Grazer.

☐ **Aquarium behaviour:** Usually peaceful with other fish, but will eat corals.

Above: Chaetodon lunula
Juvenile forms of the Racoon Butterflyfish have an 'eye-spot' in the rear part of the dorsal fin, but this fades with age. Another difference between young and adult fishes is that the adult fish has more yellow on the snout. This peaceful fish is relatively easy to keep and can have a fairly long lifespan in the aquarium.

☐ **Invertebrate compatibility:** No, not to be trusted.

A white-edged black bar runs down over the eye and immediately sweeps up again into the mid-dorsal area. A white bar crosses this bar immediately behind the eye. A black blotch appears on the caudal peduncle, a feature missing in the otherwise similar species, *C. fasciatus*. *C. lunula* is fairly long-lived in the aquarium and will readily accept most foods.

Chaetodon melannotus
Black-backed Butterflyfish

☐ **Distribution:** Indo-Pacific.

☐ **Length:** 150mm/6in (wild).

☐ **Diet and feeding:** Crustaceans, coral polyps, algae. Grazer.

☐ **Aquarium behaviour:** A peaceful, but difficult, fish.

☐ **Invertebrate compatibility:** No, not to be trusted.

The white body is crossed by diagonal thin black stripes and bordered by yellow dorsal, anal and caudal fins. A black eye bar divides the yellow head.

Chaetodon punctofasciatus
Spot-banded Butterflyfish

☐ **Distribution:** A wide distribution throughout the Pacific.

☐ **Length:** 100mm/4in (wild).

☐ **Diet and feeding:** Coral polyps. Frozen foods. Grazer.

☐ **Aquarium behaviour:** May be intolerant of other members of its own, or other, species.

☐ **Invertebrate compatibility:** No, not to be trusted.

The vertical black stripes end halfway down the body, and turn into many spots. A black spot appears immediately in front of the dorsal fin. The eye bar is yellow with a black edge.

Top right: Chaetodon melannotus
For hobbyists requiring absolute accuracy in fish descriptions, this species should perhaps be renamed the Black-sided Butterflyfish.

Right: Chaetodon punctofasciatus
This species has a wide distribution, ranging from the China Sea in the north, out to the Philippines, and south as far as Australia's Great Barrier Reef.

Chaetodon semilarvatus

Addis Butterflyfish; Lemonpeel Butterflyfish; Golden Butterflyfish

☐ **Distribution:** Indian Ocean, Red Sea.

☐ **Length:** 200mm/8in (wild).

☐ **Diet and feeding:** Crustaceans, coral polyps, algae. Grazer.

☐ **Aquarium behaviour:** May be intolerant of other members of its own, or other, species. Difficult.

☐ **Invertebrate compatibility:** No, not to be trusted.

The yellow body is crossed by thin orange vertical lines. A blue-black inverted teardrop patch covers the eye. Red Sea specimens are difficult to obtain and thus costly.

Below: Chaetodon semilarvatus
The bright coloration and distinctive markings of this species make it a perfect photographic subject. Specimens from the Red Sea are difficult to obtain and therefore very expensive.

Chaetodon striatus
Banded Butterflyfish

☐ **Distribution:** Tropical Atlantic Ocean.

☐ **Length:** 150mm/6in (wild).

☐ **Diet and feeding:** Crustaceans, coral polyps, algae. Grazer.

☐ **Aquarium behaviour:** May be intolerant of other members of its own, or other, species.

☐ **Invertebrate compatibility:** No, not to be trusted.

An easily recognizable species with four dark bands across its body. A continuous dark band passes through the outer edges of the dorsal, caudal and anal fins to connect with the ends of the third vertical band. The juvenile form has a white-ringed black spot on the soft dorsal fin. A good community fish.

Take care not to shock or otherwise stress this easily frightened fish during transportation and introduction into the aquarium.

Chaetodon vagabundus
Vagabond Butterflyfish; Criss-cross Butterflyfish

☐ **Distribution:** Indo-Pacific.

☐ **Length:** 200mm/8in (wild).

☐ **Diet and feeding:** Crustaceans, coral polyps, algae. Grazer.

☐ **Aquarium behaviour:** Peaceful.

☐ **Invertebrate compatibility:** No, not to be trusted.

Diagonal lines cross the white body in two directions. Black bars cross the eye and fringe the rear part of the body. The rear part of the dorsal and anal fins are gold-yellow edged with black; the yellow caudal fin has two black vertical bars. A similar species is *C. pictus*, often classified as a subspecies or a colour variant. *C. vagabundus*, like all marines, appreciates good water conditions.

Above: Chaetodon striatus
The black and white stripes of the Banded Butterflyfish make it easily recognizable and are reminiscent of the markings of the freshwater angelfish.

Below: Chaetodon vagabundus
The Vagabond Butterflyfish is a long-standing favourite; a peaceful fish that would make a good introduction to the marine aquarium.

Chelmon rostratus
Copper-band Butterflyfish

☐ **Distribution:** Indo-Pacific, Red Sea.

☐ **Length:** 170mm/6.7in (wild).

☐ **Diet and feeding:** Frozen foods, small animal foods, algae. Picks in between coral heads.

☐ **Aquarium behaviour:** Aggressive towards members of its own species. Difficult.

☐ **Invertebrate compatibility:** No, not to be trusted.

The yellow-orange vertical bands on the body have blue-black edging. These distinctive colours, coupled with the false 'eye-spot' at the rear of the upper body, make this fish difficult to confuse with any other. It may take time to acclimatize to aquarium foods. It is very sensitive to deteriorating water conditions.

Above: Chelmon rostratus
The outstanding colours and attractive (and unusual) shape of this fish should be all the inducement you need to keep it in the best aquarium conditions.

Above: Forcipiger flavissimus
Note how the black triangle on this species camouflages the eye.

Far right: Heniochus acuminatus
The Wimplefish is often confused with the Moorish Idol (Zanclus sp.).

Forcipiger flavissimus
Long-nosed Butterflyfish

- **Distribution:** Indo-Pacific, Red Sea.

- **Length:** 200mm/8in (wild), 100-150mm/4-6in (aquarium).

- **Diet and feeding:** Small animal foods, algae. Picks in between coral heads.

- **Aquarium behaviour:** Not as aggressive as *C. rostratus.*

- **Invertebrate compatibility:** No, not to be trusted.

The body, dorsal, anal, and pelvic fins are bright yellow, but a black triangle disrupts the contours of the head. The lower jaw is white, and the caudal fin is clear. A false 'eye spot' on the anal fin confuses predators. It is similar in its aquarium requirements and treatment to *C. rostratus.* The Big Long-nosed Butterflyfish (*F. longirostris)* is similar, but less common, with a longer snout.

Heniochus acuminatus
**Wimplefish; Pennant Coralfish
Poor Man's Moorish Idol**

- **Distribution:** Indo-Pacific, Red Sea.

- **Length:** 180mm/7in (wild).

- **Diet and feeding:** Most frozen marine foods and greenstuff. Grazer.

- **Aquarium behaviour:** Companionable.

- **Invertebrate compatibility:** No, not to be trusted.

Two wide, forward-sloping black bands cross the white body. The rear parts of the dorsal, pectoral and anal fins are yellow while the pelvic fins are black. The front few rays of the dorsal fin are much extended. These fish appreciate plenty of swimming room and young specimens act as cleanerfishes. Other species, such as the Horned Coralfish (*H. chrysostomus*) are rarely imported. *H. acuminatus* varies little in juvenile and adult coloration, but other species in the genus differ; some adults develop protuberances on the forehead.

This is an excellent substitute for the Moorish Idol (*Zanclus canescens*), which has a very poor survival rate in captivity.

Family: CIRRHITIDAE

Hawkfishes

Family chacteristics

Despite their attractive appearance and friendliness, Hawkfishes are predators – albeit of small prey. They rest or 'perch' on a piece of coral and wait for food to pass by, at which point they dash out and seize it. Reproduction in these fishes is by demersal eggs, i.e. the eggs are laid and fertilized on a firm surface, where they subsequently hatch. In the nine genera there are some 32 species, most of which are not commonly kept in the aquarium. The largest species described has a maximum length of 55cm (22in).

Diet and feeding

In the wild, these fishes will take small invertebrates – shrimps, etc. – and smaller fishes. In the aquarium, provide live foods and suitable meaty frozen foods.

Aquarium behaviour

Hawkfishes will appreciate plenty of 'perching places' in the aquarium. Despite their predatory nature, they appear not to harm sedentary invertebrates such as tubeworms, soft corals, etc.

Neocirrhites armatus
Scarlet Hawkfish

☐ **Distribution:** Central and Western Pacific.

☐ **Length:** 75mm/3in (wild)

☐ **Diet and feeding:** Accepts most marine frozen and live foods of a suitable size. Once settled in the aquarium, it will also accept marine flake.

☐ **Aquarium behaviour:** Peaceful.

☐ **Invertebrate compatibility:** Yes, generally well behaved.

This extremely attractive fish is much sought after by aquarists and usually commands quite a high price as a consequence. It is not a good swimmer and spends much of its time resting on rocks and Gorgonians waiting to ambush small crustaceans and plankton.

Fortunately, it is not a threat in the home aquarium and makes an ideal community addition.

Below: Neocirrhites armatus
The aptly named Scarlet Hawkfish is regularly imported, but not in large numbers. It is a very popular fish in short supply and, although quite expensive, is not difficult to keep. N. armatus seems to fare best in a mixed fish and invertebrate system, where it also tends to look more natural.

Oxycirrhites typus
Longnosed Hawkfish

☐ **Distribution:** Indian Ocean mainly.

☐ **Length:** 100mm/4in (wild).

☐ **Diet and feeding:** Most frozen marine foods. Sits on coral or a rock, then dashes out to grab food.

☐ **Aquarium behaviour:** Peaceful. Can be kept in small groups.

☐ **Invertebrate compatibility:** Yes, generally well behaved.

The elongate body of this hardy fish is covered with a squared pattern of bright red lines. The snout is very long and suited to probing the coral crevices for food. There are small cirri, or crestlike growths, at the end of each dorsal fin spine and on the nostrils. The female is larger than the male and the male has darker red lower jaws. There are black edges to the pelvic and caudal fins.

In nature, spawnings occur from dusk onwards. Reports of aquarium spawnings suggest that the female lays patches of adhesive eggs after courtship activity.

Below: Oxycirrhites typus
The Long-nosed Hawkfish depends on its natural coloration and markings to provide camouflage as it perches on a suitable sea whip or sea fan waiting to ambush its next unsuspecting meal, which will usually consist of a shrimp or other unfortunate crustacean!

Family: DIODONTIDAE

Porcupinefishes

Family characteristics

Porcupinefishes are very similar to the other 'inflatable' fishes, the puffers (see pages 298-301). They can be distinguished from the pufferfishes, however, by the spines on their scales and by their front teeth, which are fused together. Hard crustacean foods present little problem to them. The pelvic fins are absent. Normally the spines are held flat, but in times of danger they stand out from the body as the fish inflates itself. The appearance and inflatability of the porcupinefishes make them interesting subjects for the aquarium.

Diet and feeding

Earthworms, shrimps, crab meat and other meaty foods can be given, but cut food into pieces for smaller specimens.

Aquarium behaviour

Although they grow very large in the wild, porcupinefishes rarely grow as big in captivity. Nevertheless, be sure to keep them alone in a large aquarium. You may find that specimens become hand tame. It may be a temptation to encourage these fish to inflate, but this should be resisted as it is very stressful for the individual.

Chilomycterus schoepfi
Spiny Boxfish; Striped Burrfish; Burrfish

☐ **Distribution:** Tropical Atlantic, Caribbean.

☐ **Length:** 300mm/12in (wild).

☐ **Diet and feeding:** Crustaceans, molluscs, meat foods. Bold.

☐ **Aquarium behaviour:** Aggressive among themselves. Do not keep them with small fishes.

☐ **Invertebrate compatibility:** Not suitable.

The undulating dark lines on its yellow body provide *C. schoepfi* with excellent camouflage, and a first line of defence; then the short spines –

fixed and usually held erect – may be called upon to play their part. Not as liable to inflate itself as other related species.

Below: Chilomycterus schoepfi
The short spines of this species are kept erect – a constant defence against predation or capture. In addition, its coloration provides good camouflage in its natural environment.

Diodon holacanthus
Long-spined Porcupinefish;
Balloonfish

☐ **Distribution:** All warm seas.

☐ **Length:** 500mm/20in (wild),
150mm/6in (aquarium).

☐ **Diet and feeding:** Crustaceans,
molluscs, meat foods. Bold.

☐ **Aquarium behaviour:** Do not
keep with small fishes.

☐ **Invertebrate compatibility:**
Not suitable as invertebrates are
its main food.

The colour patterning on this
species is blotched rather than
lined, but it serves the same
excellent purpose – to disguise the
fish as part of the surrounding
underwater scenery.

Above: Diodon holacanthus
The Long-spined Porcupinefish is one of
the most commonly found species, being
native worldwide. It feeds on
crustaceans, so do not include it in an
aquarium containing invertebrates.

Diodon hystrix
Common Porcupinefish;
Porcupine Puffer

☐ **Distribution:** All warm seas.

☐ **Length:** 900mm/36in (wild),
considerably smaller in the
aquarium.

☐ **Diet and feeding:** Crustaceans,
molluscs, meat foods. Bold.

☐ **Aquarium behaviour:**
Generally peaceful with other
fishes.

☐ **Invertebrate compatibility:**
Not suitable.

This species has longer spines held
more flatly against the body. It is not
constantly active, remaining at rest
for long periods until hunger or
some other action-provoking event

Above: Diodon hystrix
Despite its formidable adult size, the
Common Porcupinefish poses little threat
to smaller fishes, though, again, it should
not be kept in a mixed fish and
invertebrate set-up and its aquarium
will need good filtration.

stirs it. You will need to keep the
tank efficiently filtered to cope with
these fishes, which can sometimes
prove 'messy eaters'.

Family: GOBIIDAE

Gobies

Family characteristics

Comparatively little is known about this very large family, which – paradoxically – contains one of the smallest known vertebrates (*Pandaka* sp). In the wild, gobies are found in several different locations: tidal shallow beaches; on the coral reef itself; and on the muddy seabed. Some species are found in fresh water. All rely on having a secure bolt hole in which to hide when danger threatens. Such 'bolt holes' may be located within sponges, caves, crevices or may simply be burrows in the seabed.

The gobies are endearing little characters for the aquarium. Unlike the blennies, the colours of the gobies can be quite brilliant. Their bodies are elongate, the head blunt with high-set eyes. Gobies can be further distinguished from blennies, with whom they share a similar habitat, by the presence of a 'suction-disc' formed by the fusion of the pelvic fins.

Sexing gobies can be difficult, although females may become distended with eggs at breeding time and there are the typical differences in the size and shape of the genital papillae – if you can see them! (The genital papillae are breeding tubes that extend from the vent of each fish; usually longer in females than in males.) In some species, the male may change colour or develop longer fins during the breeding period. Spawning occurs in burrows or in sheltered areas, with the eggs being guarded by the male. Several gobies have been spawned in captivity. *Gobiosoma oceanops* and *Lythrypnus dalli*, for example, will breed willingly in the aquarium, but rearing the young fry is not easy. The young of *G. oceanops* have been reared with more success than the smaller fry of *L. dalli*.

Diet and feeding

Gobies are carnivorous fishes that will eat brineshrimp, finely chopped meat foods, frozen foods and *Daphnia*.

Aquarium behaviour

Many reef-dwelling species provide cleaning services for larger fishes, the cleaning sequence following the pattern described for the Cleaner Wrasse (see pages 216-217).

Gobiodon citrinus
Lemon Goby

☐ **Distribution:** Indo-Pacific.

☐ **Length:** 30mm/1.2in (wild)

☐ **Diet and feeding:** Once settled, will accept most marine foods of a suitable size. Particularly fond of live foods.

☐ **Aquarium behaviour:** Very peaceful.

☐ **Invertebrate compatibility:** Good.

In common with many gobies,

Below left: Gobiodon citrinus
Beginners and experienced aquarists alike will find this species an absolute delight. It is not a difficult species to keep but should ideally be kept in a mixed aquarium – it looks rather out of place in a fish-only tank.

G. citrinus is not a good swimmer and spends much of its time perched in the branches of gorgonians or on rocks with a good vantage point. The mucus of this species is known to be poisonous and a protection against being eaten by other fish. However, it is not known to cause a toxic problem.

Gobiodon okinawae
Yellow Goby

☐ **Distribution:** Pacific Ocean.

☐ **Length:** 30mm/1.2in (wild).

☐ **Diet and feeding:** Will readily accept small particles of most marine fare, including frozen, live and flake foods. Live foods are particularly relished, especially when the fish is first settling in.

☐ **Aquarium behaviour:** Very peaceful, except with the same species.

☐ **Invertebrate compatibility:** Excellent.

Although small, the Yellow Goby is highly territorial and, once settled into an aquarium, will very rarely tolerate other individuals of the same species, even in a fairly large tank. Sometimes a pair may be identified and introduced at the same time, in which case a spawning may be possible but, so far, none of these very desirable little fish have been raised to adulthood. Poor water quality is not tolerated, and the fish will refuse to eat and may turn a dirty brown colour as a result. If no improvement is made, death usually follows.

Below: Gobiodon okinawae
The Yellow Goby is, in common with many other gobies, a very poor swimmer and prefers to remain perched on favourite vantage points. Once settled, it shows little concern for other inhabitants that pose no threat.

Gobiosoma oceanops
Neon Goby

- ☐ **Distribution:** Western Atlantic, especially Florida, and the Gulf of Mexico.

- ☐ **Length:** 60mm/2.4in (wild), 25mm/1in (aquarium).

- ☐ **Diet and feeding:** Parasites, small crustaceans and plankton. Bold nibbler; bottom feeder.

- ☐ **Aquarium behaviour:** Peaceful and uninhibited.

- ☐ **Invertebrate compatibility:** Ideal.

Two characteristics distinguish this most familiar goby: the electric blue coloration of the longitudinal line on the body and the cleaning services it offers to other fishes. The species can be positively identified by the gap visible between the two blue lines on the snout when the fish is seen from above; other species have connected lines or other markings between the ends of the lines.

G. oceanops has been bred in the aquarium and is now regularly bred on a commercial basis. Pairing occurs spontaneously. Before spawning, the male's colour darkens and he courts the female with exaggerated swimming motions, assuming a position on the aquarium floor until the female takes notice of him. Spawning activity occurs in a cave or other similar sheltered area. The fertilized eggs hatch after 7-12 days. In the wild, the fry feed on planktonic foods for the first few weeks. In the aquarium, start the fry off with cultured rotifers followed by newly hatched brineshrimp.

Unfortunately, these eminently suitable (and practicable) aquarium subjects are relatively shortlived – perhaps only a year or two, but their breeding possibilities should enable you to propagate them for several generations if given the ideal conditions. Hopefully, one day, all Neon Gobies will be tank raised.

Above: Gobiosoma oceanops
This colourful goby is peaceful, doesn't need a large tank, and will spawn quite readily given the correct conditions.

Below: Lythrypnus dalli
A number of Blue-banded Gobies could share a reasonably sized tank without too much quarrelling.

Lythrypnus dalli
Blue-banded Goby; Catalina Goby

- ☐ **Distribution:** Californian Pacific coast.

- ☐ **Length:** 60mm/2.4in (wild), 25mm/1in (aquarium).

- ☐ **Diet and feeding:** Small crustaceans and other small marine organisms. Live foods preferred. Bottom feeder.

- ☐ **Aquarium behaviour:** Peaceful with small fishes but may be bullied by larger fishes.

- ☐ **Invertebrate compatibility:** Ideal.

The red body is crossed by brilliant blue vertical lines and the first dorsal fin has an elongated ray. The male has longer dorsal fin spines than the female. Despite territorial requirements several fish will share even a reasonably small aquarium quite happily, but they are not naturally a longlived species. This species does not require the same high water temperatures as other marine fish. This fish has been spawned successfully in the aquarium.

Valenciennea puellaris
Orange-spotted Goby

☐ **Distribution:** Indo-Pacific.

☐ **Length:** 150mm/6in (wild), 100mm/4in (aquarium).

☐ **Diet and feeding:** Once settled, will accept most marine fare of a suitable size, including frozen, live and flake foods.

☐ **Aquarium behaviour:** Very peaceful. An excellent fish for the community aquarium.

☐ **Invertebrate compatibility:** Generally good, but may upset sessile invertebrates with its digging activities.

The Orange-spotted Goby is a common fish throughout the Indo-Pacific. The patterning on the body may vary widely from region to region, sometimes making identification a little difficult. This species is an excellent aquarium subject and a very good choice for the beginner. It requires a sandy substrate in which to dig and burrow.

Valenciennea strigata
Blue-cheek Goby

☐ **Distribution:** Western Pacific and Indian Oceans.

☐ **Length:** 180mm/7in (wild).

☐ **Diet and feeding:** Thrives on nearly all marine frozen, live and flake foods.

☐ **Aquarium behaviour:** Very peaceful. A good community fish.

☐ **Invertebrate compatibility:** Generally good, but may upset sessile invertebrates with its digging activities.

V. strigata, which is closely related to the previous species, *V. puellaris*, constantly digs and sifts the sand for food particles. This can be a boon for aquarists with undergravel filtration as the activity helps to keep the substrate loose and less likely to compact. *V. strigata* also enjoys the company of its own species and may generally be considered safe to keep in pairs or small groups where space and filtration capacity allows. This easy-to-keep fish makes an ideal introduction to gobies.

Top: Valenciennea puellaris
The Orange-spotted Goby is a popular fish with many aquarists, who like to observe its interesting digging habits.

Above: Valenciennea strigata
The Blue-cheek Goby enjoys the company of its own species and a pair are often observed following each other.

Family: GRAMMIDAE

Fairy Basslets; Pygmy Basslets

Family characteristics

This small group of fish are confined to the Caribbean and only three species are known to date – the ever popular *Gramma loreto*, the Royal Gramma; *Gramma melacara*, the Black-Cap Gramma; and lastly, the very rare *Gramma linki* (not featured here).

In common with their closely related cousins, the Pseudochromids (see pages 280-283), these fish are shy and secretive. They spend most of their lives within the maze of reef crevices and are highly territorial in nature. They lead largely solitary lives, coming together only at breeding times. The Royal Gramma (*Gramma loreto*), although often aggressive with its own kind, has been regularly spawned in the aquarium.

Diet and feeding

The wild diet of these fish is largely made up of drifting planktonic animals and crustaceans. Most frozen and live marine foods are accepted within the aquarium. The Royal Gramma may also accept flake foods. *Gramma melacara* will need a little more care and may need to be tempted with live brineshrimp and *Mysis* initially.

Aquarium behaviour

Gramma melacara is highly territorial and may not be kept with its own or similar species. The Royal Gramma may be tempted, in a large enough aquarium, to form a breeding unit of one male to several females. Aprroach this with care, however, as two males will fight, usually to the death.

Above: Gramma loreto
As far as coloration goes, if big is beautiful then small can be simply stunning – as this Royal Gramma shows. A cave-dwelling fish, it is often very possessive, positively resenting any intrusion by other fishes into its chosen home. It is equally intolerant of others of its own kind (though you may be able to keep a pair in a large tank). An aquarium stocked with many soft corals and hideaways suits it perfectly.

Gramma loreto
Royal Gramma

☐ **Distribution:** Western Atlantic.

☐ **Length:** 130mm/5in (wild), 75mm/3in (aquarium).

☐ **Diet and feeding:** Eats a wide range of foods, including chopped shrimp and live brineshrimp.

☐ **Aquarium behaviour:** This cave-dwelling fish should be acclimatized gradually to bright light. It may resent other cave-dwelling species, particularly the Yellow-headed Jawfish, (see page 231). Aggressive towards its own kind, but pairings are possible in a large tank.

☐ **Invertebrate compatibility:** Yes, ideal.

The main feature of this species is its remarkable colouring. The front half of the body is magenta, the rear half bright golden-yellow. A thin black line slants backwards through the eye. These somewhat secretive cave dwellers should not be kept with boisterous species. An almost identical species, *Pseudochromis paccagnellae* (see page 282) has a narrow white line dividing the two main body colours.

Spawning activity has been observed – paradoxically, not in nature but in captivity. Four fish grouped themselves into two 'pairs', each comprising one small and one large fish. The larger fish lined a pit in the sand with strands of algae glued together with a glandular secretion and the pair spawned 'stickleback' fashion. The smaller fish was enticed into the pit several times, closely followed by the larger of the two, who then stood guard over the pit full of eggs. Unfortunately, the fry were not raised successfully, but this does shed light on the possible reproductive methods practised by this species. The dissimilar sizes of the fishes making up the 'pairs' seems to bear out other reports that the male fish is usually larger than the female where both fishes are of the same age.

Above: Gramma melacara
The Black-cap Gramma positively thrives in a mainly invertebrate aquarium. Although never on constant show, it will make regular appearances.

Gramma melacara
Black-Cap Gramma

☐ **Distribution:** Caribbean.

☐ **Length:** 100mm/4in.

☐ **Diet and feeding:** Prefers live foods but will usually accept frozen marine fare after an initial settling-in period.

☐ **Aquarium behaviour:** Shy and secretive by nature. Requires a comprehensive arrangement of rockwork in which to hide.

☐ **Invertebrate compatibility:** Ideal.

The common name Black-Cap is certainly apt in this case, for the purple body is only disturbed by a black patch over the crown of the head. This most attractive fish is extremely territorial and should be kept in the absence of its own, or similar, species. Overly aggressive specimens have been known to bite their keeper's hand! Having said that, these fish make ideal additions to the invertebrate aquarium.

Family: HAEMULIDAE

Grunts

Family characteristics
Grunts can be distinguished from the similar-looking snappers by differences in their dentition. Many grind their pharyngeal teeth, the resulting sound being amplified by the swimbladder. Juveniles often perform cleaning services for other fishes.

Diet and feeding
Members of this family of fishes eat well, enjoying a diet of small fishes, shrimps and dried foods.

Aquarium behaviour
Grunts may grow too quickly for the average aquarium.

Above: Anisotremus virginicus
Provided they are given a suitably large aquarium, these juvenile Porkfish will live quite happily together. They are not difficult to keep and will accept a wide variety of foods.

Anisotremus virginicus
Porkfish

☐ **Distribution:** Caribbean.

☐ **Length:** 300mm/12in (wild).

☐ **Diet and feeding:** In the wild, brittle starfish, crustaceans, etc. In the aquarium, chopped meaty foods etc. Nocturnal feeder.

☐ **Aquarium behaviour:** Keep juvenile specimens only. A large tank will suit them well.

☐ **Invertebrate compatibility:** Voracious predators on all forms of invertebrates and so should be kept in a fish-only aquarium.

The body is triangular, the highest part being just behind the head. The steep forehead is fairly long and the eyes are large. The yellow body is streaked with bright blue lines and two black bars cross the head region, one through the eye and one just behind the gill cover. The juvenile coloration is different; the cream body has black horizontal stripes and a black blotch on the caudal peduncle. The head is yellow and the larger fins have red marks on their edges.

The common collective name of grunts comes from the noise these fishes make when they are taken from the water. They are very similar in appearance to the Majestic Snapper, (see page 223).

Family: HOLOCENTRIDAE

Squirrelfishes

Family characteristics

Squirrelfishes are large-eyed nocturnal fishes that hide by day among crevices in the coral reefs of the Indo-Pacific and Atlantic Oceans. They usually have red patterning on their elongate bodies, spines on the gill covers and sharp rays on the fins. The dorsal fin looks as if it has two separate parts: a long-based spiny part at the front and a high triangular softer rayed section at the back. The scales and gill cover are extremely rough; take great care when netting or handling, as this can easily damage the fish.

Diet and feeding

In the aquarium, squirrelfishes rapidly adjust to a daytime eating routine and a diet consisting of chopped worm foods and small fish.

Aquarium behaviour

Squirrelfishes are very active and need a sufficiently large aquarium to accommodate their energetic way of life. Remember that other small fishes may not welcome such boisterous companions.

Holocentrus diadema
Common Squirrelfish

☐ **Distribution:** Indo-Pacific.

☐ **Length:** 300m/12in (wild).

☐ **Diet and feeding:** All foods. Bold.

☐ **Aquarium behaviour:** Do not keep with small fishes.

☐ **Invertebrate compatibility:** Yes, but may be unsafe with small crustaceans.

Horizontal red lines and a red dorsal fin make this common fish a very colourful addition to any sufficiently large aquarium. In nature, it is a nocturnal species and therefore needs some hiding places in which to rest during the day. It does adapt to aquarium life, however, and will swim around in daylight hours.

Below: Holocentrus diadema
In the wild, these strikingly attractive fishes swim in large shoals among the coral reefs. Their relatively large size and constant activity render them less suited to the average aquarium, but if you can provide adequate swimming space and plenty of companions, this fish will prove a good addition to the tank.

Holocentrus rufus
White-tip Squirrelfish

☐ **Distribution:** Western Atlantic.

☐ **Length:** 200mm/8in (wild).

☐ **Diet and feeding:** Crustaceans and meaty foods. Bold.

☐ **Aquarium behaviour:** Shoaling fish for a large aquarium.

☐ **Invertebrate compatibility:** Yes, except with small crustaceans.

Holocentrus rufus is a similar colour to other squirrelfishes, but it has a white triangular mark on each dorsal fin spine. Reasonably fearless, predatory squirrelfishes can hold their own with grunts and even moray eels; they will meet any threats to their safety with grunting noises and quivering actions.

Above: Holocentrus rufus
A feature of the White-tipped Squirrelfish is the long rear part of the dorsal fin; the upper part of the caudal fin is larger than the lower.

Myripristis murdjan
Big-eye Squirrelfish; Blotcheye

☐ **Distribution:** Indo-Pacific.

☐ **Length:** 300mm/12in (wild).

☐ **Diet and feeding:** Crustaceans and meaty foods. Bold.

☐ **Aquarium behaviour:** Peaceful.

☐ **Invertebrate compatibility:** Will eat small crustaceans.

The Big-eye Squirrelfish clearly lives up to its popular name! The organs in question are used to good advantage at night, when the fish is active. The red edge on each scale gives this fish a reticulated appearance. There is a dark red vertical area behind the gill cover.

It lacks the spine on the rear of the gill cover that is carried by members of the genus *Holocentrus*. Although the squirrelfishes are peaceful, do not be tempted to include smaller fishes in their tank.

Above: Myripristis murdjan
Although the aptly named Big-Eyed Squirrelfish is generally peaceful, it would be wise to avoid keeping it with smaller fishes, which might provide an easy and tasty meal. As its huge eyes suggest, this is a nocturnal species.

Family: LABRIDAE

Wrasses

Family characteristics

The Labridae is a very large family, encompassing about 400 species. It is not surprising, therefore, that the body shape varies; some wrasses are cylindrical, while others are much deeper bodied. Like many other marine fishes, wrasses swim without making much use of the caudal fin, which is mainly used for steering or held in reserve for emergencies. The main propulsion comes from the pectoral fins.

Sex reversal is quite common in wrasses, the necessary change occurring in single-sexed groups as required. (The female stage always precedes the male one.) Reproductive activity can occur between pairs or collectively in groups. In both cases, the fishes spiral upwards towards the surface to spawn. Occasionally, this activity is based around a preselected territory. Coastal species take advantage of the outgoing tide to sweep the fertilized eggs away from the reef to safety. Species from temperate zones in Europe and the Mediterranean build spawning nests of algae or sand.

Diet and feeding

Feeding habits vary from species to species, but most relish molluscs and crustaceans. They will take brineshrimps, shrimps, and most frozen marine fare.

Aquarium behaviour

Juvenile forms are quite suitable for the aquarium. Wrasses and rainbowfishes are interesting for several reasons: the juvenile coloration patterns are different from those of adult fishes. Many bury themselves in the sand for long periods during the day as well as at night time, so a sandy substrate is essential. Others spin mucus cocoons in which to rest; and a number of fishes perform 'cleaning services' on other species, removing parasites in the process. Fishes in this group are usually quite active and therefore may disturb more sedate fishes in the aquarium.

Bodianus pulchellus

Cuban Hogfish; Spotfin Hogfish

☐ **Distribution:** Western Atlantic.

☐ **Length:** 250mm/10in (wild), 150mm/6in (aquarium).

☐ **Diet and feeding:** Crustaceans, shellfish meat. Bold bottom feeder.

☐ **Aquarium behaviour:** Generally peaceful although small fishes in the aquarium may not be entirely safe.

☐ **Invertebrate compatibility:** Safe when young, but become destructive with age and size.

The front of the body is red, divided by a white horizontal band; the upper part of the back is bright yellow. There is a black spot at the end of the pectoral fins. Juveniles are yellow with a dark spot on the front of the dorsal fin, a colour

Above: Bodianus pulchellus
This smart Cuban Hogfish from the western Atlantic is easy to acclimatize to aquarium life. It is best kept with fishes of equal or larger size.

scheme similar to that of the juvenile form of *Thalassoma bifasciatum* (see page 220). It is likely that both these species have evolved similar markings to signal their cleaning services. Usually it is quite easy to acclimatize these fishes to aquarium foods.

Bodianus rufus
Spanish Hogfish

☐ **Distribution:** Western Atlantic.

☐ **Length:** 600mm/24in (wild), 200mm/8in (aquarium).

☐ **Diet and feeding:** Crustaceans, shellfish meat. Bold bottom feeder.

☐ **Aquarium behaviour:** Peaceful.

☐ **Invertebrate compatibility:** Safe when young, but become destructive with age and size.

Juvenile specimens are yellow with an area of blue along the upper body. Adult fishes show the standard red and yellow coloration, although the proportions may vary according to the habitat and depth of water. Like other members of the genus, juveniles perform cleaning actions on other fishes.

Cirrhilabrus rubriventralis
Dwarf Parrot Wrasse; Sea Fighter

☐ **Distribution:** Indian Ocean.

☐ **Length:** 75mm/3in (wild)

☐ **Diet and feeding:** Will accept most marine fare, including frozen, live and flake foods.

☐ **Aquarium behaviour:** Peaceful community fish.

☐ **Invertebrate compatibility:** Good.

An extremely attractive species with distinct male and female coloration. The male is much more intensely coloured than the female and has an elongated leading edge to the dorsal fin. In suitably sized tanks it is possible to keep a pair, or one male and several females, together. In this arrangement, the male will intensify his colours and display to the females by dashing in front of them flaring his fins. Optimum water conditions are essential for this fish.

Above: Bodianus rufus
A brightly coloured juvenile specimen of this wrasse from the Caribbean. This is a peaceful community fish.

Below: Cirrhilabrus rubriventralis
The male of a pair will show off his splendid markings to the female, dashing in front of her, flaring his fins.

Above: Coris angulata
The adult Twin-spot Wrasse not only loses the spots seen in the juvenile, but also outgrows most home aquariums.

Coris angulata
Twin-spot Wrasse

☐ **Distribution:** Indo-Pacific, Red Sea.

☐ **Length:** 1200mm/48in (wild), 200-300mm/8-12in (aquarium).

☐ **Diet and feeding:** Small marine animals, live foods. Bottom feeder.

☐ **Aquarium behaviour:** Peaceful but grows very quickly.

☐ **Invertebrate compatibility:** Safe when young, but become destructive with age and size.

A spectacularly coloured species, both as a juvenile and as an adult. When young, this fish is white with two prominent orange spots on the dorsal surface. The front of the body, together with the fins, is covered with dark spots and there are two white-edged black blotches on the dorsal fin. The adult is green with yellow-edged purple fins and is referred to as the Napoleon Wrasse.

Coris formosa
African Clown Wrasse

☐ **Distribution:** Indian Ocean.

☐ **Length:** 300mm/12in (wild), 200mm/8in (aquarium).

☐ **Diet and feeding:** Small marine animals, live foods. Bottom feeder.

☐ **Aquarium behaviour:** Peaceful but grows large.

☐ **Invertebrate compatibility:** Safe when young, but become destructive with age and size.

The juvenile fish is dark brown with a thick vertical white band crossing the body and dorsal fin just behind the gill cover. Two shorter bands cross the head and two more appear on the rear of the dorsal fin and caudal peduncle. The caudal fin is white. In the adult, the head and body are green-brown and two green-blue stripes run in front of and behind the gill cover. The dorsal fin is red with an elongated first ray, the anal fin is green and purple, and the caudal fin is red, bordered with white.

Below: Coris formosa
Adults lack these white markings.

Gomphosus caeruleus
Birdmouth Wrasse

☐ **Distribution:** Indo-Pacific.

☐ **Length:** 250mm/10in (wild).

☐ **Diet and feeding:** Small animal life gathered from coral crevices.

Coris gaimardi
Clown Wrasse; Red Labrid

☐ **Distribution:** Indo-Pacific.

☐ **Length:** 300mm/12in (wild), 150mm/6in (aquarium).

☐ **Diet and feeding:** Crustaceans, shellfish meat. Bold bottom feeder.

☐ **Aquarium behaviour:** May quarrel among themselves.

☐ **Invertebrate compatibility:** Safe when young, but become destructive with age and size.

Juveniles are similar to *C. formosa* but are orange rather than brown; the middle white band does not

Above: Coris gaimardi (adult)
Here, the final adult colours of the Clown Wrasse are established. The blue facial markings and spotted body bear little resemblance to the appearance of the young fish shown below.

extend right down the body and the dorsal fin lacks a spot. Adults are brown-violet with many blue spots. The dorsal and anal fins are red, the caudal fin is yellow. There are blue markings on the face. These can be nervous fishes that dash about when first introduced into the aquarium, so try not to shock them.

Below: Coris gaimardi (juvenile)
The first colour stage of the Clown Wrasse is orange, with white markings on the body similar to, but less extensive than, those of C. formosa.

In the aquarium, they will take brineshrimp, *Mysis* shrimp, krill and chopped fish meats, plus some green foods. Grazer.

☐ **Aquarium behaviour:** Peaceful.

☐ **Invertebrate compatibility:** Safe when young, but become destructive with age and size.

The body in adult males is blue-green; younger males, and females, are brown. The snout is elongated. A very active fish that is constantly on the move around the aquarium. This species looks and swims like a dolphin. Juveniles act as cleaners.

Below: Gomphosus caeruleus
The Birdmouth Wrasse is a very active species, always on the move around the aquarium. Generally, this species is very peaceful, and minds its own business, but its constant movement may annoy more leisurely or smaller species and specimens often become more destructive with age and size.

Halichoeres chrysus
Banana Wrasse; Golden Rainbowfish.

☐ **Distribution:** Western Pacific.

☐ **Length:** 100mm/4in (wild).

☐ **Diet and feeding:** Easily fed with most marine frozen, live and flake foods.

☐ **Aquarium behaviour:** Peaceful. A good community fish.

☐ **Invertebrate compatibility:** Very good.

Like most wrasses, *H. chrysus* needs a substrate of soft coral sand under which to retreat during the hours of darkness and when frightened. This very attractive species has a rich egg-yellow body and a clear caudal fin. In adulthood, the facial area develops green markings, making this an even more desirable fish.

Halichoeres trispilus
Four-spot Wrasse; Banana Wrasse

☐ **Distribution:** Indian Ocean.

☐ **Length:** 100mm/4in (wild).

☐ **Diet and feeding:** Accepts most marine frozen and live foods.

☐ **Aquarium behaviour:** A peaceful community fish.

☐ **Invertebrate compatibility:** Very good.

At first glance, *H. trispilus* and *H. chrysus* would appear to be one and the same fish. However, they are found in separate oceans and *H. trispilus* is white in the lower half of the body. Its common name of Four-Spot Wrasse derives from the small black spots, three of which are found on the dorsal fin and one on the caudal peduncle. This is an ideal fish for the beginner.

Labroides dimidiatus
Cleaner Wrasse

☐ **Distribution:** Indo-Pacific.

☐ **Length:** 100mm/4in (wild).

☐ **Diet and feeding:** Skin parasites of other fishes in the wild; in captivity, finely chopped meat foods make an excellent substitute. Bold.

☐ **Aquarium behaviour:** Peaceful.

☐ **Invertebrate compatibility:** Safe.

This is the most familiar of the wrasses because of its cleaning activities. This cleaning process, also practised by some gobies and Cleaner Shrimps, is almost ritualistic. When approached by a Cleaner Wrasse, the subject fish – or host – often remains stationary with fins spread, in a head-up or head-

Above left: Halichoeres chrysus
Many species possess the common name Banana Wrasse; perhaps this completely yellow fish deserves it more than most.

Above: Halichoeres trispilus
An ideal fish for beginners, the Four-spot Wrasse is colourful, readily available and disease resistant.

Right: Labroides dimidiatus
A true asset to the aquarium, the Cleaner Wrasse provides a service much appreciated by the other tank inmates.

down attitude. Sometimes the colours of the host fish fade, maybe so that the Cleaner Wrasse can see any parasites more clearly.
The elongate blue body of the Cleaner Wrasse has a horizontal dark stripe from snout to caudal fin. The mouth is terminal, and it is this feature that distinguishes *Labroides dimidiatus* from the predatory lookalike *Aspidontus taeniatus*, the so-called False Cleanerfish.

Lienardella fasciata
Harlequin Tuskfish

☐ **Distribution:** Western Pacific.

☐ **Length:** 600mm/24in (wild), 350mm/14in (aquarium).

☐ **Diet and feeding:** Generally quite easily fed on meaty marine foods, including shrimps, mussels and squid.

☐ **Aquarium behaviour:** A peaceful community subject if kept with fish of the same size or larger.

☐ **Invertebrate compatibility:** Definitely not.

The Harlequin Tuskfish is an extremely attractive species with a set of menacing teeth that belie its peaceful nature – at least with fish it is incapable of swallowing! Although

Left: Lienardella fasciata
Despite its large size, the Harlequin Tuskfish is very peaceful.

Below: Novaclichthys taeniourus
Keep this Wrasse in a species tank or with equal-sized tankmates.

small specimens are available, it should be noted that this fish, when fed correctly, is capable of reaching a fairly large size in the aquarium and adequate room will be required.

Novaculichthys taeniorus
Dragon Wrasse

☐ **Distribution:** Indo-Pacific.

☐ **Length:** 200mm/8in (wild); 60mm/2.5in (aquarium).

☐ **Diet and feeding:** Crustaceans, meaty foods. Bottom feeder.

☐ **Aquarium behaviour:** Peaceful when young.

☐ **Invertebrate compatibility:** Safe.

The blotchy green coloration and elongated first rays of the dorsal fin are features of the juvenile; adult fishes are brown with marks radiating from the eye; these fade as the fish grows older. Keep it with similarly sized tankmates as this species is intimidated by larger, more aggressive, fish.

Left: Pseudocheilinus hexataenia
The Pyjama Wrasse is a constantly busy fish, always on the look out for small pieces of food missed by its tankmates. Once settled, it shows little regard for other fish, as long as they are not of the same species or with similar markings, otherwise fighting will occur.

Right: Thalassoma bifasciatum
Adult, fully coloured male Bluehead Wrasse make a colourful and peaceful addition to a suitable fish community aquarium. The juveniles, in common with several other species, are called Banana Wrasse but tend to be more sensitive than the adults and require water of the highest quality to do well.

Below right: Thalassoma lunare
A distinguishing feature of the adult Moon Wrasse is the bright yellow central section to the caudal fin and its green/ blue edging at the top and bottom. Note the radiating facial pattern around the eye; many species share this striking characteristic.

Pseudocheilinus hexataenia
Pyjama Wrasse; Six-line Wrasse

☐ **Distribution:** Indo-Pacific.

☐ **Length:** 50mm/2in (wild), 75mm/3in (aquarium).

☐ **Diet and feeding:** Will readily accept most marine frozen, live and flake foods.

☐ **Aquarium behaviour:** Peaceful, except with the same, or similar, species.

☐ **Invertebrate compatibility:** Ideal.

P. hexataenia has many of the attributes of an ideal aquarium fish: it rarely bothers invertebrates or other fish (as long as they are not of its own kind, or similar), it is colourful, easily fed, interesting in its activities and generally very inexpensive. This disease-resistant species can be heartily recommended to any newcomer to the hobby.

Thalassoma bifasciatum
Bluehead Wrasse

☐ **Distribution:** Caribbean.

☐ **Length:** 140mm/5.5in (wild).

☐ **Diet and feeding:** Crustaceans, meaty foods. Bottom feeder. Most specimens have a very healthy appetite but may need weaning off live brineshrimp on to other, more convenient, meaty foods.

☐ **Aquarium behaviour:** Peaceful.

☐ **Invertebrate compatibility:** Become destructive with age and size.

This fish undergoes remarkable colour changes: juveniles are yellow (the shallow water types are white) with a dark spot on the front of the dorsal fin and/or a dark horizontal stripe along the body. Dominant males develop the characteristic blue head and green body separated by contrasting black and white bands. This colour pattern acts as the focus for their harem formation.

Thalassoma lunare
Moon Wrasse; Lyretail Wrasse; Green Parrot Wrasse

☐ **Distribution:** Indo-Pacific, Red Sea.

☐ **Length:** 330mm/13in (wild).

☐ **Diet and feeding:** All meaty foods. Greedy bottom feeder.

☐ **Aquarium behaviour:** Active all day, then sleeps deeply at night. Its constant daytime movement may disturb smaller fishes and it may attack new additions to the tank, regardless of their size. Needs plenty of room.

☐ **Invertebrate compatibility:** Become destructive with age and size.

Adult specimens lose the dark blotches of the juvenile and are bright green with red and blue patterns on the head. The centre of the caudal fin is bright yellow, while the 'lyretail' effect is given by the green/blue edging.

Family: LUTJANIDAE

Snappers

Family characteristics

The snappers are a large and varied group of fishes, many of which are caught commercially for food purposes. Of those species that may be kept in the aquarium, most are only suitable as juveniles, as growth is quick and ultimate sizes can be quite unmanageable, even for the most enthusiastic aquarist. In many cases, the attractive coloration of the juveniles fades quite badly with age. Spawning is virtually unknown in captivity.

Diet and feeding

In the wild, these fishes feed on small fish and invertebrates. In the aquarium, a good, varied diet of meaty foods should be provided. This should include live river shrimp, frozen prawns, mussels, squid and lancefish.

Aquarium behaviour

Snappers are predatory fishes and should not be trusted with smaller species. They make fine specimen fish when young and can become very tame, although most will outgrow the average hobbyist's tank quite quickly. Snappers are best kept singly but are usually peaceful with other fish of the same, or larger, size. Plenty of room is needed if they are to achieve their full potential.

Lutjanus sebae
Emperor Snapper

☐ **Distribution:** Indo-Pacific.

☐ **Length:** 900mm/36in (wild).

☐ **Diet and feeding:** Animal and meaty foods. Bold.

☐ **Aquarium behaviour:** Although peaceful, do not keep this species with smaller fishes. Despite its attractive coloration, it will soon outgrow the tank, and is therefore not really suitable for anything but the largest public aquarium.

☐ **Invertebrate compatibility:** Unsuitable.

The white body has three red-brown transverse bands: the first runs from the snout up the forehead and the second is 'L-shaped' running vertically down from the dorsal fin to the pelvic fins and along the ventral surface into the front half of the anal fin. The third band is crescent shaped, beginning in the rear of the dorsal fin and crossing the caudal peduncle into the lower half of the caudal fin, which has a similarly coloured brown bar on the top edge.

Below: Lutjanus sebae
The Emperor Snapper is popular not only with marine aquarists – it also makes good eating! Sadly, the striking coloration fades with age.

Symphorichthys spilurus
Majestic Snapper

☐ **Distribution:** Pacific.

☐ **Length:** 320mm/12.6in (wild).

☐ **Diet and feeding:** Meaty foods. Bold feeder.

☐ **Aquarium behaviour:** Do not keep with smaller fishes.

☐ **Invertebrate compatibility:** Unsuitable.

The yellow body has horizontal blue lines and the fins are yellow. Two vertical black bars cross the head and there is a distinguishing white-ringed black blotch on the caudal peduncle. The dorsal fin develops extremely long filamentous extensions. The anal fin also has long extensions, which virtually mirror those of the dorsal.

Above: Symphorichthys spilurus
Being of a peaceful temperament, the Majestic Snapper makes a good community fish with other species that are as large as, or larger than, itself. Tankmates should, in any event, not be inclined to nip fins!

Family: MICRODESMIDAE

Wormfishes

Family characteristics

Firefishes were originally included in the family Blenniidae (blennies), subsequently placed in with the family Gobiidae (gobies) and have now found a home with the Microdesmids (Hoese, 1986, and Birdsong, 1988). This family consists of approximately 36 species of eel-like fishes, hence the common name of wormfishes (a not very flattering but, nonetheless, very apt description).

In the wild, firefishes may be seen hovering in groups close to the reef, head into the current, feeding on small planktonic animals that drift their way. All possess a favourite 'bolthole' in the reef into which they rapidly retreat at the first sign of danger.

Three species are represented in the genus: *Nemateleotris magnifica* (Firefish), *N. decora* (Purple Firefish) and the rarely seen, but superbly coloured, *N. helfrichi*. The first dorsal ray is greatly extended in all three species, particularly so in *N. magnifica*. The function of this ray is to act as a signalling device to other firefish and, additionally, to provide a 'locking' mechanism when the fish is positioned in its rocky retreat.

Aquarium spawnings are extremely rare and, to date, no larvae have been raised.

Diet and feeding

Small planktonic animals are taken in the wild. In the aquarium, firefishes will accept nearly all live, frozen and flake foods of a suitable size.

Aquarium behaviour

These are fairly nervous fishes, requiring plenty of rockwork crevices into which to retreat when disturbed and as a night-time shelter. In the absence of suitable rockwork; or in the company of boisterous companions, firefishes have a tendency to jump out of the water; therefore a tight-fitting cover glass is recommended.

Firefishes may be kept in small groups, but each fish will need a small territory of 10-20 square centimetres (1½-3 sq in) vertically. A hierarchy is usually established, the most dominant fish taking the most desirable position.

Nemateleotris decora
Purple Firefish

☐ **Distribution:** Indo-Pacific.

☐ **Length:** 75mm/3in (wild).

☐ **Diet and feeding:** Most marine frozen, live and flake foods.

☐ **Aquarium behaviour:** Peaceful.

☐ **Invertebrate compatibility:** Ideal.

This species is closely related to *N. magnifica,* but has a splendid purple coloration and a shorter first dorsal ray. Like its cousin, it is generally very peaceful, but squabbles may occur if the fish are kept in cramped conditions where required territory is at a premium. Provide plenty of rockwork so that this Firefish has somewhere to hide. Under active lighting the purple/blue colour at the margins of the fins is shown to its best effect.

Left: Nemateleotris decora
The Purple Firefish is far less common than N. magnifica *and consequently commands a higher price. Many aquarists do not see this as a deterrent, however, as this colourful fish makes a very desirable addition to the mixed aquarium.*

Above: Nemateleotris magnifica
The beautifully coloured Firefish never ventures too far from its bolthole. If kept in a sufficiently large aquarium, several will live quite happily together.

Nemateleotris magnifica
Firefish

☐ **Distribution:** Indo-Pacific.

☐ **Length:** 60mm/2.4in (wild).

☐ **Diet and feeding:** Small crustaceans, plankton and other small live foods. Bottom feeder.

☐ **Aquarium behaviour:** Peaceful once established.

☐ **Invertebrate compatibility:** Yes, ideal.

This beautifully coloured fish settles into aquarium life much more successfully if plenty of boltholes in the rockwork are available to it. The blue, green and yellow on the body gradually give way to a magnificent fiery red on the rear of the body and on the dorsal, anal and caudal fins. The elongated first dorsal fin is used to signal to other firefish, as well as providing a 'locking' device.

Family: MONOCANTHIDAE

Filefishes

Family characteristics
Filefishes, like the triggerfishes (family Balistidae, see pages 174-179) to whom they are related (and with whom they are classified by some authorites), have two dorsal fins, the first being a rudimentary spine which can be locked into position. The ventral fins are reduced to a single spine. The skin is rough in texture, which has given rise to the fishes' alternative common name of Leatherjackets. The teeth are developed for nibbling. Most species occur around Australia.

Diet and feeding
In the wild, these fishes use their tiny mouths to feed mainly on polyps and algae. This may cause initial problems in the aquarium until they can be persuaded to accept alternative foods.

Aquarium behaviour
Despite their similarity to triggerfishes, filefishes are generally smaller, less active and – once acclimatized and feeding well – make good additions to a community tank.

Chaetodermis pencilligerus
Tassel Filefish

☐ **Distribution:** Indo-Pacific.

☐ **Length:** 180mm/7in (wild).

☐ **Diet and feeding:** May have to be tempted with live foods initially, but thereafter should accept most marine fare.

☐ **Aquarium behaviour:** Peaceful.

☐ **Invertebrate compatibility:** No.

The Tassel Filefish is quickly becoming a firm favourite with many marine aquarists owing to its interesting, not to say unusual, shape and its obliging nature. Once settled in the aquarium, rate of growth is rapid (the length given above does not take into account the odd specimen that has been documented at 250mm/10in), and a large aquarium will quickly become necessary.

Below: Chaetodermis pencilligerus
If you are looking for something a little different, the Tassel Filefish could well be the answer. Its unusual appearance provides camouflage in its natural habitat among floating seaweeds.

Family: MONOCENTRIDIDAE

Pine-cone Fishes

Family characteristics
This small family contains some interesting fishes known to have existed millions of years ago. Their bodies are enclosed in the rigid covering of a few large scales fused together. They are deepwater fishes and it is thought that they use their light-generating organs to attract prey and for recognition purposes. These fishes are only occasionally available through dealers and therefore command a high price. The dorsal fin spines alternate to right and left rather than in the normal straight row.

Diet and feeding
Because of their deepwater origins, little is known of their natural diet. It probably consists mainly of small marine animals attracted to them by their light-generating organs. In the aquarium, offer live foods and then try to wean them on to other suitable foods.

Aquarium behaviour
You may find it advisable to keep these fishes in a species tank so that you can study them more closely.

Monocentrus japonicus
Pine-cone Fish

☐ **Distribution:** Indo-Pacific.

☐ **Length:** 160mm/6.3in (wild), 100-150mm/4-6in (aquarium).

☐ **Diet and feeding:** Provide chopped white fish meat or shellfish meat, such as mussel. Also supply frozen meaty foods.

☐ **Aquarium behaviour:** Keep in a dimly lit species aquarium.

☐ **Invertebrate compatibility:** Not recommended.

The head forms about one third of the body size. The large brass-coloured scales have dark edges and spiny centres. Spiny dorsal rays alternate from left to right. The pelvic fins are restricted to strong spines. May not thrive if kept at temperatures above 23°C(74°F) for long periods.

Below: Monocentrus japonicus
The Pine-cone Fish, one of the more unusual fishes for the aquarium, can boast an ancestry going back some millions of years. Although not commonly encountered in retail outlets, it is worth considering if you are looking for something different.

Family: MONODACTYLIDAE

Fingerfishes

Family characteristics

These silver rhomboidal fishes are reminiscent of the freshwater angelfishes. They are found in coastal waters, particularly estuaries, often entering brackish or even fresh water. Although young specimens will thrive in slightly brackish water, they do even better in full strength salt water. Along with *Scatophagus* spp, they are scavengers, frequently found in dirty waters, where they appear to thrive in the conditions! Their scientific name 'Monodactylus' means 'one finger' and refers to the shape and colour of the dorsal fin.

Diet and feeding

These fishes will eat any foods, including *Tubifex* worms.

Aquarium behaviour

Fast-moving shoaling fishes that may reach up to 150mm(6in) long in captivity.

Monodactylus argenteus

Fingerfish; Malayan Angel; Silver Batfish

☐ **Distribution:** Indo-Pacific.

☐ **Length:** 230mm/9in (wild).

☐ **Diet and feeding:** Will eat anything. Bold scavenger.

☐ **Aquarium behaviour:** Peaceful but constantly active. shoaling fishes.

☐ **Invertebrate compatibility:** Not recommended.

One or two black bars cross the front part of the silver body and the fins are yellow. The pelvic fins are rudimentary. These fishes are very

Above: Monodactylus argenteus
Although young specimens may be kept successfully in freshwater or brackish water tanks, adults really thrive in full-strength sea water. These active fishes need plenty of swimming space.

fast swimmers when disturbed, and can be difficult to catch in the aquarium. They appear to thrive better when kept in shoals.

Monodactylus sebae
Striped Fingerfish

☐ **Distribution:** Eastern Atlantic, West African coast.

☐ **Length:** 200mm/8in (wild).

☐ **Diet and feeding:** Will eat anything. Bold scavenger.

☐ **Aquarium behaviour:** Peaceful, but restless.

☐ **Invertebrate compatibility:** Not recommended.

Monodactylus sebae is slightly darker than the previous species and the body is much taller. Two additional dark vertical stripes cross the body, one connecting the tips of the dorsal and anal fins, the other crossing the extreme end of the caudal peduncle. The pelvic fins are rudimentary. They appear to be less hardy than *M. argenteus*.

Below: Monodactylus sebae
The less common relative of M. argenteus *is also an active shoaling fish.*

Family: MURAENIDAE

Moray Eels

The tropical eels are often splendidly marked and all grow very long. Many are nocturnal and are hardly seen during the day, since they hide among caves and crevices. They detect their food by smell and are usually quite undemanding in captivity, providing that they have sufficient room, refuges and food. Keeping them in the company of small 'bite-sized' fishes is tempting providence a little too much. Needless to say, the aquarium should be tightly covered – and beware your fingers when feeding!

Breeding moray eels in captivity is unlikely because sexual maturity is reached only when the eels attain a large size. This stage is not normally reached in the confines of a domestic aquarium, and so successful breeding is doubtful even if a sufficient number of specimens are kept together to allow natural pairings to take place.

The Romans valued moray eels very highly; they were kept in captivity and favourite fish were bedecked with jewels! Wealthier Romans fed spare slaves to their morays!

Echnida nebulosa
Snowflake Moray

☐ **Distribution:** Indo-Pacific.

☐ **Length:** 1000mm/39in (wild), 600m/24in (aquarium).

☐ **Diet and feeding:** Requires meaty marine foods and relishes live river shrimp.

☐ **Aquarium behaviour:** Generally peaceful if kept with other fishes too large to swallow.

☐ **Invertebrate compatibility:** Possible, but not with crustaceans.

The Snowflake Moray is usually seen on sale as relatively small, juvenile specimens, but given the correct diet and aquarium conditions, it can grow very quickly. However, unlike many other moray species, it is unlikely to reach an unmanageable size in the aquarium. Although this species can become hand-tame, it is still capable of giving a nasty bite and should be treated with respect.

Above right: Echnida nebulosa
Although the Snowflake Moray is usually bought as a small specimen, it is fast growing and requires spacious quarters.

Right: Gymnothorax tesselatus
Concentrating on the 'business end' of this species is very wise. The bite can be painful, even leading to infection – and that's only for humans.

Gymnothorax tesselatus
Reticulated Moray; Leopard Moray

☐ **Distribution:** Indo-Pacific.

☐ **Length:** 1500mm/60in (wild).

☐ **Diet and feeding:** Will eat anything it can swallow. Predatory ambushers.

☐ **Aquarium behaviour:** Do not keep with anything small.

☐ **Invertebrate compatibility:** Not safe with crustaceans.

This species can inflict a very painful bite, even leading to infection (and that's only for humans); just imagine encountering this predator if you were a fish! Only for the 'big tank' hobbyist or public aquariums.

The dark body is covered with a reticulated pattern of pale markings, producing an effect very similar to that of a giraffe's markings. The nostrils are tubular.

Family: OPISTOGNATHIDAE

Jawfishes

Family characteristics

Behind the large mouth and eyes of these fish is a tapering cylindrical body. The main appeal of jawfishes is their habit of constructing tunnels or burrows in the substrate into which they retreat when threatened. At night they use a small pebble or shell to cover the entrance.

While the commonly kept Yellow-headed Jawfish (*Opistognathus aurifrons*) shows no sexual dimorphism, males of other less familiar species, such as the Atlantic Yellow Jawfish (*O. gilberti*) and the Pacific Blue-Spotted Jawfish do undergo colour changes at breeding times. Perhaps the best guide to distinguishing the sexes is the fact that males take on the oral incubation of the eggs.

Diet and feeding

Hovering close to their burrowed tunnels, Jawfishes wait for their prey of crustaceans, small fishes and plankton. Some species are not so adventurous, preferring to remain in their holes waiting for small live foods to pass by.

Aquarium behaviour

Providing sufficient 'accommodation' is available, a number of these fishes can be kept together. They are easily frightened, retreating into their holes like lightning. These tolerant fishes bother nobody, but may be harassed by other cave-dwelling species. Jawfishes are excellent jumpers, so ensure that the tank has a closely fitting cover.

Opistognathus aurifrons
Yellow-headed Jawfish

☐ **Distribution:** Western Atlantic.

☐ **Length:** 125mm/5in (wild).

☐ **Diet and feeding:** Finely chopped shellfish meat. Makes rapid lunges from a vertical hovering position near its burrow to grab any passing food.

☐ **Aquarium behaviour:** Peaceful and rarely disturbed by other fishes.

☐ **Invertebrate compatibility:** Yes, ideal.

The delicately coloured yellow head is normally all you see of this fish, but the rest of the body is an equally beautiful pale blue. The eyes are large. It needs a reasonably soft substrate in which to excavate a burrow, entering the hole tail first at any sign of trouble. Like other jawfishes, it is a good jumper.

Right: Opistognathus aurifrons
On the rare occasions when it is seen outside its hiding place, the Yellow-headed Jawfish reveals that its slim body has the most delicate coloration. How the fish manages to retreat into its bolthole tail-first at such speed almost defies belief.

Family: OSTRACIIDAE

Boxfishes and Trunkfishes

Family characteristics

These fishes have a rigid body made up of bony plates covered with a sensitive skin that may be damaged by cleanerfishes. The only flexible part is the caudal peduncle, where the most obvious growth occurs rearwards. The pelvic fins are missing, although bony stumps may appear at the corners of the body box in some species. They are slow moving – some have been designated 'hovercraft fishes' by imaginative authors – and they do indeed have a similar form of locomotion, making rapid movements of the dorsal, anal and pectoral fins to good effect. When buying these fishes, avoid any with concave looking sides, as these never recover from this probable semi-starved state. Regular feeding is strongly urged.

Most are poisonous, releasing a poison into the water when threatened. In the confines of the aquarium, or in the transportation container, this often proves fatal both to the Boxfish and to other fishes. Some authorities advocate introducing these fish into the aquarium in advance of other fishes to reduce the chances of fatal consequences should the boxfishes become frightened.

Diet and feeding

These fishes will try anything, but appear to particularly relish live foods.

Aquarium behaviour

Some reports suggest that these fish resent the attentions of Cleaner Gobies or any other inquisitive fishes perhaps attracted by their slow swimming action.

Lactoria cornuta
Cowfish

☐ **Distribution:** Indo-Pacific.

☐ **Length:** 500mm/20in (wild); 400mm/16in (aquarium).

☐ **Diet and feeding:** Algae and crustaceans. These are shy bottom feeders.

☐ **Aquarium behaviour:** Intolerant of each other.

☐ **Invertebrate compatibility:** Generally yes, but may peck at tubeworms.

The two 'horns' on the head and two more projections at the bottom rear of the body make for easy identification. The body is brilliant

Above: Lactoria cornuta
This fish's large horny projections make the fish hard to swallow by predators, although its poisonous flesh would, in any case, inflict retribution.

yellow with bright blue spots in the centre of each segment of body 'armour plating'. Specimens in domestic aquariums do not usually reach very large sizes.

Ostracion meleagris
Spotted Boxfish; Pacific Boxfish

☐ **Distribution:** Indo-Pacific.

☐ **Length:** 200mm/8in (wild).

☐ **Diet and feeding:** Crustaceans and greenstuff. Bottom feeder.

☐ **Aquarium behaviour:** Peaceful.

☐ **Invertebrate compatibility:** Generally yes, but may peck at tubeworms.

It now seems certain that *O. meleagris* appears in a distinct male and female form. The male is the more colourful of the species. The top of the body is black with white spots and the lower flanks are violet with yellow spots. The two sections are separated by a yellow line. The eye is yellow-gold. By contrast, the female is almost entirely black with white spots.

In a large enough tank, it is possible to keep a pair, although breeding is rarely known in captivity. This is a difficult fish that is prone to bacterial infections of the skin and eyes. Initial feeding may be a problem and usually demands all the skill of an experienced hobbyist. Excellent water quality at all times is essential for this species.

Below: Ostracion meleagris
Although often confused with young specimens of pufferfish, especially Arothron meleagris*, boxfishes rely on poisonous secretions to ward off predators rather than inflating their bodies, as do the pufferfishes. This is a difficult species to keep.*

Ostracion tuberculatum
Blue-spotted Boxfish

☐ **Distribution:** Indo-Pacific.

☐ **Length:** 450mm/18in (wild).

☐ **Diet and feeding:** Crustaceans and greenstuff. Bottom feeder.

☐ **Aquarium behaviour:** Peaceful; best left undisturbed.

☐ **Invertebrate compatibility:** Generally yes, but may peck at tubeworms.

The almost cube-shaped body of juveniles is light cream or yellow with dark blue spots; it is easy to imagine that they are animated dice, slowly swimming around looking for food. Adult fishes develop a more elongate body and the colour changes to a yellowy green, while the armoured plates on the body become more clearly defined. The fins are tinted yellow.

Tetrosomus gibbosus
Hovercraft Boxfish

☐ **Distribution:** Indo-Pacific.

☐ **Length:** 400mm/16in (wild), 100mm/4in (aquarium).

☐ **Diet and feeding:** Can be tempted to eat most marine frozen, live and flake foods of a suitable size.

☐ **Aquarium behaviour:** Peaceful, except with its own, or similar, species.

☐ **Invertebrate compatibility:** No, except when very small.

T. gibbosus is usually seen on sale as a juvenile specimen some 25mm/1in in length and will rarely exceed 100mm/4in in the aquarium. Its curious swimming habit – it appears to hover in mid-air – has given rise to its common name of Hovercraft Boxfish. It should be kept with other non-boisterous fish, as it is a prime target for fin-nipping species.

Above: Ostracion tuberculatum
This species really lives up to its popular name, especially when viewed from the angle captured here. A side view would reveal yellow fins, a pointed snout and longish caudal peduncle.

Below: Tetrosomus gibbosus
If you are looking for something different – say a fish that doesn't look look a fish, doesn't swim like a fish, but is full of charm – the Hovercraft Boxfish would make an excellent choice!

Family: PLATACIDAE

Batfishes

Family characteristics

The oval-bodied, high-finned Batfish is unmistakable. It is found in coastal and brackish waters and in mangrove swamps. It often lies on its side 'playing dead', floating like a leaf to avoid capture or detection. It now appears fairly certain that there are four species of batfish; *Platax orbicularis*, *P. pinnatus*, the rarer *P. tiera* (Longfinned Batfish) and *P. batavianus* (Marbled Batfish). The latter two are rarely available to the hobbyist. Adult fishes are less colourful than juveniles. Some leading authorities clarify batfishes as belonging to the subfamily Platacinae of the family Ephippididae (the spadefishes).

Diet and feeding

These fishes may be difficult to accustom to the usual commercial marine foods and patience will be needed, especially with *P. pinnatus*. Try to purchase juvenile specimens that are already feeding.

Aquarium behaviour

The Batfish usually adapts to captivity well, not quarrelling with similarly sized tankmates. It does need a spacious tank, however, as it grows very quickly. The shape of the body dictates the need for a very deep tank to enable the fish to develop proportionately.

Platax orbicularis
Batfish; Orbiculate Batfish; Round Batfish

☐ **Distribution:** Indo-Pacific.

☐ **Length:** 500mm/20in (wild), 380mm/15in (aquarium).

☐ **Diet and feeding:** Will eat anything. Scavenger.

☐ **Aquarium behaviour:** Peaceful, but grows fast. Keep away from boisterous fin-nipping species.

☐ **Invertebrate compatibility:** Usually well behaved.

The body is round, with large rounded fins. There are one or two dark stripes on the head and front part of the body, but these fade with age. Young specimens have more elongated fins and also more red coloration.

Right: Platax orbicularis
This is a young specimen, showing the typical dark red-brown stripes; adults are less colourful. Allow a generous depth of water and plenty of swimming space for these tall-finned fishes.

Platax pinnatus
Red-faced Batfish

☐ **Distribution:** Indo-Pacific.

☐ **Length:** 500mm/20in (wild), 450mm/18in (aquarium).

☐ **Diet and feeding:** Can be difficult to start feeding. Live brine shrimp should be offered regularly.

☐ **Aquarium behaviour:** Peaceful, but grows fast. Keep away from boisterous fin-nipping species.

☐ **Invertebrate compatibility:** Usually well behaved.

This is a difficult fish to keep, demanding plenty of room and excellent water conditions. Do not choose this species unless you are prepared to invest the time and patience needed. Only juveniles from a reliable source (the best possible handling) will do well in the aquarium. This is definitely not a fish for the beginner. The body shape is much shorter and higher than in *P. orbicularis* and the fins are very elongate. The colour is much darker, with a red outline to the body and fins. It is a pity that it should lose such magnificent colours and gracefulness with advancing age.

Below: Platax pinnatus
The red edges of the fins outline this splendid fish to perfection.

Family: PLECTORHYNCHIDAE

Sweetlips

Family characteristics
Fishes in this family are often classified in the Haemulidae, alternatively known as the Pomadasydae. They resemble grunts or snappers, but differ from them in dentition details. The coloration of juveniles and adults differs quite dramatically. The sweetlips are confined to the Indo-Pacific Ocean areas.

Diet and feeding
Crustaceans, live animal and meaty foods. Shy slow eaters.

Aquarium behaviour
Juveniles are excellent subjects for a large quiet tank. They are gentle and independent natured and will usually totally ignore the other fish in their aquarium.

Plectorhynchus albovittatus

Yellow Sweetlips; Yellow-lined Sweetlips

☐ **Distribution:** Indo-Pacific, Red Sea.

☐ **Length:** 200mm/8in (wild).

☐ **Diet and feeding:** Crustaceans, animal and meaty foods. Bottom feeder.

☐ **Aquarium behaviour:** Hardy, but keep with non-boisterous fishes.

☐ **Invertebrate compatibility:** Yes, when small. Progressively more destructive with age.

In juveniles the body is yellow, with two white-bordered dark bands running the length of the body. The lower band is level with the terminal mouth and centre line of the fish. The patterning of the body extends into the rear of the yellow dorsal fin and into the caudal fin. Adult fishes lose this interesting coloration and become brown.

Below: Plectorhynchus albovittatus
Young fish, such as this, look very appealing, but they lose these colours with age; adults are brown. Juvenile Yellow Sweetlips are peaceful and make ideal subjects for a large, quiet aquarium.

Plectorhynchus chaetodonoides

Harlequin Sweetlips; Clown Sweetlips; Polka-dot Grunt

☐ **Distribution:** Pacific.

☐ **Length:** 450mm/18in (wild).

☐ **Diet and feeding:** Crustaceans, animal and meaty foods. (Small live or frozen shrimps will often get them feeding in the aquarium.) Bottom feeder.

☐ **Aquarium behaviour:** Shy; keep with non-boisterous fishes.

☐ **Invertebrate compatibility:** Yes, when small. Progressively more destructive with age.

Juveniles have a dark brown body covered with well-defined white blotches and this pattern is repeated on the fins. Adult fishes are a drab

brown with dark dots. Feeding requires special attention; be sure to offer only small portions.

Below and right: Plectorhynchus chaetodonoides
There is a striking difference between young and adult Harlequin Sweetlips.

Plectorhynchus orientalis
Oriental Sweetlips

☐ **Distribution:** Indo-Pacific.

☐ **Length:** 400mm/16in (wild).

☐ **Diet and feeding:** Crustaceans, animal and meaty foods. Bottom feeder.

☐ **Aquarium behaviour:** Shy; keep with non-boisterous fishes.

☐ **Invertebrate compatibility:** Yes, when small. Progressively more destructive with age.

Juvenile fishes have large cream-yellow patches on a dark background. Adults may sometimes be confused with young *P. albovittatus*, although there are more stripes on *P. orientalis* and the coloration is not quite so yellow. In the aquarium, this species rarely exceeds 300mm (12in) in length.

Below: Pletorhynchus orientalis
Members of the sweetlips group are surprisingly shy; another even more unexpected feature is their habit of taking small morsels of food. This specimen is a juvenile; adults have stripes and are similar in coloration to the juvenile Yellow Sweetlips.

Family: PLOTOSIDAE

Catfishes

Family characteristics

Two features make it easy to identify the marine catfishes: the second dorsal and anal fins merge with the caudal fin, and there are barbels around the mouth. The spines preceding the dorsal and pectoral fins are venomous.

These fishes are very gregarious when young – species grouping together in a tight ball for safety – but this habit is lost (along with any colour pattern) when adult. In the wild, adult fishes may enter river systems. Aquarium spawnings have been reported, but so far the fry have not been reared successfully.

Diet and feeding

Chopped shellfish meats form an ideal food for these fishes in the aquarium.

Aquarium behaviour

Only juvenile specimens are suitable for the aquarium, as the adult fishes not only outgrow their juvenile coloration but also can become dangerous to inexperienced handlers. If a fairly large number are kept together in the aquarium, it is possible to witness the collective defence behaviour – 'balling' – as when threatened in the wild.

Plotosus lineatus
Saltwater Catfish

☐ **Distribution:** Indo-Pacific.

☐ **Length:** 300mm/12in (wild).

☐ **Diet and feeding:** Chopped shellfish meats. Bottom feeder.

☐ **Aquarium behaviour:** Peaceful. Prefers to be in a small shoal.

☐ **Invertebrate compatibility:** Yes.

Two parallel white lines run along the length of the dark body. The second dorsal and anal fins are very long-based and merge with the caudal fin. In the wild, young specimens shoal together, forming a tight, ball-like clump when threatened. In the aquarium, they are best kept in small shoals, since solitary specimens seem to pine away. The spines are venomous, so handle these fishes with care. (For action if stung, see page 287 – introduction to *Scorpaenidae*.)

This species presumably spawns in the same way as the freshwater Plotosid *Tandanus*, which constructs a nest of debris, sand or gravel. The male (usually identified simply because it does not lay the eggs) is said to guard the eggs after spawning has taken place.

Below: Plotosus lineatus
The habit of these smartly striped young specimens congregating together suggests that the catfish may be a good community subject. However, this fast-growing species soon loses its stripes and its sociable nature as it matures.

Poeciliidae
Black Mollies

☐ **Distribution:** 'Man-made' tank hybrids.

☐ **Length:** Male: 70mm/2.8in; female: 95mm/3.7in.

☐ **Diet and feeding:** Bold feeder. Appreciates most live foods, including *Daphnia* and midge larvae. Frozen foods, flake and some greenstuffs should also be fed.

☐ **Aquarium behaviour:** Peaceful.

☐ **Invertebrate compatibility:** Generally satisfactory, although some macro algae and delicate invertebrates may suffer.

Black Mollies are not normally associated with the marine aquarium; they are usually seen as freshwater community fishes. However, in the wild some species of molly live in brackish or fully marine conditions. Being hardy and nitrite tolerant, they can be used either as a non-aggressive fish to mature a new aquarium or as an unusual addition to graze on algae.

Black Mollies cannot, however, be taken straight from a freshwater environment and introduced into the marine aquarium; they must first be 'converted'. This is easily achieved by using an airline and valve to drip marine water at a rate of one drop per second into their container. As the container fills, the water should be discarded. The whole process will take 6-12 hours. (Converting back is the

Above: Black Mollies
Normally thought of as freshwater fish, many species of molly live in brackish or sea water in the wild. Freshwater specimens must be 'converted' to marine conditions; black hybrids, such as these, are the most suitable.

procedure in reverse.) Once 'converted' to a fully marine environment, it has been noted that colours intensify and general disease resistance improves. Breeding possibilities remain unaffected.

WARNING: Not all species of molly are suitable for convertion to marine conditions. Black hybrids and Hi-fins are generally fine but true species are not. If you are in any doubt, stick with the pure black hybrids.

Family: POMACANTHIDAE

Angelfishes

Family characteristics

Despite the close similarities between angelfishes and butterflyfishes (see pages 186-197), the surest way of distinguishing between the two groups is to look for the spine found on a the gill cover of the former. As a general rule, angelfishes are also thicker set in the body and less ovoid in shape.

The Pomacanthids are an exceedingly attractive group of fish that vary widely in terms of size and colour. Some species may grow to exceed 600mm (24in) in the wild, whereas the dwarf species (see pages 244-249) may never grow any larger than 100mm (4in).

As juveniles, many of the larger angelfish have a different coloration pattern from the adults, and it is not always possible to identify these species with certainty; many are very similar at this stage – blue with white markings, or, sometimes, black with yellow markings. These juveniles commonly act as 'cleaner fish' to larger fish in the wild, and it is thought that the markings not only advertize this fact, but also distract the aggression that could be directed at them by older and fully coloured fish of the same species.

As with so many marine species, sexual differences are barely, if at all, distinguishable, although during the spawning season, females may be observed to be noticeably swollen with eggs. Spawning always occurs at dusk as a response to falling light levels and takes place between a single pair of fish only, although males of some species may mate with more than one female at different times. As with butterflyfishes, angelfishes ascend the water column and release eggs and sperm simultaneously. The fertilized eggs then float, hatch and develop in the planktonic layers for about one month, after which the juvenile fishes settle to the bottom. Although extremely difficult, several species of angelfishes have been spawned and reared successfully to adulthood in the aquarium.

Diet and feeding

Most angelfishes are omnivorous, feeding on a wide variety of smaller animals, algae, sponges and corals. Although members of the dwarf species (*Centropyge* sp.) rarely damage corals and sponges, the same cannot be said of large Pomacanthids and, indeed, their diet may consist entirely of sponges and corals. Adults of these species may be very reluctant to accept the normal foods available to the marine aquarist during the initial settling-in period.

Angelfishes should be offered a variety of foods, including live and frozen brineshrimp and *Mysis*, frozen squid, mussels, prawns, sponge-based foods, as well as some algae.

Aquarium behaviour

Only a few angelfishes can be recommended to the beginner, as most require some previous feeding and water management skills. Many species are particularly sensitive to deteriorating water conditions and quickly display their unhappiness by contracting various diseases, as well as losing interest in food.

The highly territorial nature of angelfishes means that mixing fish of the same, or similar, species within the confines of an average-sized aquarium is usually out of the question. Having said that, these are fish that can become very tame and accept food from the hand once their trust has been gained. The more difficult species should ideally be purchased as juveniles, as they seem to adapt much better to aquarium conditions.

Suitable rockwork should always be provided in which the fish can shelter at night and retreat into when disturbed. A comprehensive rockwork arrangement is essential for dwarf species, as this is where they spend much of their time. Make sure that larger species have plenty of swimming space in the aquarium.

Apolemichthys trimaculatus

Three-spot Angelfish; Flagfin Angelfish

☐ **Distribution:** Indo-Pacific.

☐ **Length:** 250mm/10in (wild).

☐ **Diet and feeding:** Mainly algae, but offer freeze-dried foods and greenstuff. Grazer.

Left: Apolemichthys trimaculatus
This fish's name refers to the three prominent spots on the body.

□ **Aquarium behaviour:**
Territorial, keep individual specimens only. This is not the easiest angelfish to maintain in captivity.

□ **Invertebrate compatibility:**
Yes, when small, but adults may do some damage.

The three 'spots' that give the fish its common name are around its head – one on top and one on each side of the body behind the gill covers. The lips are bright blue, and the anal fin is black with a broad white area immediately next to the body.
   This species is fussy about water conditions and may also be difficult to acclimatize to aquarium life. Provide a variety of foods, particularly sponge-based.

Arusetta asfur
Purple Moon Angel

□ **Distribution:** Indian Ocean, Persian Gulf and Red Sea.

□ **Length:** 150mm/6in (wild).

□ **Diet and feeding:** Meat foods and plenty of greenstuff. Grazer.

□ **Aquarium behaviour:**
Territorial, and should not be kept with the same, or similar, species.

□ **Invertebrate compatibility:**
Yes, when small. Progressively unsuitable with age and size.

This species has a yellow vertical bar across the blue body in front of the

Above: Arusetta asfur
Compare the extended dorsal and anal fin outlines of this Indo-Pacific Angelfish with those of the smaller 'dwarf' species of the genus Centropyge.

anal fin. In this respect it differs from a similar-looking species, *Pomacanthus maculosus*, whose yellow bar begins well into the anal fin. The dorsal and anal fins are elongated and the yellow colour is repeated on the caudal fin.
   Arusetta asfur is regarded by some as a 'dwarf' angelfish, and this might be correct while the fish is still young, but as it grows into adulthood, the Purple Moon Angel behaves more like larger angelfish becoming progressively more destructive and quite aggressive in the mixed aquarium.

Genus Centropyge
Dwarf Angelfishes

Species in the genus *Centropyge* deserve a special introduction, although we have maintained their position in the A-Z sequence of the angelfish section.

Most species are ideal aquarium fishes, principally because they are miniature versions of the larger angelfishes. Most specimens make ideal subjects for the invertebrate aquarium, being almost invariably well behaved, readily accepting aquarium conditions and eating all the usual commercial marine foods. In the wild, they are found more commonly at the base of the reef rather than among the coral polyps, although they are never far away from a safe retreat. Unlike some other angelfishes, *Centropyge* species more often than not associate in pairs, with several pairs sharing the same area. Their main diet appears to be algae, which they graze from the reef surfaces. Should treatment for White Spot, Oodinium or other protocoal diseases be required, these fishes will accept lower doses of a copper-based remedy for a longer time than normal, but may not tolerate the higher doses given to larger related species.

Centropyge acanthops
African Pygmy Angelfish; Fireball Angelfish

☐ **Distribution:** Indian Ocean, along the eastern seaboard of Africa.

☐ **Length:** 75mm/3in (wild).

☐ **Diet and feeding:** Meat foods and plenty of greenstuff. Grazer.

☐ **Aquarium behaviour:** Peaceful.

☐ **Invertebrate compatibility:** Generally well behaved in the invertebrate aquarium.

A blue fish with a yellow head and dorsal area, plus a pale yellow caudal fin. An ideal, peaceful aquarium subject. It is similar to *C. aurantonotus* (Flame-backed Angelfish) from the West Indies, which can be distinguished by its blue tail. The spine on the gill cover distinguishes this species as an angelfish, despite having a body more like a damselfish.

Below: Centropyge acanthops
Its diminutive size, attractive coloration, harmless disposition and interesting habits make the African Pygmy Angelfish an ideal choice for the mixed fish and invertebrate aquarium.

Centropyge argi
Pygmy Angelfish; Cherubfish; Purple Fireball

☐ **Distribution:** Western Atlantic.

☐ **Length:** 75mm/3in (wild).

☐ **Diet and feeding:** Meat foods and plenty of greenstuff. Grazer.

☐ **Aquarium behaviour:** Usually peaceful.

☐ **Invertebrate compatibility:** Generally well behaved in the invertebrate aquarium.

A deeper water fish, which lives around the base of the reef rather than at the top. The colour patterns around the head may vary in detail

from one specimen to another and there is no difference in the juvenile colour form, as in other angelfishes. It is possible to keep compatible pairs in the aquarium, since natural territories are not particularly large.

Above: Centropyge argi
This attractive species is generally peaceful, and if two fishes appear to keep each other's company consistently, the result may be a spontaneous spawning in the aquarium.

Centropyge bicolor
Bicolor Cherub; Oriole Angel

☐ **Distribution:** Pacific.

☐ **Length:** 125mm/5in (wild).

☐ **Diet and feeding:** Meat foods and plenty of greenstuff. Grazer.

☐ **Aquarium behaviour:** Peaceful, providing plenty of hiding places are available.

☐ **Invertebrate compatibility:** Generally well behaved in the invertebrate aquarium.

The rear part of the body, from behind the head as far as the caudal fin, is bright purple-blue. The small bar across the head over the eye is the same bright shade, while the head and caudal fin are yellow. In groups, a solitary male will dominate a 'harem' of females. If the male is removed from the group or

Above: Centropyge bicolor
Literally, a two-colour angelfish. Juveniles are found in shallower waters than the adults. Despite its wide distrubtion in the Pacific, C. bicolor is not found in Hawaii.

– as in nature – dies, then one of the females will change sex to replace him. This procedure occurs every time the group becomes 'male-less'. This fish is susceptible to disease. Use copper remedies with care.

Centropyge bispinosus
Coral Beauty

☐ **Distribution:** Indo-Pacific.

☐ **Length:** 120mm/4.7in (wild).

☐ **Diet and feeding:** Meat foods and plenty of greenstuff. Grazer.

☐ **Aquarium behaviour:** Will settle down if retreats are close at hand.

☐ **Invertebrate compatibility:** Generally well behaved in the invertebrate aquarium.

In young specimens, the head and body are outlined in deep purple; red flanks are vertically crossed by many thin purple lines. The adult fish has much larger areas of gold/yellow on the flanks, again crossed by dark vertical stripes. The pattern is very variable, however; specimens from the Philippines, for example, have more purple and red coloration than those from Australasian waters.

Above: Centropyge bispinosus
The colour patterns of this species are very variable and depend on the native home of the individual specimen.

Centropyge eibli
Eibl's Angelfish

☐ **Distribution:** Indo-Pacific.

☐ **Length:** 150mm/6in (wild), 100mm/4in (aquarium).

☐ **Diet and feeding:** Most foods. Grazer.

☐ **Aquarium behaviour:** Peaceful.

☐ **Invertebrate compatibility:** Generally well behaved in the invertebrate aquarium.

The pale grey-gold body is crossed with gold and black lines, and some gold patterning appears in the anal fin. The rear part of the dorsal fin, the caudal peduncle and caudal fin are black, edged in pale blue. The eye is ringed with gold.

Above: Centropyge eibli
The combination of subtle colours of Eibl's Angelfish come as a pleasant surprise when compared to the more vivid – sometimes even gaudy – hues of other dwarf angelfishes.

Centropyge flavissimus
Lemonpeel Angelfish

☐ **Distribution:** Indo-Pacific.

☐ **Length:** 100mm/4in (wild).

☐ **Diet and feeding:**
Predominantly greenstuff, but
might be persuaded to take meaty
foods. Grazer.

☐ **Aquarium behaviour:** Peaceful.

☐ **Invertebrate compatibility:**
Generally well behaved in the
invertebrate aquarium.

A plain yellow fish except for the
blue outlines around the eye,
bottom lip and gill cover edge.

Right: Centropyge flavissimus
The species may be easily distinguished
from Centropyge heraldi *by the presence*
of blue rings around the eyes.

Centropyge heraldi
Herald's Angelfish

☐ **Distribution:** Indo-Pacific.

☐ **Length:** 100mm/4in (wild).

☐ **Diet and feeding:** Mainly
greenstuff, but also takes meaty
foods. Grazer.

☐ **Aquarium behaviour:** Peaceful.

☐ **Invertebrate compatibility:**
Generally well behaved in the
invertebrate aquarium.

C. heraldi is also plain yellow,
lacking even the blue details of *C.
flavissimus*. Fijian specimens have a
black edge to the dorsal fin.

Above: Centropyge heraldi
In the world of marine fishes, mono-
coloration is quite exceptional. This
makes the totally yellow Herald's
Angelfish an extremely easy fish to
identify, and a popular choice.

Centropyge loriculus
Flame Angelfish

☐ **Distribution:** Pacific.

☐ **Length:** 100mm/4in (wild).

☐ **Diet and feeding:** Meat foods and plenty of greenstuff. Grazer.

☐ **Aquarium behaviour:** Peaceful.

☐ **Invertebrate compatibility:** Generally well behaved in the invertebrate aquarium.

The fiery red-orange body has a central yellow area crossed by vertical dark bars. The dorsal and anal fins are similarly dark-tipped. The Flame Angelfish is not difficult to keep but does require excellent water conditions and a varied diet. Although generally expensive, this is a most rewarding fish to keep.

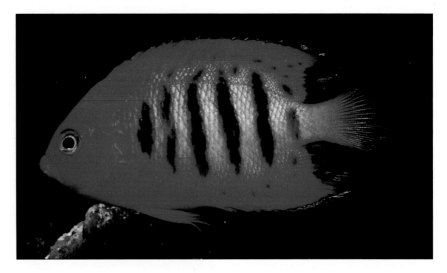

Above: Centropyge loriculus
The vivid coloration of this species clearly distinguishes it from any other dwarf angelfish.

Centropyge potteri
Potters Angel

☐ **Distribution:** Hawaiian Islands only.

☐ **Length:** 100mm/4in (wild).

☐ **Diet and feeding:** May initially be a difficult feeder, which should be tempted with live foods. When settled, it will usually accept the usual frozen marine food. Algae should be available to graze on as they make up an important part of this species' diet.

☐ **Aquarium behaviour:** Peaceful.

☐ **Invertebrate compatibility:** Very good.

C. potteri should not be regarded as an easy fish to keep; it demands excellent water quality and a varied

Above: Centropyge potteri
A peaceful, interesting species that can be kept in pairs as it is found in the wild around its native Hawaiian Islands.

diet with plenty of algae in it. However, this is a peaceful fish, which can often be kept with its own kind, much in the same way as C. resplendens. C. potteri is rather larger than most Centropyge species, in the middle size range between dwarf angels and the larger angels.

Centropyge resplendens
Resplendent Angel

☐ **Distribution:** Eastern Atlantic, mainly Ascension Island and St. Helena Islands.

☐ **Length:** 40mm/1.5in (wild), 60mm/2.5in (aquarium).

☐ **Diet and feeding:** Easily fed on most marine frozen, live and vegetable flake foods. Appreciates a good growth of algae on which to browse.

☐ **Aquarium behaviour:** Peaceful.

☐ **Invertebrate compatibility:** Excellent.

The Resplendent Angel is one of the very few *Centropyge* species that not only tolerates its own species but even appreciates being kept in pairs or small groups in a larger aquarium. If maintained correctly, aquarium specimens usually attain a far larger size than those found in the wild, adapting very well to aquarium life. Altogether an excellent choice for the fish-only or living reef type tank.

Below: Centropyge resplendens
Unlike many other dwarf Angelfishes, this species can often be kept in pairs.

Centropyge vroliki
Pearl-scaled Angelfish

☐ **Distribution:** Pacific.

☐ **Length:** 120mm/4.7in (wild).

☐ **Diet and feeding:** Provide meat foods and plenty of greenstuff for this grazer.

☐ **Aquarium behaviour:** Territorial.

☐ **Invertebrate compatibility:** Ideal.

At first glance this species could be confused with *C. eibli*, but it lacks the vertical lines. The pale body is edged with dusky black dorsal, anal and caudal fins. The eye is ringed in gold, and the rear edge of the gill cover and the base of the pectoral fin are also gold.

Although widely available, this species may be less popular with

Above: Centropyge vroliki
Like its relative, C. eibli, *the Pearl-scaled Angelfish is not endowed with striking colours, but it deserves to be more popular due to an adaptable nature and ideal compatibility with all species of invertebrates. In addition, feeding usually presents no problem.*

fishkeepers simply because of its muted colours. It requires an elaborate rockwork structure in the aquarium to feel fully at home.

Chaetodontoplus conspicillatus
Conspicuous Angelfish

☐ **Distribution:** Pacific.

☐ **Length:** 250mm/10in (wild).

☐ **Diet and feeding:** Crustaceans, coral polyps, algae. Grazer.

☐ **Aquarium behaviour:** Little is known about how well this species adapts to life in the aquarium.

☐ **Invertebrate compatibility:** Suitable when very young, but become destructive with age. Overall, not recommended.

The brown body is ringed by blue-edged dorsal and anal fins. The clearly defined eyes are a 'conspicuous' feature of the vivid yellow face; the mouth is a contrasting blue. Further areas of bright yellow appear at the base of the pectoral and caudal fins; the pelvic fins are blue.

This rare species – considered by many marine fishkeepers to be the 'Jewel of the Angelfishes' – is found mainly around Lord Howe Island, about 640km(400 miles) off the east coast of Australia.

Above: Chaetodontoplus conspicillatus
Sadly, the 'Jewel of the Angelfishes' is both rare and difficult to keep.

Right: Chaetodontoplus duboulayi
A rare, but generally peaceful, species.

Chaetodontoplus duboulayi
Scribbled Angelfish

☐ **Distribution:** Pacific.

☐ **Length:** 220mm/8.5in (wild).

☐ **Diet and feeding:** Crustaceans, coral polyps, algae. Grazer.

☐ **Aquarium behaviour:** Generally peaceful.

☐ **Invertebrate compatibility:** Suitable when very young, but become destructive with age.

The rear portion of the body from the gills, together with the anal and dorsal fins, is dark blue with scribbled markings. A vertical yellow bar behind the white gill cover is joined to the yellow caudal fin by a narrow yellow stripe along the top of the body. A dark bar covers the eye and the mouth is yellow. A nitrate-free, but algae-covered tank is ideal.

Chaetodontoplus septentrionalis
Blue Striped Angelfish

☐ **Distribution:** Western Pacific.

☐ **Length:** 210mm/8.25in (wild).

☐ **Diet and feeding:** Crustaceans, coral polyps, algae. Grazer.

☐ **Aquarium behaviour:** Not well documented.

☐ **Invertebrate compatibility:** Suitable when very young, but become destructive with age.

The brown body, dorsal and anal fins are covered with horizontal wavy blue lines. All the other fins are yellow. Juveniles are differently marked, being black with a yellow black-based caudal, and yellow margins to the dorsal and anal fins. The size at which the adult colour is assumed can be very variable.

Euxiphipops navarchus
Blue-girdled Angelfish; Majestic Angelfish

☐ **Distribution:** Pacific.

☐ **Length:** 250mm/10in (wild).

☐ **Diet and feeding:** Meat foods and greenstuff. Grazer.

☐ **Aquarium behaviour:** Young specimens adapt better to aquarium life.

☐ **Invertebrate compatibility:** Suitable when very young, but become destructive with age. Overall, not recommended.

Blue-edged dark areas on the head and caudal peduncle of this fish are connected by a dark ventral surface. The rest of the body is rich orange flecked with fine blue iridescent spots. The plain orange anal and blue dorsal fins are both edged in pale blue, as are the dark pelvic fins. Like other large angelfishes, the juvenile form is dark blue with vertical white stripes.

Above: Chaetodontoplus septentrionalis
This species' brown body makes a contrasting background for the wavy blue stripes. It is rarely kept.

Below: Euxiphipops navarchus
When seen underwater, the dark blue areas, yellow saddle-back patch and yellow caudal fin help to disrupt the 'fish shape' outline of this species.

Euxiphipops xanthometapon

Blue-faced Angelfish; Yellow-faced Angelfish; Blue-masked Angelfish

☐ **Distribution:** Indo-Pacific.

☐ **Length:** 380mm/15in (wild), 300mm/12in (aquarium).

☐ **Diet and feeding:** Meat foods and greenstuff. Grazer.

☐ **Aquarium behaviour:** Young specimens adapt better to aquarium life.

☐ **Invertebrate compatibility:** Suitable when very young, but become destructive with age. Overall, not recommended.

Despite the inclusion of 'xantho' (the Greek word for yellow) in the specific name, this fish is usually known as the Blue-faced Angelfish. Do not allow the attractive colouring and majestic appearance of this fish to tempt you unless you are an experienced fishkeeper; it needs special care. Juveniles are dark blue with white markings.

Below: Euxiphipops xanthometapon
This fish is difficult to keep, although juveniles are often found to be easier. Note the false eye on the dorsal fin.

Holacanthus bermudensis
Blue Angelfish

☐ **Distribution:** Western Atlantic.

☐ **Length:** 450mm/18in (wild).

☐ **Diet and feeding:** Meat foods and greenstuff. Grazer.

☐ **Aquarium behaviour:**
Aggressive when young; grows large.

☐ **Invertebrate compatibility:**
Suitable when very young, but become destructive with age. Overall, not recommended.

When adult, these fish are blue-grey in colour with yellow tips to the dorsal and anal fins. The juveniles of this striking species can be distinguished from those of the

Queen Angelfish, *H. ciliaris*, by the straight blue vertical lines on their dark blue bodies. This species was formerly known as *H. Isabelita*.

Above: Holacanthus bermudensis
The beautiful blue body hues of this spectacular angelfish appear to change under varied lighting conditions.

Holacanthus ciliaris
Queen Angelfish

☐ **Distribution:** Western Atlantic.

☐ **Length:** 450mm/18in (wild).

☐ **Diet and feeding:** Meat foods and greenstuff. Grazer.

☐ **Aquarium behaviour:**
Aggressive when young; grows large. Intolerant of its own and similar species and can be very territorial.

☐ **Invertebrate compatibility:**
Suitable when very young, but become destructive with age.

Above: Holocanthus ciliaris (adult)
When faced with this fish, who could deny its claim to be the 'Queen' of angelfish? Adult coloration may also be adopted at half the fish's maximum size, given the proper diet and spacious tank.

A very beautiful fish in the aquarium. Variations in colour pattern occur, but generally this species has more yellow than *H. bermudensis*. Hybrids between this species and *H. bermudensis* are classified as '*H. townsendi*' but this is a non-valid name. Young specimens of *H. ciliaris* have more curving blue vertical lines on the dark blue body than the young of *H. bermudensis*.

Queen Angelfish are reported to be prone to outbreaks of white spot disease, but can be successfully treated with copper-based remedies. They are quite resistant to such treatment, but proper management should reduce the likelihood of such outbreaks.

Below: Holocanthus ciliaris (juvenile)
This beautifully coloured and highly prized species just seems to go on improving with age. If you purchase it as a juvenile (shown here), and have a large enough tank, you will have the pleasure of witnessing its various colour progressions until it reaches the majestic adult form shown below left.

Holacanthus passer
King Angelfish

☐ **Distribution:** Pacific.

☐ **Length:** 450mm/18in (wild).

☐ **Diet and feeding:** Mainly greenstuff. Grazer.

☐ **Aquarium behaviour:** Aggressive; grows large.

☐ **Invertebrate compatibility:** Suitable when very young, but become destructive with age. Overall, not recommended.

The body is a dark brownish gold colour with a single vertical white stripe. The caudal fin is yellow. The dorsal and anal fins show gold patterning and edging. Young specimens have extra blue stripes on the rear of the body. These are extremely aggressive fish.

Above: Holocanthus passer
Only the yellow caudal fin and white stripe remain from the gold juvenile coloration. The blue stripes also merge.

Holacanthus tricolor
Rock Beauty

☐ **Distribution:** Western Atlantic.

☐ **Length:** 600mm/24in (wild), 300mm/12in (aquarium).

☐ **Diet and feeding:** In nature, sponges. Will eat meat foods and algae but may not thrive. Grazer.

☐ **Aquarium behaviour:** Progressively aggressive with age.

☐ **Invertebrate compatibility:** Suitable when very young, but become destructive with age. Overall, not recommended.

Juvenile forms of this fish are yellow with a blue-edged dark spot on the body, but this enlarges to cover two-thirds of the body as the fish matures into adult coloration.

This good-looking but aggressive species will offer a challenge to the experienced fishkeeper with a spacious aquarium. Feeding can be a problem, requiring patience and care; even when they appear to be feeding well, these fishes miss their normal diet of marine sponges. If the aquarium conditions are good, however, and you feed good-quality frozen foods, then you may well achieve success with this fish. Recently introduced sponge-based foods may prove helpful in the successful upkeep of this species. Again, juvenile specimens tend to adapt far better to aquarium life. Fish less than 25mm (1in) long are pure yellow with a neon-blue ocellus on the hind part of the body.

Pomacanthus annularis
Blue Ring Angelfish

☐ **Distribution:** Indo-Pacific.

☐ **Length:** 400mm/16in (wild), 250mm/10in (aquarium).

☐ **Diet and feeding:** Meat foods and greenstuff. Grazer.

☐ **Aquarium behaviour:** Territorial.

Above: Holacanthus tricolor
The Rock Beauty is an extremely attractive fish. However, it can be a hard species to acclimatize to aquarium diets and tends to be aggressive.

Below: Pomacanthus annularis
It is easy to see how the common name of 'Blue Ring Angelfish' was inspired. Unfortunately, this fish becomes territorial and destructive with age.

☐ **Invertebrate compatibility:** Suitable when very young, but become destructive with age.

Blue lines run from either side of the eye diagonally across the brown body. The lines rejoin at the top of the rear portion of the body. A dominant blue ring lies behind the gill cover. Juveniles are blue with a distinctive pattern of almost straight transverse white lines. Juveniles and adults were once considered to belong to different species.

Pomacanthus imperator
Emperor Angelfish

☐ **Distribution:** Indo-Pacific.

☐ **Length:** 400mm/16in (wild), 300mm/12in (aquarium).

☐ **Diet and feeding:** Animal foods and greenstuff. Grazer.

☐ **Aquarium behaviour:** Generally peaceful.

☐ **Invertebrate compatibility:** Suitable when very young, but become destructive with age. Overall, not recommended.

The yellow body is crossed with diagonal blue lines. The dark blue of the anal fin extends into a vertical wedge behind the gill cover. The eye is hidden in a blue-edged dark band. The caudal fin is yellow. Juveniles are dark blue with white semicircular or oval markings. Like all angels, this species requires the very best water conditions possible. Many fishkeepers have grown this species on from juvenile to adult.

Above: Pomacanthus imperator (adult)
Despite the colour changes apparent in this adult, the eye remains hidden beneath a dark bar, giving it protection against attack. Many juveniles have been successfully grown on to adulthood.

Below: Pomacanthus imperator (juvenile)
The beautiful pattern of distinctive white markings on a blue background provides excellent disruptive camouflage for the juvenile Emperor Angelfishes in the dappled light of the coral reef.

Pomacanthus paru
French Angelfish

☐ **Distribution:** Western Atlantic.

☐ **Length:** 300mm/12in (wild).

☐ **Diet and feeding:** Meat foods and greenstuff. Grazer.

☐ **Aquarium behaviour:** Young specimens may be 'nippy', since they act as cleaner fishes.

☐ **Invertebrate compatibility:** Suitable when very young, but become destructive with age. Overall, not recommended.

The young fish is black with bright yellow vertical bands. The adult fish is predominantly grey with bright speckles. Juvenile members of the Atlantic pomacanthids exhibit cleaning tendencies towards other fishes, and each angel's territory is recognized as a cleaning station.

Above and right: Pomacanthus paru
The impressive juvenile French Angelfish (above) has a pattern of yellow stripes on a black background. As the fish develops into a sub-adult (top right), the stripes begin to fade and the adult coloration appears (seen fully developed at below right).

Pomacanthus semicirculatus
Koran Angelfish

- ☐ **Distribution:** Indo-Pacific, Red Sea.

- ☐ **Length:** 400mm/16in (wild), 380mm/15in (aquarium).

- ☐ **Diet and feeding:** Meat foods and greenstuff. Grazer.

- ☐ **Aquarium behaviour:** Territorial.

- ☐ **Invertebrate compatibility:** Suitable when very young, but become destructive with age. Overall, not recommended.

The body is golden brown with blue speckles and the fins are outlined in blue. There are vertical blue lines on the head. The white markings on the juvenile are in the shape of semicircles rather than straight lines. During the colour change to adulthood, the markings on the caudal fin often resemble Arabic characters in the Koran – hence the common name.

Above: Pomacanthus semicirculatus (adult)
This near-adult form has almost lost the juvenile white markings.

Below: Pomacanthus semicirculatus (juvenile)
As in most angelfishes, the juvenile's coloration differs from the adult's.

Pygoplites diacanthus
Regal Angelfish; Royal Empress Angelfish

☐ **Distribution:** Indo-Pacific and Red Sea.

☐ **Length:** 250mm/10in (wild), 180mm/7in (aquarium).

☐ **Diet and feeding:** Sponges and algae in nature. In the tank, they will take frozen bloodworm, frozen *Mysis* shrimp, mussel meat, etc; even flake sometimes. Grazer.

☐ **Aquarium behaviour:** Shy, requires plenty of hiding places.

☐ **Invertebrate compatibility:** Suitable when very young, but become destructive with age. Overall, not recommended.

Dark-edged, bright orange slanting bands cross the body and extend into the dorsal and anal fins. The caudal fin is plain yellow. Feeding can be a problem, and may make it difficult to acclimatize this species to aquarium life. It requires a low nitrate level in the tank. Definitely a species for the experienced hobbyist only.

Specimens from the Philippines are relatively pale in colour, and consequently less desirable, and may be virtually impossible to feed.

Those from Sri Lanka, the Maldives and the Red Sea will eat in good water conditions and are brighter in colour. The new sponge-based foods may prove helpful in facilitating the upkeep of this species. As some encouragement, one documented specimen of this species has grown 125mm (5in) during seven years in captivity.

Below: Pygoplites diacanthus
The Regal Angelfish is a species for the experienced aquarist; but if you can keep it in the very best of conditions and provide it with just the right diet, then over a period of years it will repay you by displaying amazing colours and, indeed, a truly regal manner.

Family: POMACENTRIDAE

Anemonefishes (Clownfishes) and Damselfishes

Family characteristics

Fishes within this family are usually divided into two distinct groups: the anemonefishes and the damselfishes.

Anemonefishes have come to represent the very essence of tropical marine fishkeeping; they are brilliantly coloured, full of character and relatively easy to keep. Many hobbyists would not consider an aquarium complete without one.

The curious 'waddling' swimming action of the anemonefishes, together with their appropriate markings, has given them their other collective common name of clownfishes. As their principal common name suggests, anemonefishes live in close association with sea anemones, especially *Heteractis* and *Stoichactis* species. The anemonefish was thought to be immune to the stinging cells of the sea anemone, but it now seems certain that the mucus on the fish prevents the stinging cells from being activated by masking the fish as potential prey.

The clownfish/anemone relationship is usually referred to as being symbiotic; in fact, it is commensal, which means that the fishes and anemones live in close proximity to one another, often to their mutual benefit. There may be some doubt about the benefit derived by the anemones in return for not stinging the fish. The favoured theory is that the anemonefishes, being highly territorial by nature, chase away fishes that might eat the host anemone.

Despite their close relationship in the wild, many anemonefishes will live quite happily in the aquarium without an anemone (and vice-versa), but then you will not see the fish behave as it does in nature. (It is worth noting that sea anemones are sensitive creatures and need special care and attention, see pages 317-322.) In the wild, large sea anemones can accommodate a whole group of anemonefish. Of these, only a dominant pair will breed, the rest remaining as male fish. If any disaster should remove the dominant female though, all is not lost; the dominant male will change sex and become a breeding female that will then pair off with the next dominant male.

Many anemones are only big enough to house a pair of fish, and all other prospective 'tenants' are seen off. This is more often than not the case in the aquarium and it is recommended that each anemone should host a pair of anemonefish only; any other fish may be remorselessly attacked and, if they are not killed, their lives will be made miserable. Very large tanks may accommodate several pairs of anemonefish without too much trouble, but smaller tanks should only house one pair, otherwise serious territorial disputes could ensue.

Anemonefish are fairly easy to pair, spawn and raise (see *Breeding fish*, pages 138-143); damselfish are slightly more difficult. In many countries, commercially bred anemonefish are now readily available and these should be purchased wherever possible to preserve wild stocks under pressure from pollution and habitat destruction.

Damselfishes are small busy fishes that bob constantly around the coral heads, using them as their territory and retreating into them when threatened. They are agile fishes – particularly when you are trying to catch them!

There are a number of similar-looking 'Electric-blue' or 'Blue Devil' fishes and also several blue-and-yellow coloured species. Because of these similarities, the nomenclature of this large group of fishes is, quite literally, a scientific minefield. Many authorities have endeavoured to reclassify the damsels, but this has only served thoroughly to confuse hobbyists, authors and scientists alike. Fortunately, for the purposes of the average hobbyist, all this matters little, as most damselfish share the same behaviour patterns.

Damselfishs' colours sometimes fade, which may be a response that helps the fish to blend in more effectively with the surroundings, or it may indicate undue stress in captivity. Because damselfishes are considered to be hardy, they are much abused, often suffering high ammonia and nitrite levels to mature a filter system. Apart from this being bad aquarium management, these species are no less likely to contract White Spot or Oodinium diseases than any other marine fish subject to adverse conditions, which could spell long-term disaster for a newly set-up aquarium.

Normally, there are no clear distinctions between the sexes. There is one method of determining sex by external observation – a technique similar to that used for determining sex in freshwater cichlids – and that is by looking at the genital papillae (often called the ovipositor). The male genital papilla is narrower and more pointed than the female's. Observations are best delayed until breeding activity is noticed, when the papillae are easier to see. Spawning in damselfishes entails the selection of a site and the laying and subsequent guarding of eggs.

Diet and feeding

In the aquarium, these fishes will take live foods, algae, frozen foods, flake etc. All species of damselfishes will take dried foods readily.

Aquarium behaviour

Anemonefishes are eminently suitable for keeping in the aquarium, where they will naturally associate with suitable anemones. In this way, you can keep a pair of Anemonefishes in a relatively small tank.

Damselfishes, despite their enormous popularity as a result of being colourful and hardy, do have a negative side: they are often aggressive and intolerant of their own, and other, species.

Abudefduf cyaneus
Blue Damsel

☐ **Distribution:** Indo-Pacific.

☐ **Length:** 60mm/2.4in (wild).

☐ **Diet and feeding:** Finely chopped meats and dried food. Bold feeder.

☐ **Aquarium behaviour:** May squabble with members of its own species. Keep singly or in shoals.

☐ **Invertebrate compatibility:** Yes, good choice.

Although some specimens may have yellow markings on the caudal fin and ventral area, the predominant colour of this fish is a stunning royal blue. There is some confusion about the positive classification of this species, since some authorities also refer to it as *Chrysiptera*.

Below: Abudefduf cyaneus
One of many blue damselfishes, this fish has more synonyms than most.

Abudefduf oxyodon
Blue-velvet Damselfish; Black Neon Damselfish; Blue-streak Devil

☐ **Distribution:** Pacific.

☐ **Length:** 110mm/4.3in (wild), 75mm/3in (aquarium).

☐ **Diet and feeding:** Finely chopped meats, algae and greenstuff. Bold.

☐ **Aquarium behaviour:** Aggressive.

☐ **Invertebrate compatibility:** Yes, good choice.

A vertical yellow stripe crosses the deep blue-black body just behind the head. The electric blue wavy lines on the head and upper part of the body fade with age.

Above: Abudefduf oxyodon
This handsome fish from the Pacific Ocean may not adapt well if it is in less than first-class condition.

Abudefduf saxatilis
Sergeant Major

☐ **Distribution:** Indo-Pacific, tropical Atlantic.

☐ **Length:** 150mm/6in (wild), 50mm/2in (aquarium).

☐ **Diet and feeding:** Finely chopped meats, algae and greenstuff. Bold grazer.

☐ **Aquarium behaviour:** Juveniles are very active; adults can become very aggressive.

☐ **Invertebrate compatibility:** Yes, good choice.

Five vertical dark bars cross the yellow/silvery body. Depending on the geographic location of the individuals, the caudal fin may be a dusky colour. Juvenile forms in the Atlantic have bright yellow upper parts on an otherwise silver body. However, it may lose its colours when disturbed. This is a hardy fish, and a good choice for the beginner. A shoal in a large tank is impressive.

Above: Abudefduf saxatilis
The Sergeant Major is widely distributed throughout the tropics. The male changes colour and an ovipositor appears from the vent during spawning. Eggs are laid on rocks, shells or coral.

Amphiprion akallopisos
Skunk Clown;

☐ **Distribution:** Indo-Pacific.

☐ **Length:** 75mm/3in (wild), 40-50mm/1.6-2in (aquarium).

☐ **Diet and feeding:** Small crustaceans, small live foods, algae, vegetable-based foods. Bold feeder.

☐ **Aquarium behaviour:** Peaceful but anemone-dependent.

☐ **Invertebrate compatibility:** Yes, ideal.

A white line runs along the very top of the brown-topped golden body, from a point level with the eye to the caudal peduncle. Some anemonefishes do not live up to the family's reputation as sea anemone dwellers, but the Skunk Clown seems to need the association more than other species.

Left: Amphiprion akallopisos
These yellow Skunk Clowns are doing what comes naturally, resting among the tentacles of their host sea anemone.

Amphiprion clarkii
*Two-banded Anemonefish;
Banded Clown; Clark's
Anemonefish*

☐ **Distribution:** Indo-Pacific.

☐ **Length:** 120mm/4.7in (wild),
75mm/3in (aquarium).

☐ **Diet and feeding:** Small
crustaceans, small live foods,
algae, vegetable-based foods.
Bold feeder.

☐ **Aquarium behaviour:** Peaceful.

☐ **Invertebrate compatibility:**
Yes, ideal.

This species is highly variable in
coloration depending on its
location, but generally the body is
predominantly dark brown but for
the ventral regions, which are
yellow. All the fins, with the
exception of the paler caudal fins,
are bright yellow. Two tapering
white vertical bars divide the body

into thirds; in juvenile forms, there
is a third white bar across the rear
of the body.
    Clark's Anemonefish is peaceful
and makes an ideal choice for a
mixed aquarium.

Below: Amphiprion clarkii
*Clark's Anemonefish can be found in
many colour variations and
identification is therefore sometimes
difficult. This bold fish is excellent for the
community aquarium with little need
for a host anemone in most cases.*

Left: Amphiprion ephippium
This Tomato Clown is but one species so named, many others having the similar red body coloration with a dark patch.

Amphiprion ephippium
Tomato Clown; Fire Clown; Red Saddleback Clown

☐ **Distribution:** Indo-Pacific.

☐ **Length:** 120mm/4.7in (wild), 75mm/3in (aquarium).

☐ **Diet and feeding:** Small crustaceans, small live foods, algae, vegetable-based foods. Bold feeder.

☐ **Aquarium behaviour:** Can be aggressive.

☐ **Invertebrate compatibility:** Yes, ideal.

This fish is often confused with *A. frenatus*, since both are a rich tomato-red with a black blurred blotch on the body rearwards of the gill cover. Juveniles have a white vertical bar just behind the head disappearing as the fish matures.

Above: Amphiprion frenatus
Unlike the previous species, adult specimens of this Fire Clown retain the vertical white stripe into adulthood.

Amphiprion frenatus
Tomato Clown; Fire Clown; Bridled Clownfish

☐ **Distribution:** Pacific.

☐ **Length:** 75mm/3in (wild).

☐ **Diet and feeding:** Small crustaceans, small live foods, algae, vegetable-based foods. Bold feeder.

☐ **Aquarium behaviour:** May be quarrelsome in confined spaces.

☐ **Invertebrate compatibility:** Yes, ideal.

A. frenatus is very similar to *A. ephippium*, but it has the white stripe behind the head (sometimes two in juveniles) and the body blotch is often larger. The confusion is not helped by the fact that some authorities call this fish *A. ephippium* or *A. melanopus*.

Amphiprion nigripes
Black-footed Clownfish

☐ **Distribution:** Indian Ocean.

☐ **Length:** 80mm/3.2in (wild), 50mm/2in (aquarium).

☐ **Diet and feeding:** Eats plankton and crustaceans in the wild, but finely chopped foods are ideal in captivity. Bold feeder.

☐ **Aquarium behaviour:** Shy.

☐ **Invertebrate compatability:** Yes, ideal.

Although it is similar to the two previous species, *A. nigripes* is much more subtly coloured. It is a soft golden brown with a white stripe just behind the head. The pelvic fins are black, but the anal fin is not always so, hence the common name.

Above right: Amphiprion nigripes
A native of the Maldive Islands.

Amphiprion ocellaris
Common Clown; Percula Clown

☐ **Distribution:** Indo-Pacific.

☐ **Length:** 80mm/3.2in (wild), 50mm/2in (aquarium).

☐ **Diet and feeding:** Finely chopped foods. Bold feeder.

☐ **Aquarium behaviour:** Will exclude other anemonefishes from its territory.

☐ **Invertebrate compatibility:** Yes, ideal.

This is the clownfish that everyone recognizes, thanks to its bold and unforgettable coloration. There has been much debate on whether there is a similar, but distinct species called *A. percula*. In fact, this is a very colour-variable fish and the two fish are very probably one and the same species.

Right: Amphiprion ocellaris
'The essence of the coral reef.'

Amphiprion perideraion
Salmon Clownfish; Pink Skunk Clownfish

☐ **Distribution:** Pacific Ocean.

☐ **Length:** 80mm/3.2in (wild), 38mm/1.5in (aquarium).

☐ **Diet and feeding:** Finely chopped foods. Not quite as bold as other species.

☐ **Aquarium behaviour:** Shy.

☐ **Invertebrate compatibility:** Yes, ideal.

This species is very similar in colour to *A. akallopisos*, but can be easily distinguished from it by the vertical bar just behind the head. The body colour is perhaps a little more subdued and a white stripe reaches the snout. *A. perideraion* is rather more sensitive than other species and is best kept in a species tank with adequate space for sea anemones. Males have orange edging to the soft-rayed part of the dorsal fin and at the top and bottom of the caudal fin.

Below: Amphiprion perideraion
The Salmon Clownfish has a vertical bar and pink hue, which help distinguish it from similar white-backed species.

Amphiprion polymnus
Saddleback Clownfish

☐ **Distribution:** Pacific.

☐ **Length:** 120mm/4.7in (wild), 100mm/4in (aquarium).

☐ **Diet and feeding:** Small crustaceans, small live foods, algae, vegetable-based foods. Bold feeder.

☐ **Aquarium behaviour:** Can be territorial, and aggressive.

☐ **Invertebrate compatibility:** Yes, ideal.

The dark red-brown body is marked by two white bands; one broad band lies just behind the head, the other begins in the middle of the body and curves upwards into the rear part of the dorsal fin. There is also a dash of white along the top of the caudal fin.

Above: Amphiprion polymnus
The white markings on the dorsal area and the dark brown body make recognition of this species quite easy.

Chromis caerulea
Green Chromis

☐ **Distribution:** Indo-Pacific, Red Sea.

☐ **Length:** 100mm/4in (wild), 50mm/2in (aquarium).

☐ **Diet and feeding:** Chopped meats. Shy.

☐ **Aquarium behaviour:** Generally peaceful.

☐ **Invertebrate compatibility:** Yes, good choice.

This hardy colourful shoaling species has a brilliant green-blue sheen to the scales. The caudal fin is more deeply forked than in some damselfishes. Keep these fishes in shoals; individuals may go into decline. Ideally, a shoal should consist of at least six fishes.

Above: Chromis caerulea
Like many damselfishes, the peaceful Green Chromis has a gregarious nature and appreciates being kept in a small shoal. It is a lively and attractive species and can safely be kept in a mixed fish and invertebrate set-up.

Chromis cyanea
Blue Chromis; Blue Reef Fish

☐ **Distribution:** Tropical Atlantic.

☐ **Length:** 50mm/2in (wild).

☐ **Diet and feeding:** Chopped meats. Dried foods.

☐ **Aquarium behaviour:** A peaceable shoaling fish that prefers to be with some of its own kind to feel at home.

☐ **Invertebrate compatibility:** Yes, good choice.

This species thrives in vigorously aerated water. The body colour is brilliant blue with some black specks, topped with a black dorsal surface. There are black edges to the

dorsal and caudal fins. The eye is also dark. In shape and size (but not colour) *C. cyanea* closely resembles *C. multilineata*, the Grey Chromis.

At breeding time, a brown ovipositor extends from just in front of the anal fin in a similar manner to that of freshwater cichlids.

Above: Chromis cyanea
This peaceful shoaling fish is best kept with others of its kind, possibly in a reef aquarium. The normally narrow black area on the top of the male Blue Chromis spreads during spawning time and a brown ovipositor appears. The male usually guards the eggs.

Chromis xanthurus
Yellow-tailed Damselfish

☐ **Distribution:** Indo-Pacific.

☐ **Length:** 100mm/4in (wild), 50mm/2in (aquarium).

☐ **Diet and feeding:** Chopped meats and dried foods. Bold feeder.

☐ **Aquarium behaviour:** Can be aggressive.

☐ **Invertebrate compatibility:** Yes, good choice.

The deep royal blue body contrasts sharply with the bright yellow caudal fin and caudal peduncle. Again, there is some confusion over the correct name of this species,

Above: Chromis xanthurus
Quite understandably, Yellow-tailed Damsels are among the most popular fish in the marine hobby; they are colourful, hardy, disease resistant and will eat almost all marine food. They are also, however, highly territorial.

both *Pomacentrus caeruleus* and *Abudefduf parasema* are given by other sources.

Dascyllus aruanus
Humbug; White-tailed Damselfish

☐ **Distribution:** Indo-Pacific.

☐ **Length:** 80mm/3.2in (wild), 75mm/3in (aquarium).

☐ **Diet and feeding:** Chopped meats. Bold feeder.

☐ **Aquarium behaviour:** Aggressive towards its own kind and very territorial.

☐ **Invertebrate compatibility:** Yes, a good choice.

This white fish has three black bars across the body. The front bar covers the eye and follows the slope of the head up into the first rays of the dorsal fin. The rear two bars extend into the pelvic and anal fins and also into the dorsal fin, where they are linked by a horizontal bar along the top part of the fin. The caudal fin is unmarked. This is the hardiest of the damsels.

Left: Dascyllus aruanus
This fish shares its common name with the similarly coloured confection. It is relatively hardy, but very territorial.

Dascyllus carneus
Cloudy Damsel; Blue-spotted Dascyllus

☐ **Distribution:** Indo-Pacific.

☐ **Length:** 80mm/3.2in (wild).

☐ **Diet and feeding:** Chopped foods. Dried foods. Bold.

☐ **Aquarium behaviour:** Aggressive towards its own kind.

☐ **Invertebrate compatibility:** Yes, a good choice.

All the fins, except the white caudal, are black and the body is greyish brown with a pattern of blue dots. There is a white patch on the top of the body, towards the front part of the dorsal fin and immediately behind a black bar, which covers the pectoral fin. A similar fish, *D. reticulatus*, is an overall grey, lacks the white patch and has a vertical black bar running from the rear of the dorsal to the rear of the anal fin.

Below: Dascyllus carneus
Of a similar size, but less starkly coloured than the Humbug, the Cloudy Damsel has more grey-brown in its body.

Dascyllus marginatus
Marginate Damselfish; Marginate Puller

☐ **Distribution:** Red Sea.

☐ **Length:** 100mm/4in (wild).

☐ **Diet and feeding:** Chopped meats. Dried foods. Bold feeder.

☐ **Aquarium behaviour:** Aggressive and territorial.

☐ **Invertebrate compatibility:** Yes, a good choice.

A brown area slopes backwards from the front of the black-edged dorsal fin to the point of the anal fin. The rest of the body is cream in colour. This active fish will shelter among coral during the night.

Below: Dascyllus marginatus
Like all Dascyllus *species, this fish occasionally makes quite audible purring or clicking sounds*

Dascyllus melanurus
Black-tailed Humbug

☐ **Distribution:** West Pacific.

☐ **Length:** 75mm/3in (wild).

☐ **Diet and feeding:** Chopped foods. Dried foods. Bold feeder. Frozen mysis and brineshrimp.

☐ **Aquarium behaviour:** Aggressive and territorial.

☐ **Invertebrate compatibility:** Yes, good choice.

This fish is very similar to *D. aruanus*, except that the black bars are more vertical and a black bar crosses the caudal fin.

Above: Dascyllus melanurus
This black and white damsel is, like its almost lookalike relative, the Humbug, a shoaling fish. It is found over a more limited area, however, being confined to the western Pacific Ocean around the Philippines and Melanesia. It is an aggressively territorial species that is best housed with other fish that can take care of themselves.

Dascyllus trimaculatus
Domino Damsel; Three-spot Damselfish

☐ **Distribution:** Indo-Pacific, Red Sea.

☐ **Length:** 125mm/5in (wild), 75mm/3in (aquarium).

☐ **Diet and feeding:** Chopped meats and dried foods. Bold.

☐ **Aquarium behaviour:** Territorial.

☐ **Invertebrate compatibility:** Yes, good choice.

This fish is velvety black overall, including the fins. The only markings are the three spots from which one of the comon names is derived. There is one white spot on each upper flank, midway along the length of the dorsal fin; the third spot is situated on the centre of the head, just behind the eye. The spots fade with age.

Below: Dascyllus trimaculatus
This very common damselfish is instantly recognizable by the three white spots on its body, and it would be very hard to imagine a more appropriate popular name for it. Unfortunately, the spots usually fade with age.

Paraglyphidodon melanopus
Yellow-backed Damselfish

☐ **Distribution:** Indo-Pacific.

☐ **Length:** 75mm/3in (wild).

☐ **Diet and feeding:** Chopped meats. Bold feeder.

☐ **Aquarium behaviour:** May be aggressive towards its own, and smaller, species.

☐ **Invertebrate compatibility:** Yes, good choice.

An oblique bright yellow band runs from the snout to the tip of the dorsal fin above a pale violet body. The anal and pelvic fins are light blue, edged with black. The caudal fin is edged with yellow. A spacious

Above: Paraglyphidodon melanopus
The combination of black-edged pelvic fins and brilliant colours of this damselfish has inspired several alternative common names, including Bow-tie Damsel, Bluefin Damsel and Royal Damsel. It is found over a wide area of the Indo-Pacific.

tank with plenty of hiding places suits this brilliantly coloured fish very well.

Above: Pomacentrus coeruleus
The brilliant electic blue colour of the Blue Devil makes it an instant eye-catcher in the dealer's tanks, and it will certainly add an extra splash of colour to the home aquarium.

Pomacentrus coeruleus
Blue Devil; Electric-blue Damsel

☐ **Distribution:** Indo-Pacific.

☐ **Length:** 100mm/4in (wild), 50mm/2in (aquarium).

☐ **Diet and feeding:** Chopped meats. Dried food. Bold feeder.

☐ **Aquarium behaviour:** Extremely pugnacious.

☐ **Invertebrate compatibility:** Yes, good choice.

The bright blue coloration of this fish really makes it stand out in the aquarium. There may be some black facial markings. It is a hardy species and lives peacefully in small groups when young but may turn aggressive when adult.

Pomacentrus melanochir
Blue-finned Damsel

☐ **Distribution:** Pacific.

☐ **Length:** 80mm/3.2in (wild).

☐ **Diet and feeding:** Chopped meats. Dried foods.

☐ **Aquarium behaviour:** Pugnacious.

☐ **Invertebrate compatibility:** Yes, good choice.

This rare blue damsel can be identified by the dark edge to each scale and defined blue patterning on the head. The dorsal, anal and caudal fins are more extended than on other species. Pectoral fins are yellowish.

Below: Pomacentrus melanochir
The extended finnage of this damselfish make it easy to identify.

Pomacentrus violascens
Yellow-tailed Demoiselle

☐ **Distribution:** Pacific.

☐ **Length:**80mm/3.2in (wild), 50mm/2in (aquarium).

☐ **Diet and feeding:** Chopped meats. Dried food. Bold feeder.

☐ **Aquarium behaviour:** Pugnacious.

☐ **Invertebrate compatability:** Yes, good choice.

The markings on *Pomacentrus violascens* resemble those of *P. melanochir*, but the tips of the dorsal and anal fins are yellow and the yellow of the caudal fin does not spread quite so far onto the body.

Left: Pomacentrus violascens
This beautiful damsel is slightly more sensitve than others in this family.

Left: Premnas biaculeatus
An attractive deep-red clownfish.

Above: Stegastes leucostictus
Common, and suitable for the beginner.

Premnas biaculeatus
Maroon Clownfish

☐ **Distribution:** Pacific Ocean.

☐ **Length:** 150mm/6in (wild), 100mm/4in (aquarium).

☐ **Diet and feeding:** Finely chopped foods. Bold.

☐ **Aquarium behaviour:** Aggressive towards other anemonefishes and its own species, if not a mated pair.

☐ **Invertebrate compatibility:** Yes, ideal.

This larger species differs from other clownfishes by having two spines beneath the eye, as well as the usual small spines on the back edge of the gill cover. The body is a deep rich red with three narrow white bands crossing it, behind the head, midway along the body and just behind the dorsal and anal fins.

Stegastes leucostictus
Beau Gregory

☐ **Distribution:** Caribbean.

☐ **Length:** 150mm/6in (wild), 50mm/2in (aquarium).

☐ **Diet and feeding:** Animal and vegetable matter. Dried foods. Bold feeder.

☐ **Aquarium behaviour:** Aggressive.

☐ **Invertebrate compatibility:** Yes, a good choice.

The yellow body is topped by a golden brown area covered in bright blue dots. There is a dark blotch at the rear of the dorsal fin. All the other fins are yellow. This common damsel is hardy enough for the beginner, but may bully fishes with similar feeding habits; kept alongside species with different feeding habits it is not so aggressive.

Family: PSEUDOCHROMIDAE

Pygmy Basslets/Dottybacks

Family characteristics

Many of the behavioural patterns of both pygmy basslets (Pseudochromidae) and fairy basslets (Grammidae) are essentially the same, the main difference being geographical location; the former group originates from the Red Sea and parts of the Indo-Pacific, while the latter are confined to the Caribbean. The pygmy basslets are a reasonably large group of fishes spread over a wide area. The vast majority are to be found within the Red Sea and adjacent areas and, as imports are restricted from these parts of the world, numbers are at a premium and specimens usually command high prices.

Pygmy basslets are generally very shy but highly territorial in nature, spending much of their time travelling the maze of crevices within the coral reef structure, occasionally dashing out to capture a morsel of food carried in the current.

Little is known of the reproduction process of this family and reports of aquarium spawning are extremely rare. To date, no larvae have been raised successfully.

Diet and feeding

Small crustaceans and drifting plankton make up much of the wild diet of these fishes. In the aquarium, much of the normal frozen and live marine fare is readily acceptable after an initial settling-in period. Flake foods are not appreciated by most species at any time.

Aquarium behaviour

These fishes are highly territorial by nature and will usually be very intolerant of the same, or similar, species. A good arrangement of rockwork is essential to satisfy their secretive habits, although most individuals acclimatize quite well, losing a good deal of their shyness once established.

Pseudochromis diadema
Flash-back Gramma

☐ **Distribution:** Western Pacific.

☐ **Length:** 55mm/2.2in (wild).

☐ **Diet and feeding:** Will readily accept most marine frozen and live foods; even flake.

☐ **Aquarium behaviour:** Peaceful, but requires plenty of hiding places.

☐ **Invertebrate compatibility:** Ideal.

Left: Pseudochromis diadema
The Flash-back Gramma is a beautiful fish often overlooked by aquarists. Given the right environment, it is easy to keep and will thrive for many years.

P. diadema is almost totally yellow with a wedge of purple beginning at the snout, running across the back and disappearing at the caudal peduncle. Although generally a peaceful fish, it will not tolerate its own, or similar, species, but may be regarded as a good beginner's fish if this is taken into account.

Pseudochromis dutoiti
Neon-back Gramma

☐ **Distribution:** Central Indian Ocean.

☐ **Length:** 88mm/3.5in (wild).

☐ **Diet and feeding:** Live foods are recommended, but frozen marine fare is usually acceptable once the fish has settled in. Flake foods are nearly always rejected.

Above: Pseudochromis dutoiti
The Neon-back Gramma is occasionally imported and is a popular, if expensive, choice. It is money well spent, however, as you will be rewarded with a naturally inquisitive fish, full of character.

☐ **Aquarium behaviour:** Shy but territorial. Aggressive towards fish of the same, or similar, species.

☐ **Invertebrate compatibility:** Ideal.

P. dutoiti is always popular when it appears on sale and makes an ideal addition to the living reef aquarium. However, it can become highly aggressive towards its own kind, or similar fish. Like *P. flavivertex*, it is not a beginner's fish and will usually need extra care where feeding is concerned.

Pseudochromis flavivertex
Sunrise Dottyback

☐ **Distribution:** Red Sea.

☐ **Length:** 70mm/2.75in (wild).

☐ **Diet and feeding:** Favours live foods but will eventually accept frozen marine fare. Flake is nearly always rejected.

☐ **Aquarium behaviour:** Has a shy, secretive nature and requires plenty of rockwork in which to hide.

☐ **Invertebrate compatibility:** Ideal.

The Sunrise Dottyback certainly lives up to its name – its cobalt blue body has a bright yellow band running from the tip of the nose, across the back and into the tail, giving the impression of a tropical sunrise. It should be kept on its own or in the company of dissimilar fish if fighting is to be avoided. Although not a beginner's fish, *P. flavivertex* makes a highly attractive addition to the invertebrate aquarium for those with more experience.

Pseudochromis paccagnellae
False Gramma; Dottyback; Royal Dottyback; Paccagnella's Dottyback

☐ **Distribution:** Pacific.

☐ **Length:** 50mm/2in (wild).

☐ **Diet and feeding:** Finely chopped meat foods, brineshrimp.

☐ **Aquarium behaviour:** Do not keep with similar fishes. May tend to nip at other fishes.

☐ **Invertebrate compatibility:** Ideal.

This species is almost identical to *Gramma loreto,* but a thin white line – often incomplete or hard to see – divides the two body colours.

Above: Pseudochromis flavivertex
Like its cousin, P. dutoiti, *this fish is only an occasional, and expensive, import. But, such is its beauty, many cannot resist the temptation. It does best in a mixed fish/invertebrate tank.*

Below: Pseudochromis paccagnellae
The False Gramma requires similar aquarium conditions to its lookalike, the Royal Gramma. Make sure that you provide plenty of retreats in the aquarium to help it feel secure.

Pseudochromis porphyreus
Strawberry Gramma

☐ **Distribution:** Central and Western Pacific.

☐ **Length:** 55mm/2.2in (wild), 75mm/2.75in (aquarium).

☐ **Diet and feeding:** Easily fed. Will accept most marine frozen, live and flake foods.

☐ **Aquarium behaviour:** Bold. Can be very aggressive towards its own kind and similar species.

☐ **Invertebrate compatibility:** Very good.

There is always an exception to every rule and, in many respects, *P. porphyreus* is it. Most pygmy basslets are shy and secretive but, once settled, the Strawberry Gramma can be bold and

Above: Pseudochromis porphyreus
The Strawberry Gramma is a fish full of character; it is bold and generally fearless, even with much larger fish. Colourful, disease resistant and fairly inexpensive, it is a popular choice.

aggressive, intolerant of its own kind and even unrelated fish it takes a disliking to. However, it is still a desirable aquarium addition and its tough constitution makes it ideal for the beginner.

Family: SCATOPHAGIDAE
Butterfishes/Scats

Family characteristics
Like the Monodactylidae, the fishes in this family are also estuarine and can be kept with some success in brackish water or even freshwater aquariums. The family, which has only four species in two genera, is found around many of the islands of the Malay Archipelago to New Guinea and Northern Australia. In the latter region they are confined to tidal zones.

Diet and feeding
These fishes will eat anything, including greenfood, such as lettuce, spinach and green peas.

Aquarium behaviour
It is usual to keep scats in the company of fingerfishes (*Monodactylus* species, see pages 228-229). They need plenty of swimming space.

Scatophagus argus
Scat; Argus Fish

☐ **Distribution:** Indo-Pacific.

☐ **Length:** 300mm/12in (wild).

☐ **Diet and feeding:** Will eat anything, including greenstuff. Scavenger.

☐ **Aquarium behaviour:** Peaceful.

☐ **Invertebrate compatibility:** Generally not recommended.

Like *Monodactylus* sp., the Scat is almost equally at home in salt, brackish or even fresh water, but it thrives best in sea water. It frequents coastal and estuarine waters, where it is assured of a good supply of animal waste and other unsavoury material. (Its scientific name means 'excrement eater'.) The oblong body

Above: Scatophagus argus
An active fish that will eat anything.

is laterally compressed, reminiscent of butterflyfishes and angelfishes. It is green-brown with a number of large dark spots, which become less prominent on adult fishes. A deep notch divides the spiny first part and the soft-rayed rear section of the dorsal fin. Juveniles have more red coloration, especially on the fins.

Family: SCIAENIDAE

Croakers and Drums

Family characteristics

Most of the species likely to be suitable for the aquarium come from the western Atlantic, although a species from the eastern Pacific, is another possible contender.

The fishes in this family are also capable of making sounds by resonating the swimbladder. Their strikingly marked bodies are usually elongated, often with a characteristic high first dorsal fin.

Diet and feeding

Most species may pose problems in their day to day care, being somewhat fussy eaters; success in the aquarium relies upon a constant supply of small live foods.

Aquarium behaviour

Fine with peaceable tankmates. Will spawn in the aquarium if provided with ideal conditions.

Equetus acuminatus
Cubbyu; High Hat

☐ **Distribution:** Caribbean.

☐ **Length:** 250mm/10in (wild), 150mm/6in (aquarium).

☐ **Diet and feeding:** Crustaceans, molluscs, soft-bodied invertebrates; live foods preferred in captivity. Slow bottom feeder.

☐ **Aquarium behaviour:** The long fins may be tempting to other fish, so be sure to keep this species with non-agressive tankmates.

☐ **Invertebrate compatibility:** No, destructive.

The main feature of this fish is the very tall first dorsal fin, which is carried erect. The pale body is

Above: Equetus acuminatus
This is a bottom-feeding fish – note the small barbels underneath the mouth.

covered with many horizontal black bands and the black fins have white leading rays. The chin barbels are used to detect food swimming below the fish, which then snaps downward to catch its prey. This species is probably the hardiest of the genus.

Equetus lanceolatus
Jack-knife Fish; Ribbonfish

☐ **Distribution:** Caribbean.

☐ **Length:** 250mm/10in (wild).

☐ **Diet and feeding:** Crustaceans, molluscs, soft-bodied invertebrates. Slow bottom feeder.

☐ **Aquarium behaviour:** Its fins may be attacked by other fish.

Can be aggressive towards its own kind when adult. A difficult species requiring care.

☐ **Invertebrate compatibility:** No, destructive.

The high first dorsal fin of this very beautiful fish has a white-edged black line through it, that continues like a crescent through the body to the tip of the caudal fin. This gives the fish a forward sloping

Above: Equetus lanceolatus
The strikingly attractive Jack-knife Fish is unfortunately rather delicate, and succumbs easily to shock and stress. It can also be aggressive towards its own kind when adult.

appearance. Further vertical black bars cross the eye and the body just behind the head. A delicate fish in captivity. A similar-looking species, *E. punctatus*, can easily be confused with the Jack-knife Fish.

Family: SCORPAENIDAE

Dragonfishes, Lionfishes, Scorpionfishes and Turkeyfishes

Family characteristics

Here are the exotic 'villains' of the aquarium. They are predatory carnivores that glide up to their prey and engulf it with their large mouths. The highly ornamental fins are not just there for decoration either, since they have venomous stinging cells and will inflict a very painful wound. HANDLE THESE FISHES WITH CARE. If you are stung, bathing the affected area in very hot water will alleviate the pain and help to 'coagulate' the poison.

During spawning, the pair of fishes rises to the upper levels of the water and a gelatinous ball of eggs is released. When they are 10-12mm (about 0.5in) long, the fry sink to the bottom of the aquarium.

Diet and feeding

These fish will very often only take live foods in the initial stages of captivity – usually guppies or mollies – but, with a little patience, nearly every specimen can be weaned onto a diet of dead Lancefish and other meaty frozen foods. The ethics of feeding one live fish to another must, of course, be carefully considered by aquarists.

Aquarium behaviour

Members of the Scorpaenidae are usually peaceful in captivity, but do not put temptation their way by keeping them with small fishes. Lionfishes need plenty of room in which to manoeuvre.

Dendrochirus brachypterus
Turkeyfish; Short-finned Lionfish

☐ **Distribution:** Indo-Pacific, Red Sea.

☐ **Length:** 170mm/6.7in (wild), 100mm/4in (aquarium).

☐ **Diet and feeding:** Small fishes, meat foods. Sedentary, engulfs passing prey.

☐ **Aquarium behaviour:** Keep in a species aquarium or together with larger fish.

☐ **Invertebrate compatibility:** Generally yes, but will eat crustaceans.

A very ornate fish. The red-brown body has many white-edged vertical bars. The dorsal fin is multirayed and tissue spans the elongated rays. When spread, the fins have more

Above: Dendrochirus brachypterus
Camouflage and a venomous sting protect this species in the wild.

obvious transverse patterning. The male has a longer pectoral fin and larger head than the female. At breeding time, the male darkens in colour; females become paler. This species does not grow as large as *Pterois* spp and hence it has attracted the alternative popular name of Dwarf Lionfish.

Pterois antennata
Scorpionfish; Spotfin Lionfish

☐ **Distribution:** Indo-Pacific, Red Sea.

☐ **Length:** 250mm/10in (wild), 100-150mm/4-6in (aquarium).

☐ **Diet and feeding:** Generally live foods such as small fishes, but all Lionfishes can be acclimatized to take frozen shrimps and similar items. A slow-swimming fish that takes sudden gulps of food.

☐ **Aquarium behaviour:** Predatory.

☐ **Invertebrate compatibility:** Generally yes, but will eat crustaceans.

The red bands on the body are wider and less numerous than on *P. volitans*. The white rays of the dorsal and pectoral fins are very elongated.

Above: Pterois antennata
The stationary lurking Scorpionfish is often dismissed by unsuspecting victims as a harmless piece of floating seaweed.

Below: Pterois radiata
The dark bars across the head and body of this graceful fish are accentuated by thin white borders on each side.

Pterois radiata
White-fin Lionfish; Tail-bar Lionfish

☐ **Distribution:** Indo-Pacific, Red Sea.

☐ **Length:** 250mm/10in (wild), 150mm/6in (aquarium).

☐ **Diet and feeding:** Smaller fishes and meaty foods as described for *P. antennata*. Slow-swimming sudden gulper.

☐ **Aquarium behaviour:** Predatory.

☐ **Invertebrate compatibility:** Generally yes, but will eat crustaceans.

Again, the red bands on the body are wider and less numerous than on *P. volitans*. The white rays of the dorsal and pectoral fins are very elongated and graceful. At breeding time, males of all *Pterois* species darken, while females become paler and have larger abdomens.

Pterois volitans
Lionfish; Scorpionfish

Distribution: Indo-Pacific.

Length: 350mm/14in (wild).

Diet and feeding: Smaller fishes and suitable meaty foods. Slow-swimming sudden gulper.

Aquarium behaviour: Keep with fish too large to be eaten.

Invertebrate compatibility: Generally yes, but will eat crustaceans.

This is the most well-known fish in this group. The dorsal fin rays are quite separate and the pectoral fins are only partially filled with tissue. The pelvic fins are red, and the anal and caudal fins are fairly clear. Thick and thin red bands alternate across the body and there are tentacle-like growths above the eyes.

Below: Pterois volitans

This species can be very colour variable and even an extreme black form exists. Such highly prized variants appear on the market from time to time, but, as expected, command very high prices. Lionfish should be encouraged to accept 'dead' foods as soon as possible.

Family: SERRANIDAE

Sea Basses and Groupers

Family characteristics
Many juvenile forms of this large family of predatory fishes have become aquarium favourites. Most of the species within this group are hermaphrodite, and some therefore lack any clear sexual dimorphism. Even so, many species undergo colour changes during breeding, turning darker, paler, or taking on a bicolour pattern. Not surprisingly, 'females' become distended with eggs – another clue to their likely functional sex. This is a large and varied family with over 370 species represented in tropical and temperate seas worldwide.

Diet and feeding
You should include crustaceans and meaty foods in the diet of these fishes.

Aquarium behaviour
The majority of species need a large aquarium.

Anthias squamipinnis
Wreckfish; Orange Sea Perch;
Lyre-tail Coralfish; Anthias

☐ **Distribution:** Indo-Pacific.

☐ **Length:** 125mm/5in (wild).

☐ **Diet and feeding:** Preferably live foods, or meat foods. Bold and prefers moving foods.

☐ **Aquarium behaviour:** Peaceful.

☐ **Invertebrate compatibility:** Yes, ideal.

This very beautiful orange-red fish has elongated rays in the dorsal fin, a deeply forked caudal fin and long pelvic fins. It is a shoaling species that needs companions of the same species. Males have an elongated third dorsal spine are usually larger and more conspicuously coloured than females. Dominant males are quite happy for a harem to follow them. Although these fish perform courtship behaviour in captivity, they have not yet been bred.

Left: Anthias squamipinnis
The Wreckfish is a shoaling species;
dominant males often have a harem.

Calloplesiops altivelis
Marine Betta; Comet Grouper

☐ **Distribution:** Indo-Pacific.

☐ **Length:** 150mm/6in (wild).

☐ **Diet and feeding:** Small fishes, meaty foods. Predatory.

☐ **Aquarium behaviour:** Err on the side of caution, and do not keep with small fishes.

☐ **Invertebrate compatibility:** Yes, ideal.

A very beautiful and deceptive fish: the trick is to decide which way it is facing, since the dorsal fin has a 'false-eye' marking near its rear edge. The dark brown body is covered with light blue spots and all

Above: Calloplesiops altivelis
The fins of this fish are very similar to those of the freshwater Siamese Fighting Fish, Betta splendens, *hence the popular name. Avoid keeping smaller fishes in the same aquarium.*

the fins are very elongated. This species spends much of its time in a 'head down' hunting position. Its tail allegedly resembles the head of the moray eel – a useful defence against predators.

Above: Cephalopholis miniatus
Ranging from the Red Sea to the mid-Pacific, the Coral Trout inhabits the coral reefs, looking for a meal of smaller fishes. In the aquarium, it often hides away in caves or under ledges, denying the fishkeeper a view of its spectacular colouring. An alternative, and very apt, common name for this species is Jewel Bass. The Coral Trout grows quite large in the aquarium and so is generally beyond the scope of the average hobbyist.

Cephalopholis miniatus
Coral Trout; Red Grouper; Coral Rock Cod

☐ **Distribution:** Indo-Pacific.

☐ **Length:** 450mm/18in (wild).

☐ **Diet and feeding:** Smaller fishes and meaty foods. Predatory.

☐ **Aquarium behaviour:** Do not keep with small fishes.

☐ **Invertebrate compatibility:** Yes, ideal.

The body and the dorsal, anal and caudal fins of *C. miniatus* are bright red and covered with bright blue spots. However, the pectoral and pelvic fins are plain red. Other fishes bear a resemblance to this species, but they do not have the distinguishing rounded caudal fin. A large, well-filtered aquarium is essential for this fast-growing fish.

Chromileptis altivelis
Panther Grouper; Polka-Dot Grouper

☐ **Distribution:** Indo-Pacific.

☐ **Length:** 500mm/20in (wild), 300mm/12in (aquarium).

☐ **Diet and feeding:** Live foods. Bold feeder.

☐ **Aquarium behaviour:** It is better not to keep this species with smaller fishes. However, its smallish mouth makes it the least harmful of all the grouper fishes.

☐ **Invertebrate compatibility:** Not recommended.

Juveniles have black blotches on a white body – effective disruptive camouflage. As the fish matures, these blotches increase in number but decrease in size. The result is a very graceful fish, and one that is constantly on the move in the tank.

Below: Chromileptis altivelis
These splendid juveniles in fine colour live up to their common name. As is often the case with beautiful fishes, however, they are very predatory.

Right: Grammistes sexlineatus
Introduce this grouper to a large aquarium ahead of suitably sized tankmates and it may become tame.

Grammistes sexlineatus
Golden-stripe Grouper; Sixline Grouper

☐ **Distribution:** Indo-Pacific.

☐ **Length:** 250mm/10in (wild).

☐ **Diet and feeding:** Animal and meaty foods. Bold.

☐ **Aquarium behaviour:** Do not keep with smaller fishes.

☐ **Invertebrate compatibility:** Not recommended.

Alternate black and white horizontal stripes cover the body. Although a good aquarium subject, it can give off toxic secretions when frightened, annoyed or even in the process of dying. It is unlikely to reach its full size when in captivity, unless kept in an extremely large aquarium.

Family: SIGANIDAE

Rabbitfishes

Family characteristics

The rabbitfishes have deep oblong bodies and are fairly laterally compressed. The mouth is small and equipped for browsing on algae and other vegetation. The spines on the dorsal and anal fins are venomous, so be sure to handle these fishes extremely carefully. Their alternative common name is 'Spinefoot', a reference to the fact that unsuspecting waders who disturb grazing fish risk a wound on the foot caused by the fishes' spines. Juveniles are often more brightly coloured than adults.

Only a dozen or so species belong to this family, but they have an economic significance in the tropics, where they are caught for food. The one species that is especially familiar to hobbyists, *Lo vulpinus*, has a tubular mouth, which contrasts with the normal rabbit-shaped mouth of species in this family.

Some reports of spawning in captivity – albeit of species not featured here – indicate that changing some of the water, or even decreasing its depth, may trigger spawning.

Diet and feeding

Rabbitfishes must have vegetable matter in their diet, although they will adapt to established dried foods and live foods in the aquarium.

Aquarium behaviour

Rabbitfishes are active, fast-growing fishes that need plenty of swimming space.

Lo vulpinus

Foxface; Fox-fish; Badgerfish

☐ **Distribution:** Pacific.

☐ **Length:** 250mm/10in (wild).

☐ **Diet and feeding:** Most foods, but must have vegetable matter.

Bold grazer that adopts a typical 'head-down' feeding attitude.

☐ **Aquarium behaviour:** Lively but peaceable, although it may be aggressive towards its own kind.

☐ **Invertebrate compatibility:** Not recommended.

Below: Lo vulpinus
The Foxface, or Badgerfish, has a tubular mouth, which is rather at variance with the more rabbitlike shape characteristic of other members of this family. When first introduced into a new aquarium, it may lose its distinctive coloration in favour of a temporary blotched appearance.

The white head has two broad black bands: one runs obliquely back from the snout, through the eye and up the forehead; the second band is triangular, beginning below the throat and ending behind the gill cover. This coloration obviously gave rise to the common name of Badgerfish among European hobbyists, more familiar with the badger than other hobbyists, who, for some reason, feel the fish's face looks more like that of a fox. Although superficially similar to the surgeonfishes, it has no spine on the caudal peduncle, and the pelvic fins are not very well developed, having only a few rays.

Above: Siganus virgatus
Remember that the dorsal and anal spines of this species are venomous.

Siganus virgatus
Silver Badgerfish; Double-barred Spinefoot

☐ **Distribution:** Pacific.

☐ **Length:** 260mm/10.2in (wild).

☐ **Diet and feeding:** Live foods, meat foods and plenty of greenstuff. Bold grazer.

☐ **Aquarium behaviour:** Lively but peaceable, although it may be aggressive towards fellow members of its own species.

☐ **Invertebrate compatibility:** Not recommended.

The silvery yellow body is more oval and the head more rounded than in the previous species. Again, the head has two badger-like black bars, the second of which begins narrowly just below the pectoral fins and broadens as it runs up to the top of the body. The head and forepart of the body are covered with blue lines, producing an intricate pattern.

Family: SYNGNATHIDAE

Pipefishes and Seahorses

Family characteristics

Every fishkeeper loves the seahorse, and the equally appealing pipefish, which could be described as a 'straightened out' version of the seahorse. Pipefishes are found among crevices on coral reefs, whereas seahorses, being poor swimmers, anchor themselves to coral branches with their prehensile tails. Many pipefishes are estuarine species, and are therefore able to tolerate varying salinities, even entering fresh water.

When seahorses reproduce, the female uses her ovipositor tube to deposit the eggs into the male's abdominal pouch, where they are fertilized and subsequently incubated. Incubation periods range from two weeks to two months, depending on the species.

Diet and feeding

Seahorses and pipefishes have small mouths and require quantities of small live foods to thrive; brineshrimp and rotifers are suitable, even *Daphnia* would be satisfactory if other live foods are in short supply.

Aquarium behaviour

Pipefishes and seahorses do best in a quiet aquarium. Many hobbyists have great difficulty in keeping seahorses for any length of time, for two main reasons: firstly, these are sensitive fish requiring excellent water quality all of the time; secondly, they are constant feeders and three or four good feeds a day is highly recommended.

Collection from the wild usually occurs by accident, when the fish are caught up in shrimp nets. This is extremely traumatic for the fish and may give some clue as to their unwillingness to adapt well to aquarium conditions. Always make sure seahorses and pipefishes are feeding well before you buy them.

Above: Doryrhamphus excisus
Pipefish are close cousins of the seahorse and should be treated in much the same way: a quiet tank, plenty of small livefoods and good water quality.

Doryrhamphus excisus
Bluestripe Pipefish

☐ **Distribution:** Indian and Western Pacific Oceans.

☐ **Length:** 70mm/2.75in (wild).

☐ **Diet and feeding:** Prefers live brine shrimp but may accept small frozen shrimp once settled.

☐ **Aquarium behaviour:** Very peaceful; needs a quiet aquarium.

☐ **Invertebrate compatibility:** Ideal.

Given the optimum conditions, this is one of the few pipefish that do well in the aquarium environment. An ideal situation for this fish – a close relative of the seahorse – would be a quiet invertebrate or species tank. Take care that food portions are small enough for it to swallow. It is important to long-term success that newly purchased specimens are slowly acclimatized.

Hippocampus erectus
Florida Seahorse; Northern Seahorse

☐ **Distribution:** Western Atlantic.

☐ **Length:** 150mm/6in (wild).

☐ **Diet and feeding:** Small animal foods. Browser.

☐ **Aquarium behaviour:** Needs quiet, non-boisterous companions.

☐ **Invertebrate compatibility:** Ideal.

The pelvic and caudal fins are absent, and the anal fin is very small. The tail is prehensile. The coloration of this species is variable; individuals may be grey, brown, yellow or red. The male incubates the young in the abdominal pouch. This species is also frequently referred to as *H. hudsonius*.

Right: Hippocampus erectus
A pale individual of this elegant species. Seahorses adopt a vertical position when at rest. When swimming, they lean forward, propulsion being provided by the fanlike dorsal fin.

Below: Hippocampus kuda
Apart from the fascination of its unusal body shape, with its equine appearance, and amusing activity among the coral branches, the seahorse also displays a very different method of reproduction.

Hippocampus kuda
Yellow Seahorse; Pacific Seahorse

☐ **Distribution:** Indo-Pacific.

☐ **Length:** 250mm/10in (wild) – measured vertically.

☐ **Diet and feeding:** Plenty of live foods, very small crustaceans, brineshrimp. *Daphnia* etc. Browser.

☐ **Aquarium behaviour:** Best kept in a species tank.

☐ **Invertebrate compatibility:** Ideal.

Newly imported specimens may be grey, but once they have settled into the aquarium, the body may take on a yellow hue. The colour of specimens can vary widely, however. An irresistible fish with a fascinating method of reproduction. The male incubates the fertilized eggs in his pouch for four to five weeks before they hatch. This species needs anchorage points in the aquarium, such as marine algae and other suitably branched decorations. The male's brood pouch is the end of an evolutionary process that started with the seahorses' relatives glueing the eggs to the underside of the body for protection.

Family: TETRAODONTIDAE

Puffers

Family characteristics

Puffers are generally smaller than porcupinefishes and smooth scaled. Their jaws are fused, but a divided bone serves as front teeth. '*Tetraodon*' means four toothed (two teeth at the top and two at the bottom), whereas '*Diodon*' means two teeth (one at the top and one at the bottom). These fishes use their pectoral fins to achieve highly manoeuvrable propulsion, but the pelvic fins are absent. Their inflating capabilities vary from species to species; *Tetraodon* sp. – some of which are freshwater – are 'fully inflatable', but members of the genus *Canthigaster* can only partially inflate. The flesh of all species is poisonous.

Diet and feeding

Puffers will eat readily in the aquarium, taking finely chopped meat foods. They have a bold feeding manner.

Aquarium behaviour

Generally peaceful but occasionally may be aggressive towards other fishes. Do not keep with invertebrates.

Arothron hispidus
White-spotted Blowfish; Stars and Stripes Puffer

☐ **Distribution :** Indo-Pacific, Red Sea.

☐ **Length:** 500mm/20in (wild).

☐ **Diet and feeding:** Finely chopped meat foods. Cruncher.

☐ **Aquarium behaviour:** Peaceful. Do not keep with invertebrates.

☐ **Invertebrate compatibility:** No, will eat invertebrates.

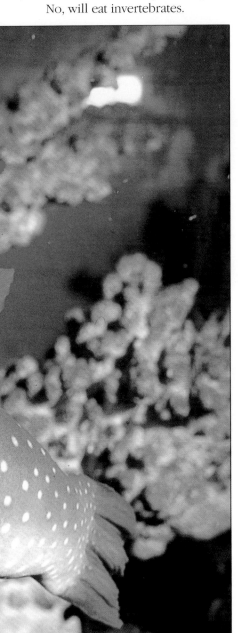

The distinctive features of this species are the number of bluish white spots over the patchy grey body. These spots are not so pronounced in adult fishes. Just behind the gill cover, and at the base of the pectoral fins, there is a dark patch surrounded by a circular yellow pattern. Like most puffers, the flesh is poisonous. The caudal fin is often seen clamped shut and plays little part in propulsion. Pufferfish have their teeth fused together to form four powerful teeth at the front of the mouth, which they use to crunch up molluscs and crustaceans.

Left: Arothron hispidus
Like all puffers, this species will inflate its body when disturbed or frightened, but do not provoke it. Just in front of the white-rimmed eyes, two tentacle-like nostrils are visible.

Below: Arothron meleagris
Because this species is a rather large and messy eater, its tank will need good filtration and you should make sure that all traces of uneaten food are removed to avoid undue pollution.

Arothron meleagris
Spotted Puffer; Guinea Fowl Puffer; Golden Puffer

☐ **Distribution:** Indo-Pacific, Red Sea.

☐ **Length:** 300mm/12in (wild).

☐ **Diet and feeding:** Finely chopped meaty foods. Cruncher.

☐ **Aquarium behaviour:** Peaceful, but do not keep with invertebrates.

☐ **Invertebrate compatibility:** No, will eat invertebrates.

Although a plain yellow colour phase occurs, normally the brown-grey body is densely covered with white spots. When kept in a spacious aquarium, it will be less likely to release its poison under stress from other fishes. Scrupulous attention to water quality must be paid, as this fish is a very messy eater. Its unconsumed portions of food will rapidly pollute the water unless great care is taken to remove them.

Canthigaster solandri
Sharpnosed Puffer; False-eye Puffer/ Toby

☐ **Distribution:** Indo-Pacific, Red Sea.

☐ **Length:** 120mm/4.7in (wild), 50mm/2in (aquarium).

☐ **Diet and feeding:** Finely chopped meat foods. Bold cruncher.

☐ **Aquarium behaviour:** Peaceful, except towards members of its own kind.

☐ **Invertebrate compatibility:** No, will eat invertebrates.

This spectacularly patterned fish has a gold-brown body and a caudal fin covered with pale spots. A blue wavy line replaces the spots on the upper part of the body and a large white-edged black spot appears at the base of the dorsal fin. The fish swims with its caudal fin folded. The pelvic fins are absent.

Canthigaster valentini
Black-saddled Puffer; Valentine Puffer

☐ **Distribution:** Indo-Pacific.

☐ **Length:** 200mm/8in (wild), 75mm/3in (aquarium).

☐ **Diet and feeding:** Finely chopped meaty foods. Bold cruncher.

☐ **Aquarium behaviour:** Peaceful, although it has a reputation for nipping the fins of species with long fins, and may not tolerate its own kind.

☐ **Invertebrate compatibility:** No, will eat invertebrates.

The lower half of the body is cream in colour and covered with small brown dots. The upper part has four saddle-shaped dark areas; the one covering the forehead also has blue lines. These blue lines also occur on the two narrow vertical bars that reach three-quarters of the way down the sides of the fish between the head and dorsal fin. There is a final plain patch on the top of the caudal peduncle. A black spot at the base of the dorsal fin may merge with the other dark markings. The bases of the fins are red, but for the caudal fin, which is yellow. The pelvic fins are absent.

Right: Canthigaster valentini
The bold Black-saddled Puffer has strong jaws with which it can crunch coral.

Below: Canthigaster solandri
This species has an attractive spotted pattern, with bright radiating stripes around the eyes and top of the body.

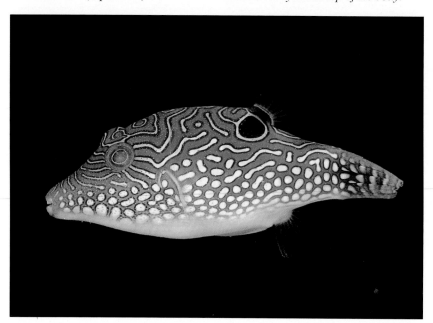

PART FIVE

TROPICAL MARINE INVERTEBRATES

There are many more species of invertebrates in the world than there are vertebrates, but within the aquarium hobby it is the fishes on which most emphasis is placed. The following section presents a selection of invertebrates suitable for the aquarium, arranged roughly in order of complexity (that is, the most biologically advanced are last). Certain species that are not recommended, either because they are dangerous to other species or to the hobbyist, or because they may not survive long in captivity are featured in Appendix Two on pages 382-383. Fortunately, few tropical marine invertebrates are threatened with extinction, but there is concern that some may be declining in number. For example, the Queen Conch and the Spiny Lobster have traditionally been important food sources for coastal people but are increasingly becoming food for rich tourists, and some popular shells have been over collected for the ornamental shell trade. On a more positive note, it is worth remembering that, compared to the number of species caught for food or destined for the jewellery and curio market, the number of animals captured for the aquarium is very small, and with the sophisticated equipment available today, many have a good chance of surviving in the aquarium at least as long as they would have done in the wild. Finally, it is worth making the point that by observing their invertebrates, recording their findings and publishing their results, aquarists can contribute to the general fund of information about these fascinating creatures.

Left: *The Red Hermit Crab may reach the size of an adult human fist with an appetite to match, but for many marine aquarists, the strange beauty and interesting behaviour of such invertebrates make them irresistible.*

Phylum: PORIFERA

Sponges

Adocia sp.
Blue Tubular Sponge

This intensely blue species is imported fairly regularly from Indonesia, but is never available in large quantities. Damaged specimens will turn progressively whiter as the living cells die, leaving behind the supporting structure. Once it has recovered from the initial transition from one tank to another, the Blue Tubular Sponge is very hardy and grows surprisingly rapidly, particularly in an aquarium with fairly slow-moving water. Given suitable conditions it will also grow quickly from small pieces. Never remove any sponges from the water; if air pockets form within them they will decline and die.

In the wild, many small animals live within sponges and, on rare occasions, small crabs, shrimps and gobies are found in imported specimens. Other animals may live on the outer surface. Try to avoid introducing 'undesirable' subjects into the tank.

Below: Adocia sp.
Few other invertebrates are blue, so this easy-to-maintain animal provides a bold splash of colour in the aquarium. In slack or slow-moving water this species will develop a branching form.

Axinellid sp.
Orange Cup Sponge

Sri Lanka, Singapore and Indonesia are the main sources of supply for this yellow-orange species with its distinctive cup or bowl shape. They often prove to be one of the hardiest species but, unfortunately, are often shipped in insufficient water, leaving the edges exposed to the air so that they may turn pale and begin to crumble. It is important to check that all sponges are intact before you buy them. Ensure that the tank is not brightly illuminated when you introduce this sponge and that there is sufficient water movement to prevent debris accumulating in the cup. Growth is slow. Several other types of yellow or orange sponges are regularly imported and most do well in the right conditions.

Right: Axinellid sp.
The Indo-Pacific Orange Cup Sponge is readily available and easy to maintain if you prevent algae from smothering it.

Haliclona compressa
Red Tree Sponge

This bright orange-red species is very common in the Caribbean Sea and regularly imported. Most specimens are about 20cm (8in) tall, but larger examples are sometimes available and, with their interesting branched habit, they provide a dramatic splash of colour. It is important to select a specimen with the base attached to a piece of rock and with no white or pale patches on the arms. This species appreciates a reasonable water flow and, like all sponges, prefers somewhat dim lighting conditions. In too bright a situation, the branches often become covered with encrusting algae that choke the sponge and eventually kill it.

Right: Haliclona compressa
This striking Red Sea species is typically beige-pink but, as with many sponges, the colour is very variable. Avoid damaged specimens and do not allow them to touch corals or anemones.

Phylum: CNIDARIA

Hard Corals, Horny Corals, Jellyfish, Polyps, Sea Anemones, Sea Pens, Soft Corals

HARD, TRUE OR STONY CORALS

Euphyllia sp.
Vase Coral; Frogspawn Coral

Considerable confusion still surrounds the nomenclature of many corals and, until recently, much classification work was based solely on the dead coral skeletons, with no consideration of the living animal. Much of this confusion is found in the genus *Euphyllia,* which includes several well-known and popular corals.

The tentacles of all these specimens are very pronounced and comparable with those of sea anemones, although they vary in shape. Vase Coral may have round or crescent-tipped tentacles, while

Above: Euphyllia divisa
A species known as Frogspawn Coral.

those of Frogspawn Coral are semi-transparent and irregularly bubblelike. An exception is Fox Coral, in which the polyps are like a row of flat toadstools. All species require optimum water and lighting conditions, otherwise lifespans will be unnaturally short.

Euphyllia picteti
Tooth Coral

Euphyllia picteti has only become widely available in the past few years, but is already firmly established as a favourite and specimens are quickly snapped up. When deflated they are fairly unprepossessing, but within a few hours of introduction to the tank they swell to four times the area of their supporting skeleton. The flesh is usually a fluorescent green, or occasionally blue, lined with anemonelike tentacles that frequently have vivid orange tips. The green colouring is due to the coral's symbiotic algae, so strong lighting is essential. This coral will take sizeable pieces of food and the tentacles pack a powerful sting; they

Above: Euphyllia picteti
This beautiful species is popular, but often more costly than related species.

are quite capable of raising a painful rash on a careless aquarist's arm, or killing any sessile invertebrates placed too close to them in the aquarium. Excellent water conditions are essential for long-term success.

Above: Favites sp.

In this striking picture, a Red Sea species demonstrates the surprising ability of many corals to fluoresce under ultraviolet light. Providing that you choose a whole specimen that is not damaged, it should do well in the invertebrate aquarium. Unfortunately, many of the specimens seen for sale have been hacked from a larger coral head and invariably die in captivity.

Favites sp.
Moon Coral

These attractive dark green corals are closely related to the Brain Corals (see page 310), but here the meanders are subdivided into numerous single pockets, each housing a polyp. The polyps are interlinked, which causes a major problem with this type of coral. All too often the specimens received by importers are merely pieces hacked from large coral heads and in these cases, even with the best care, they almost invariably die. *Favites* are among the most light-sensitive corals, but can take supplementary food. If decorative algae are growing in the tank, ensure that leaf fronds do not rob the coral of light. Moon Corals fluoresce in ultraviolet light.

Galaxea fascicularis
Star Coral

This distinctive coral has a very attractive skeleton. Small stubs arise from the central core, with polyps sitting on each of the star-shaped ends. Singapore is the main source of this brown species, but as there are often considerable losses in transport it has not achieved any great popularity.

The Star Coral is by no means impossible to keep, but it requires similar conditions to *Goniopora* (see below) and cannot be recommended for the newcomer to the hobby.

Goniopora lobata

G. lobata is common throughout the Indo-Pacific and is one of the most frequently imported corals. Its attractive feathery appearance, combined with its usually low price, ensures that this is often the first stony coral that many aquarists keep. Unfortunately, it is not necessarily a good choice, since it demands perfect water conditions and very good lighting.

Without sufficient light, the zooxanthellae will not function properly, and with too much food in the water, the polyps will not expand. Furthermore, *Goniopora* species are easily damaged. The polyps are not individual animals, rather outgrowths from the skin covering the ball-like skeleton. When the animal inflates with water to expand the polyps, the skin is stretched taut, becomes thin and is easily punctured. Sea urchins, sharp-footed crustaceans and tumbles from rocks are major causes of such punctures, which usually lead to a persistent infection that quickly engulfs the coral.

There are several species of *Goniopora*; one of the most attractive has rather thin, creamy polyps with purple centres. Treat them with great care. The *Porites* corals of the Caribbean and Pacific are closely related, but rarely imported and just as difficult to maintain in captivity.

Above: Galaxea fascicularis
Star Coral is rarely seen in good condition, but here the colour is good and polys extended. This species is not for beginners; it requires perfect water conditions and excellent lighting.

Below: Goniopora lobata
Stony corals should be erect and well expanded. Bacterial infection may result if the flesh is damaged. Shrinking polyps indicate poor lighting and water; always provide the best of both.

Heliofungia actiniformis
Plate Coral

The various species of *Heliofungia* are easily confused with sea anemones, since their circular or oval bodies are covered with long tentacles that completely hide the ridged skeleton.

Most corals are a collection of polyps, but *Heliofungia* is a solitary polyp with one central mouth. Zooxanthellae can tint them green or pink. As well as deriving nourishment from the zooxanthellae, *Heliofungia* will take chopped fish and shrimp in small quantities in the aquarium.

Heliofungia fare best when placed directly onto a coral sand substrate where they can receive good light and a moderate flow of water. Do not position them on rocks, otherwise the delicate tissue around the edge of the coral may tear and open up a path for infection. When buying specimens, check that all the tentacles are erect and that there are no bald patches.

H. actiniformis is roughly circular, as are most of the related species, but *Herpolitha limax,* which sports many short, brown tentacles, forms a long oval.

Leptoria sp.
Brain Coral

Brain corals can reach massive proportions, but small specimens are strongly recommended for an aquarium with good lighting. The optimum size for aquarium specimens is about 10-15cm (4-6in); larger ones tend to be damaged in transport. Most species are various shades of brown, but some are a vivid green and others a pale purple-pink. Brain corals receive much of their food from the action of symbiotic algae, but they will supplement this with plankton.

Occasionally, you may see small tentacles around the edges of the sinuous channels of the Brain Coral. These can be extended to sting nearby corals and are particularly prominent in the closely related Caribbean species, *Meandrina meandrites.*

Right: Leptoria sp.
It's easy to see how this species gained the common name 'Brain Coral'. One of nature's most fascinating structures.

Below: Heliofungia actiniformis
Here, the tentacles are semi extended, revealing the green-tinged body.

Plerogyra sinuosa
Bubble Coral

This accommodating species can be highly recommended as a first stony coral. The common name is a very apt description, for during the day the polyp mouths and tentacles are covered in a mass of bubbles. These may be fawn coloured in the best specimens. At night, the bubbles deflate somewhat and the coral erects flowing, 6cm (2.4in)-long stinging tentacles to capture small shrimps in the wild. In the aquarium this coral will accept whole shrimps and pieces of fish gently tucked among the bubbles. If regularly fed, a Bubble Coral can increase its expanded diameter by some 50 percent within a few weeks. Weekly feeding is sufficient and even this can be suspended if the coral is given sufficient light.

The best specimens come from Sri Lanka and Indonesia, but somewhat similar species, often greenish or light brown, are found throughout the Indo-Pacific. Do not place any coral species too close to one another.

Below: Plerogyra sinuosa
Like many other corals, this species possesses zooxanthellae algae within its tissues. Intense lighting is essential if it is to reach full potential.

Above: Tubastrea aurea
When it expands – which it does mainly at night – T. aurea *is a most dramatic coral, resembling a bunch of golden chrysanthemums.*

Tubastrea aurea
Sun Coral

This vivid orange coral lives at the mouth of, or inside, caves and crevices in Indo-Pacific reefs. It is an extremely common species and large numbers are exported annually. Living as they do in a shady habitat, they have no need of symbiotic algae and rely on trapping food particles with their abundant yellowish tentacles.

Place the coral in a shady spot in the tank and consider each polyp as an individual small anemone. If it is reluctant to open, tempt the polyps to expand and 'flower' by squeezing a shrimp head into the water. A few minutes later the polyps will open and you can feed each one a small piece of shrimp.

Newly introduced Sun Corals may not open for a week or more. From then on, if properly fed, you can expect the colony to expand by producing new polyps at the base of the mature ones. In the wild, colonies grow up to 50cm (20in) across, but 10cm (4in) is a more normal size in the aquarium.

There are several similar species, of which the Indonesian *Balanophyllia gemmifera* is a good choice for the aquarium. It is larger and even easier to feed than *T. aurea.* The related species *Dendrophyllia gracilis* has an attractive branchlike form and is only infrequently offered for sale. *Tubastrea aurea* requires optimum water conditions.

Above: Tubipora musica
The structure of its red skeleton has given rise to the common name of this attractive coral. Aquarium specimens are usually broken from larger heads.

Tubipora musica
Organ Pipe Coral

This species is much more familiar to aquarists as a dead skeleton for tank decoration than as a living animal. It is one of the few corals with a naturally red skeleton, a very attractive feature when not overgrown with algae. When it is alive, the top of the interlinked red pipelike skeleton houses a mass of short but active brown polyps. These pulse with the flow of water in similar fashion to *Anthelia* (see page 324). The polyps are all interconnected and, as specimens are usually fragments of much larger pieces, their life in the aquarium is generally limited.

— 313 —

HORNY CORALS

Gorgonia flabellum
Sea Fan

Sea Fans are very closely related to Sea Whips, but in this species one main branch grows out in a very flat plane, the myriad small offshoots linking together to form a lace-fan appearance. Sea Fans are particularly common on Caribbean reefs, and at one time many were collected for the curio trade. Fortunately, this practice is now greatly reduced, but dried specimens of Sea Fans and Sea Whips are still occasionally offered as aquarium decoration.

When properly cured (i.e. made safe for aquarium use), Sea Fans should look like black lace. Before cleaning, they are often yellow or pink and should never be used in this state, as the dried tissue will rot in the tank and pose a major pollution problem. Although Sea Fans are more difficult to maintain than Sea Whips, they can survive in captivity, so let us hope that the dried Sea Fan trade will soon be a thing of the past.

Small, 15cm (6in) specimens are best for the aquarium, as larger animals are difficult to transport without damaging the tissue.

Muricea muricata
Sea Whip

This Caribbean species is one of many that produce a cluster of 'finger', or whiplike, extensions. Sea Whips are found throughout the tropics, usually in areas of strong water movement, and appear in all the colours of the rainbow. Most are fairly easy to maintain, the thicker fingered species having proved generally the most hardy.

Like Leather Corals, the best specimens are attached to a small portion of stone or coral. This ensures that the base is not broken and allows you to position the animal in a water current without risk of the soft flesh rubbing against the rockwork. When buying specimens, check that all the fingers are intact and that there are no exposed areas of dark chitinous skeleton. Bacterial infection can easily begin at such sites.

In nocturnal Sea Whips, the mat of small polyps only emerges at night. Under ideal conditions, they may grown 2.5cm (1in) a month.

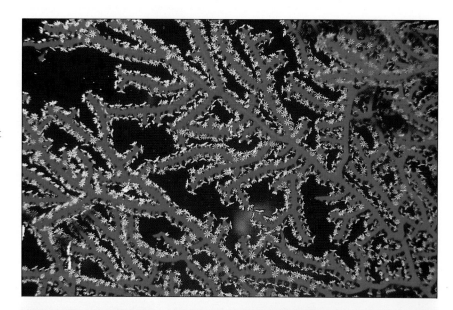

Top: Gorgonia flabellum
This close-up view clearly shows the structure of a typical Sea Fan.

Above: Muricea muricata
Feed sparingly when polyps are extended. Good water circulation in the tank is essential.

JELLYFISH

Cassiopeia andromeda
Upside-down Jellyfish

Jellyfish are familiar to beachcombers throughout the world, but virtually unknown within the hobby. *C. andromeda* is the only species ever offered for sale and the only one suitable for the aquarium. It is found throughout the Red Sea and the Indo-Pacific region and can reach up to 30cm (12in) in diameter. Only small specimens, up to 8cm (3.2in), are generally available.

Most jellyfish trail their arms behind them as they swim, but *Cassiopeia* spends much of its time lying on the substrate, with its tentacles upwards, wafting in the current. Among the tentacles are small bladderlike growths, filled with zooxanthellae that provide much of the animal's food. It therefore needs intense lighting, and you can supplement the diet with newly hatched brineshrimp. However, even with the best care, it rarely lives long in an aquarium and should therefore be left to experienced aquarists.

Below: Cassiopeia andromeda
This species needs a specialized aquarium, with intense lighting, optimum water quality, a soft sandy substrate and plenty of room.

POLYPS

Rhodactis spp.
Mushroom Polyps

Mushroom Polyps – or Mushroom Anemones, as they are sometimes known – are available as colonies clustered on small pieces of rock. They have been one of the staple additions to the invertebrate tank for many years, but are still justifiably popular with hobbyists.

There are many species; most of them measure 2-5cm (0.8-2in) in diameter, although a few giants can reach 15-20cm (6-8in), and these are very dramatic. Most species are green or brown, and often the two colours are combined in radiating stripes. Reddish brown and blue specimens are available from Indonesia but are somewhat expensive. It has been found that these red and blue pigments act as a light filter, which prevents the animal being 'sunburnt'. If the aquarium is not brightly illuminated, these attractive colours disappear, allowing more light through to the zooxanthellae, but resulting in a drab, brownish animal.

A particularly attractive species, with a rather tufted appearance to its pale bluish green polyps, is often found on living rock imported from the Caribbean.

Zoanthus sociatus
Green Polyps

There is a huge range of *Zoanthus* species and their relatives, all of which are generally sold under the title 'polyp colony'. The zoanthids mark something of a halfway house between the corals and the sea anemones. The polyps are usually found in clusters and are often interconnected at the base. However, they have many more tentacles than the true corals and no calcareous skeleton. All the 'polyp colonies' are among the easiest species in the phylum Cnidaria to keep. Most require good water quality and good lighting, but very little else. In good conditions, mature polyps often multiply by budding new, small polyps at the base so allow room for new growth.

Most species are coloured in various shades of green, brown and beige, or combinations of the three, but several bright yellow types are also regularly available.

Top: Rhodactis spp.
The many forms of Rhodactis *all have the typical mushroom shape seen here.*

Above: Zoanthus sociatus
Coral polyps are one of the mainstays of the reef tank and usually easy to keep.

SEA ANEMONES

Anthopsis koseirensis
Pink Malu Anemone

This attractive pinkish purple species is structurally very similar to *Heteractis malu* (see page 320) and is widely considered to be a colour form of the latter species – hence the common name. *A. koseirensis* is rather less common in the wild than *H. malu* and commands a higher price, but its appealing coloration, ease of maintenance and the willingness of all species of clownfishes to form a commensal relationship with it justifies its cost. It is particularly attractive if the illumination is supplemented with a red-enhancing light.

In the past, many anemones were artificially coloured by immersing them in a solution of food colouring to produce blue, green, orange and scarlet specimens. While the food dye itself appeared to cause no obvious problems, few specimens survived long after this treatment – possibly because of the effect it had on the colour of light reaching the vital zooxanthellae. Fortunately, this practice seems to have died out, but you should treat with suspicion any malu anemones in colours other than purple-pink, pale yellow and pale brown.

Below: Anthopsis koseirensis
This beautiful pink species is popular with hobbyists and clownfishes. This one houses a pair of Common Clowns.

Condylactis gigantea
Caribbean Anemone

This long-tentacled anemone is the most popular and commonly exported species from the Caribbean sea. The body can be white, brown or pink, and the tentacles are usually pink or white with a more intense pink tip.

Condylactis are very easy to keep, requiring only moderate lighting and a steady, but not vigorous, water flow. They are easily fed by dropping small pieces of fish or shrimp among the tentacles once or twice a week. As a general rule, do not feed anemones with a liquid invertebrate diet.

Despite their pleasant colouring and ease of maintenance, however, *Condylactis* are not as popular as the Pacific anemones, because the various *Amphiprion* clownfishes will only very rarely set up home within their tentacles.

Heteractis aurora
Sand Anemone

The Sand Anemone's scientific name has undergone a change (Dunn 1981), from *Radianthus simplex*. This greyish white species is easily distinguished by its tentacles, which are thickened to give a ringed appearance and have a tendency to lie flat against the surface disc. These are among the commonest anemones in their Indo-Pacific home range.

As their common name suggests, they are happiest placed on the substrate, rather than on rockwork, and will often anchor their foot through the sand and onto the undergravel filter plate or base glass of the tank. When disturbed, they can rapidly deflate and disappear beneath the substrate, thus escaping the attention of predators.

H. aurora is a small species, rarely more than 15cm (6in) in diameter, and is easy to feed with small pieces of fish or shrimp once or twice a week. Although easy to maintain for long periods, it is often thought to have only limited attraction for clownfishes. However,

several *Amphiprion* species have been recorded with this anemone and *Amphiprion clarkii* have been observed with *H. aurora* in reef areas off Borneo. It is a very suitable species for the beginner or the hobbyist on a limited budget.

Above: Condylactis gigantea
The common Caribbean Anemone is found in a variety of colours.

Below: Heteractis aurora
The Sand Anemone is common on coral gravel beds in lagoons and at reef edges.

Heteractis magnifica
Purple Base Anemone

Heteractis magnifica is one of the large anemones that regularly plays host to clownfishes. Previously known as *Radianthus ritteri,* it was one of the few whose scientific name seemed to cause no confusion! Unfortunately, the new name *Heteractis magnifica* (Dunn 1981) has been very slow to achieve the same familiarity.

This anemone is found throughout the Indian Ocean and the Indo-Pacific region. Specimens from Sri Lanka typically have a purple body with buff or light brown tentacles. A more attractive form with a scarlet body and pure white tentacles is exported from Kenya. Both are easy to keep, but in good conditions they may reach 70cm (27.5in) in diameter and quickly grow too large for all but the biggest aquarium.

At first sight, *H. magnifica,* like all anemones, appears rooted in one position, but this particular species has an annoying habit of climbing slowly up the tank glass, thus presenting a rather unattractive view to the aquarist. By gently teasing the foot loose with the ball of the thumb, you can easily move them, but take great care to avoid tearing the very delicate flesh.

Below: Heteractis magnifica
The large Heteractis magnifica *is often included in shipments from Sri Lanka. This beautiful anemone is a popular aquarium species, which, under good lighting and water conditions thrives, making an ideal host to clownfishes.*

Heteractis malu
Malu Anemone

Until recently, this species was widely known as *Radianthus malu* and will be much more familiar to experienced aquarists under this name. Malu Anemones are imported in large numbers from Singapore, Indonesia and the Philippines, and are one of the staples of the hobby. *H. malu* is very attractive to clownfishes and an ideal species for most invertebrate aquariums.

Specimens are available in sizes ranging from 10 to 40cm (4 to 16in) in diameter, but they are capable of growing even larger. All the colour forms of the Malu Anemone display the distinctive purple-red tips to the tentacles, which are regularly tapered, up to 5cm (2in) long and evenly spaced across the disc.

Given intense illumination, the lighter forms will turn brown with the development of the zooxanthellae algæ that supply much of the anemone's nutritional requirements. In most situations, you can offer *H. malu* a similar diet to *H. aurora* and, like all anemones, it will benefit from regular additions of a vitamin supplement.

This species is more likely to stay where you put it than some others, but it is quite capable of moving if your choice of site is not suitable. When introducing this species, and other anemones, to the tank, ensure that they do not rest 'face-down' on the sand. Anemones 'breathe' through the tentacles and quickly die if water movement around the tentacles is restricted.

Remember that all anemones accumulate waste products within their body cavities. To void these, they periodically collapse, pumping out the stale and polluted water within their body and often producing a stream of brown mucus at the same time. Occasionally, a sizeable anemone may shrink to the size of a golf ball. This is not a matter for concern, provided it does not happen more than once a day and the anemone does not stay closed for more than 24 hours. In this event, it may be taken as a sign that a major change of the aquarium water is overdue. Many anemones will contract when lacking in sufficient illumination or as a result of poor water conditions.

Below: Heteractis malu
Clownfishes rarely venture far from their host anemone and even lay their eggs under its protective mantle.

Heteractis sp.
Gelam Anemone

In this species we have another example of the confusion that exists in the specific names for anemones. Although it is considered here as a *Heteractis* species, Gelam Anemones are structurally different from the Malu Anemones with which they share their genus name. Gelam Anemones have shorter tentacles – which often have swollen tips –

hence the alternative common name of Bubble Anemones.

Their small size – up to 20cm (8in) – and good colouring make them justifiably popular. The most attractive Gelam Anemones have rusty red tentacles and the very best have purple bodies. They are easy to keep, feeding on small pieces of shrimp or shellfish, and most clownfishes will use them.

Gelam Anemones have a tendency to roam around the aquarium, but

can be encouraged to settle if they are placed in an opened clam shell or between two scallop shells wedged into the substrate. They usually confine themselves to the lower half of the tank, unlike *H. magnifica,* which often seeks out the highest point. This species is particularly soft bodied; check that there are no tears in the body, as these usually prove fatal. Avoid buying very pale specimens, which have lost their symbiotic algae.

Left: Heteractis sp.
A typical feature of the Gelam Anemones are the swollen tips to the tentacles. The body may be red-purple or brown and is normally hidden among the rocks.

Pachycerianthus mana
Fireworks Anemone; Tube Anemone

The cerianthid anemones are an interesting group found in all the world's warmer waters. Their chief characteristic is that they live within a tube formed of mucus and detritus gleaned from the soft substrates they inhabit. They also have many very long, thin tentacles and a very powerful sting.

P. mana is an Indo-Pacific species that often has banded tentacles, while other close relatives show colours ranging from pure white, through yellow to maroon and near-black. Here again, the strength of their sting precludes them from acting as hosts for clownfish. Their 20cm (8in)-long tentacles make them a threat to neighbouring invertebrates and fishes.

Cerianthids are night feeders, usually remaining in their tubes during daylight hours. They are easy to feed on finely chopped fish and shrimp every other day, but in view of the risk they pose to other animals, think carefully before introducing them to a well-populated aquarium.

Left: Pachycerianthus mana
The Fireworks Anemone lives up to its common name in appearance. Its tentacles are lined with very powerful poisonous stinging cells.

Above: Stoichactis gigas

This giant anemone has a powerful sting; avoid keeping it with other anemones or close to corals. It is popular with clownfishes and green and brown specimens are often available.

Stoichactis gigas
Carpet Anemone; Blanket Anemone

As the scientific name would suggest, this is one of the largest species of sea anemone, with wild specimens approaching 1m (39in) in diameter. Smaller specimens are popular with aquarists, but are no longer as readily available as in previous years, when large numbers were exported from Sri Lanka. Most were white or pale brown, but there was a steady supply of blue, purple and fluorescent green specimens. These are now rare in the hobby and command high prices. Nonetheless, this is a very hardy and long-lived species and worth seeking out.

All anemones have stinging cells with which they catch plankton, small fish and crustaceans, but this is one of the few in the hobby that can sting man. The short (1cm/0.4in), densely packed tentacles feel very sticky when touched; their effect is one of multitudes of barbed stinging cells being fired into the flesh. Stings on particularly sensitive areas, such as the wrist and inside of the forearms, can produce an annoying rash that may last several days or, on rare occasions, considerably longer.

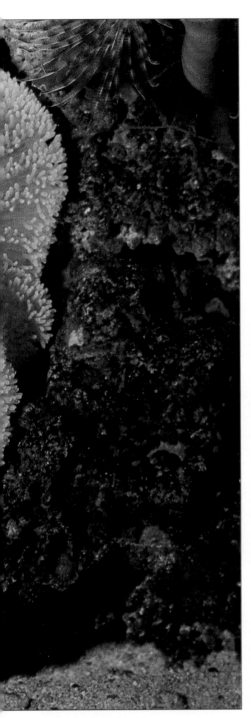

SEA PEN

Cavernularia obesa
Sea Pen

Sea Pens are an attractive and interesting group of animals, only rarely seen within the hobby. Their central tubular body is supported by a calcium 'spine' that resembles the quill of a feather, hence their common name. A distinct foot burrows into the substrate and anchors the animal in position in the fairly turbulent waters that they often inhabit.

During the day, the body, which may be orange, yellow buff or white, is contracted. Throughout the night, and occasionally during daylight hours, the body expands and previously hidden feathery polyps appear all over the animal as it starts to feed on plankton and organic detritus. Sea Pens can give off an eerie greenish blue light when disturbed. They move slowly through the sand and, in view of their habits, are not suitable for the total system aquarium with a shallow substrate.

Below: Cavernularia obesa
At night-time, the delicately branched polyps expand in the water current to capture minute food particles.

In view of this stinging potential, be very careful when servicing the aquarium – and do not house *Stoichactis* anemones with other families of anemones or allow them to come into contact with other invertebrates.

You can feed this species in the same way as the Gelam Anemones, but you may find it necessary to wriggle the food into the tentacles to elicit a stinging response before the food is passed to the central mouth and engulfed.

SOFT CORALS

Anthelia glauca
Pulse Coral

The Pulse Corals are a comparatively recent introduction to the hobby, but as Indonesian imports become more common they are justifiably increasing in popularity. *Anthelia* and the closely related *Xenia elongata* are two of the easiest corals to keep and both can be expected to spread and multiply in the aquarium.

Anthelia consists of a cluster of very feathery polyps that join at the base to form a foot anchoring the coral to a rocky substrate. *X. elongata* is similar but more treelike, the body splitting into branches that bear the polyps. The common name for both these corals comes from the continual rhythmic opening and closing of the polyps as they appear to feed. Both species require adequate lighting and appreciate a good current of water.

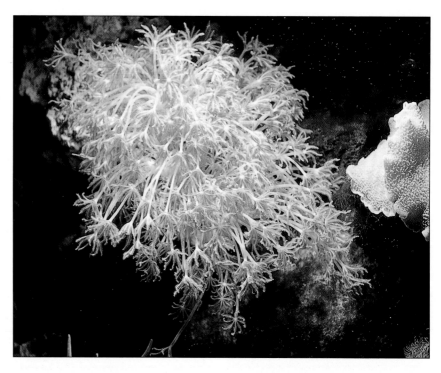

Above: Xenia elongata
Although very similar to Anthelia glauca, *this species of soft coral has a more branched appearance. Both need good lighting and water circulation.*

Below: Anthelia glauca
The attractive Pulse Coral is highly recommended for the invertebrate aquarium. It spreads quickly, colonizing neighbouring rocks.

Dendronephthya rubeola
Red Cauliflower Coral

This delicately branched species is one of the most attractive of the soft corals and the easiest of several lookalike species to maintain. *D. rubeola* lives on sand and mud sediments in the Indo-Pacific, where it anchors itself into position with a number of thick, fleshy, rootlike growths from its base. This method of attachment makes it much easier to collect this species than *Dendronephthya klunzingeri,* for example, which anchors itself to rocks and is easily damaged during collection.

During the day, these species contract into a red-and-white ball. The loose calcareous spicules, which support the flesh like a disjointed skeleton, project through the skin and make an uncomfortable handful. At night, or under subdued lighting, the animal takes in water to feed. Very large specimens reach over 1m (39in) in height, but more typical aquarium specimens are 15-20cm (6-8in) high. All the Cauliflower Corals feed on the very smallest particles.

Sarcophyton trocheliophorum
Leather Coral; Elephant Ear Coral

The Leather Corals are very widespread throughout the Indo-Pacific tropical seas and *S. trocheliophorum* is the most frequently imported. The common name comes from both the texture and colour of the animal when it retracts its polyps.

In the wild, they can form soft, undulated plates over 1m (39in) in diameter, the top surface clothed with a carpet of 1cm (0.4in)-high, delicate polyps. One of the best forms for the aquarium grows as a convoluted mushroom. Given good lighting, they fare very well and can be expected to increase in size.

When disturbed, the polyps retract and it is not unusual for them to take several days to re-open when, for example, they are transferred from one aquarium to another. This species can be strongly recommended as a first coral for the beginner. The best specimens are attached to a small piece of rock. Make sure that the base is undamaged, with no decomposing white areas. Use a small-bore siphon to remove any sediment that accumulates within the 'mushroom'.

Top: Dendronephthya rubeola
The coloured Cauliflower Corals require a constant supply of fine food and good water movement to survive. Specimens grow rapidly under good conditions.

Above: Sarcophyton trocheliophorum
The mushroom-shaped Leather Coral expands its polyps to feed. With the right conditions, it will grow steadily.

Phylum: PLATYHELMINTHES

Flatworms

Pseudoceros splendidus
Red-rim Flatworm

This vivid red, black and white species is one of the few occasionally offered for sale. Although somewhat nocturnal in its habits, it usually does well if not subject to predation. The bright colours of this group of animals are believed to serve a protective function, and few fish will eat them because of their foul-tasting mucus. However, many crabs and shrimps will quickly devour, or badly damage, flatworms. There are many small and insignificant species, one of these – reddish brown and 3-4mm (0.12-0.16in) long – can reach plague proportions (see page 384). Always remove it from a living reef aquarium, as there is no predator that will eliminate it naturally without damaging other desirable invertebrates.

Above: Thysanozoon flavomaculatum
In the past, this attractive species was occasionally imported by accident among shipments of living rock.

Below: Pseudoceros splendidus
Most flatworm species are nocturnal, but Pseudoceros splendidus *appears during the day, when its coloration, warning of a foul taste, deters predators.*

Phylum: ANNELIDA

Segmented Worms

SABELLID SPECIES (FANWORMS)

Sabellastarte magnifica; S. sanctijosephi
Fanworm; Featherduster Worm; Tubeworm

Among the huge numbers of featherduster worms and tubeworms imported each year from Singapore, Sri Lanka and Indonesia, these particular species are justifiably popular. The body of the worm is encased in a parchment tube buried in the substrate, with the feathery head extended for feeding. At the approach of danger, the feathery tentacles are very rapidly withdrawn into the tube. They are not fussy about their lighting requirements and are easily satisfied with brineshrimp *nauplii* and rotifers.

It is by no means unusual for these species to reproduce in the home aquarium. When this is about to happen, the first signs are usually evident early in the morning, when the animals emit smoky plumes of either eggs or sperm. The adult worms then often shed their feathery heads, which prevents them eating the larvae that develop very shortly after the fertilization process is complete.

Tubeworms also shed their 'feathers' if attacked by predators, as well as in response to poor water conditions or the shock of being moved. If this happens, remove the head but leave the tube in position, and after two or three weeks a short, stubby feathered head will normally reappear and eventually grow back to its former glory.

Below: Sabellastarte magnifica
Featherduster worms are imported in huge numbers, and in a wide variety of colours, from Sri Lanka, Singapore and Indonesia, and are ideal for newcomers to the marine aquarium hobby. The cilia, or 'feathers', clearly visible here, are used to trap fine particles of food, which are then channelled to the creature's central mouth.

SERPULID COLONIES (TUBEWORMS)

Serpulid worms differ from Sabellids in that they produce a stony tube and are usually considerably smaller. The most commonly available species lives in colonies, with its tubes embedded in *Porites* coral. The 1cm (0.4in) diameter worms are often very brightly coloured, with heads of red, blue, black, white and yellow – in contrast to the beige through brown to maroon heads of *Sabellastarte*. *Spirobranchus giganteus*, the Christmas Tree Worm, is common in the Caribbean and the Indo-Pacific. It is so-named on account of the two branches of spiralling tentacles that emerge from the tube. Featherduster clusters from the Caribbean are small intertwined clumps of rocky tubed Serpulid Worms, usually with red- or rust-coloured heads. Take care that these species do not become overgrown with algae.

Above: Serpulid sp.
This large Serpulid species is a fairly demanding variety and easily damaged.

Below: Serpulid sp.
This small but attractive Caribbean species lives in small aggregations.

Phylum: CHELICERATA

Horseshoe Crabs

Limulus polyphemus
Horseshoe Crab; King Crab

The Horseshoe Crab is not in fact a crab, but is more closely related to spiders, scorpions and mites than to crabs. The few living species are the only representatives of a very ancient lineage. *Limulus polyphemus* is the largest species and can reach 50cm (20in) in size. It is found on mud and sand flats on the east coast of America. Small specimens are available on the American market, but in Europe one of the three western Pacific species is more likely to be on offer. These do not grow as large and make entertaining, if somewhat clumsy, aquarium specimens. They spend a large part of their time burrowing under the substrate and can play a valuable role in keeping the sand loose in undergravel filter systems.

Worms, algae and small shellfish form the natural diet of Horseshoe Crabs. They are unlikely to find sufficient food if left to scavenge, so provide small particles of squid, cockles and shrimps.

Above: Limulus polyphemus
The Horseshoe Crab is rather a clumsy aquarium occupant and will topple delicate corals quite easily. It may be better to keep it in a species tank, where it can be given an adequate diet, and provided with room to burrow in a deep, sandy substrate, which is essential to its well-being.

Phylum: CRUSTACEA

Barnacles, Crabs, Lobsters and Shrimps

BARNACLES

Lepas anserifera
Gooseneck Barnacles

Lepas species are occasionally available, but require large quantities of fine food, which can put a strain on the filtration systems of most marine aquariums. Small conical barnacles sometimes grow up spontaneously, or may be introduced on living corals.

Left: Lepas anserifera
A small colony of Gooseneck Barnacles clustered on a piece of submerged driftwood. Although common in the tropics, they are rarely seen for sale.

Above: Lepas anserifera
Barnacles are fascinating creatures to observe, using their delicate feet to capture small, drifting particles of food and wafting them into their mouths.

CRABS

Calappa flammea
Shame-faced Crab

The Shame-faced Crab is one of the few typical crab-shaped crustaceans that you might consider for the home aquarium. A number of similar species are characterized by their over-developed but weak claws. These are normally held in front of the mouthparts with just the eyes peeping over – hence their common name.

Calappa are very efficient scavengers and will also break open and eat various types of molluscs. They are well camouflaged with algal growth on the carapace, or

Above: Calappa flammea
The Shame-faced Crab is not a retiring creature, but a powerful omnivore, capable of devouring many sessile invertebrates in its quest for food.

they may spend much of their time buried beneath the substrate – a habit that makes them interesting aquarium inhabitants.

Dardanus megistos
Red Hermit Crab

Hermit crabs vary quite considerably in size and colour, from the blue-legged hermits from Singapore and the tiny thumbnail-sized species commonly shipped from the Caribbean to the giant *Aniculus maximus,* which has attractive golden yellow legs but is a fearsome predator that will devour anything that comes within reach of its powerful claws. *D. megistos* is one of the largest species, and fist-sized specimens are by no means uncommon.

Unlike most crabs, the hermit crab's abdomen extends out from the body, with no hard protective shell on the rear. It protects itself by taking over the shells of various univalve molluscs – often by eating the previous and rightful owner. Despite the weight of some of these shells, hermit crabs are very active climbers and their inquisitive nature endears them to many hobbyists. However, they have very catholic tastes and a large specimen is capable of causing considerable damage within a well-stocked living-reef aquarium. They are useful scavengers, particularly in tanks with sizeable fishes, but will rarely fit into the average aquarium set-up.

Lybia tessellata
Boxing Crab

The small Boxing Crabs rarely grow more than 3cm (1.2in) long and make ideal aquarium occupants, particularly for small tanks, where they will not get 'lost in the crowd'.

Boxing Crabs are the only examples of invertebrates known to use tools. While the Anemone Hermit Crab, *Pagurus prideauxi,* merely shelters beneath anemones, these small crabs collect a tiny anemone in each claw and actively wave them at encroaching predators as a warning. Furthermore, even though these crabs use their first pair of walking legs to search the substrate detritus for food, they will happily collect food from the anemones. Only when they change

Above: Dardanus megistos
Given its size and large appetite, this crab is better suited to a species tank.

Below: Lybia tessellata
An attractive crab for a smaller tank with suitable benign companions.

their exoskeleton will *Lybia* deliberately release the anemones and carefully set them aside until the new shell hardens. Then they pick them up and press them into service once more.

There are several species of *Lybia,* but the nomenclature is in some confusion. All are attractive and very interesting, well worth a place in the tank. Small pieces of meaty marine fare are readily accepted.

Neopetrolisthes ohshimai
Anemone Crab

N. ohshimai is one of a small group of porcelain crabs that have evolved an immunity to anemone stings and, like *Amphiprion* clownfishes, can live among the tentacles of their venomous hosts for protection from predators. Measuring barely 2.5cm (1in) across the carapace, they are among the smallest crabs and make ideal aquarium specimens.

The crabs live in the same types of anemones as clownfishes and will use their well-developed claws on any clownfish that tries to evict them from their chosen home. For feeding purposes, however, the crab uses feathery projections on its jaw processes to trap particulate matter. As they are so small, Anemone Crabs are particularly vulnerable when changing their shells, so provide plenty of hiding places and do not house them with larger, more aggressive crustaceans.

The Indo-Pacific *N. ohshimai* has an irregularly spotted pattern that

Above: Neopetrolisthes ohshimai
An Anemone Crab in a Heteractis *anemone. This species will feel more secure housed in a fairly small tank.*

can vary depending on where it was found. *Neopetrolisthes maculatus* is densely covered with small chocolate spots on a white background, which give it an overall pinkish appearance. The very rare *N. alobatus* from East Africa has widely spaced, almost circular, dark brown spots of varying sizes, producing a polka-dot effect.

Pagurus prideauxi
Anemone Hermit Crab

Several species of *Pagurus* hermit crabs have gone one step better in their search for protection by actively encouraging certain species of stinging sea anemones to colonize their shells and ward off predators. Like all crustaceans, they periodically shed their hard outer skeleton, so be sure to include a few spare shells among the tank's decorations to provide new homes for the growing crabs.

Pagurus not only find new and suitably sized shells into which they can rapidly slip their delicate abdomens, but they also tease their anemones off the old shell and deliberately replace them on the new one. The anemones appear to accept this willingly, because as the crabs rip up their food, many small pieces drift away and into the anemones' tentacles. *Pagurus* are large, destructive crabs and will only suit the aquarist looking for a 'one-off' speciality animal.

Below: Pagurus prideauxi
This crab protects itself by encouraging anemones to colonize its shell.

Stenorhynchus seticornis
Arrow Crab

The Arrow Crab gets its common name from the distinctly triangular, arrowhead-shaped body. This feature, together with its very long thin legs, results in a creature too closely reminiscent of a spider to appeal to many tastes. This is regrettable, as Arrow Crabs are attractive, easily maintained aquarium subjects.

S. seticornis has a leg span of about 15cm (6in) and is generally well behaved, although it may pull at featherduster worms, as small, burrowing worms form a major part of its natural diet. In captivity, it will eat any meaty food and has the added advantage of being one of the few creatures that will happily consume the carnivorous, scavenging Bristleworms *(Hermodice carunculata)*, which are sometimes accidentally introduced into the aquarium.

Unless you have a very large aquarium, it is not a good idea to keep two Arrow Crabs together, as they almost invariably fight – the loser having all its legs removed before being eaten.

Below: Stenorhynchus seticornis
Within this limited family of spiderlike crabs, the Caribbean species shown here is the most readily available. Do not worry if a specimen loses a limb during transportation; it will quickly regrow. The Arrow Crab is not a swift mover, so you will have the opportunity to study it at leisure.

LOBSTERS

Enoplometopus occidentalis
Red Dwarf Lobster

This vivid red species is the most attractive of the various lobster species and, with its relatively large claws, looks like the typical

Left: Enoplometopus occidentalis
Naturally nocturnal, but usually learns to scavenge in daylight hours.

Below: Panulirus versicolor
A very popular species – but it has a healthy appetite and rapid growth rate.

fishmonger's lobster. The Red Dwarf Lobster grows to a length of about 12cm (4.7in) and looks very dramatic, but its largely nocturnal habits mean that you will generally catch only the occasional fleeting glimpse of your specimen.

Enoplometopus is highly territorial and quickly despatches any similar species and many of the more commonly available shrimps. It is also quite capable of catching and killing small fishes, particularly when these are 'dozing' at night. Think carefully before introducing *Enoplometopus* into your aquarium, since removing a particularly aggressive specimen at a later date

may involve you in a complete strip down of the tank.

The similar, but rarer, pacific *E. holthuisi* is slimmer and can be easily distinguished by a white bullseye-like mark on either side of the thorax. Two rather purplish pink species are imported from the Indo-Pacific region and both are smaller than *E. occidentalis*. *E. debelius* has a pale pink body, liberally covered with almost round, purplish red spots. *E. daumi* has a pale purple-brown body, becoming more richly purple towards the head and claws. These two species are even more shy and retiring than their red relative shown left.

Panulirus versicolor
Purple Spiny Lobster

This attractive purple and white banded species is the most attractive member of a large and commercially valuable family of animals and a worthy addition to the aquarium. There are major fisheries of its relatives in both the Caribbean and Mediterranean.

Very young specimens of *P. versicolor* are commonly imported from Singapore and Indonesia. They have a body length

of 5-7cm (2-3in), the long, rasplike antennae adding a further 10-15cm (4-6in). An adult body length of 20cm (8in) is by no means unusual. These efficient scavengers thrive in captivity on a diet of frozen fish and shrimp. Although not deliberately destructive, they can cause damage with their sharp feet, or by jerking backwards to evade a threat, either real or imagined.

Several other species of *Panulirus* are occasionally available, usually in shades of reddish brown, but these grow even larger and are suitable

only for a very large aquarium. Like the purple species, their long antennae are easily broken in confined spaces and, although they will grow with successive shell changes, the animal loses much of its appeal to the aquarist if these appendages are broken.

The Slipper Lobster, *Scyllarides nodifer,* is a close relative, but instead of antennae this species has well-developed plates around the head to dig through the substrate seeking food. It is dull-coloured and of more interest to the specialist.

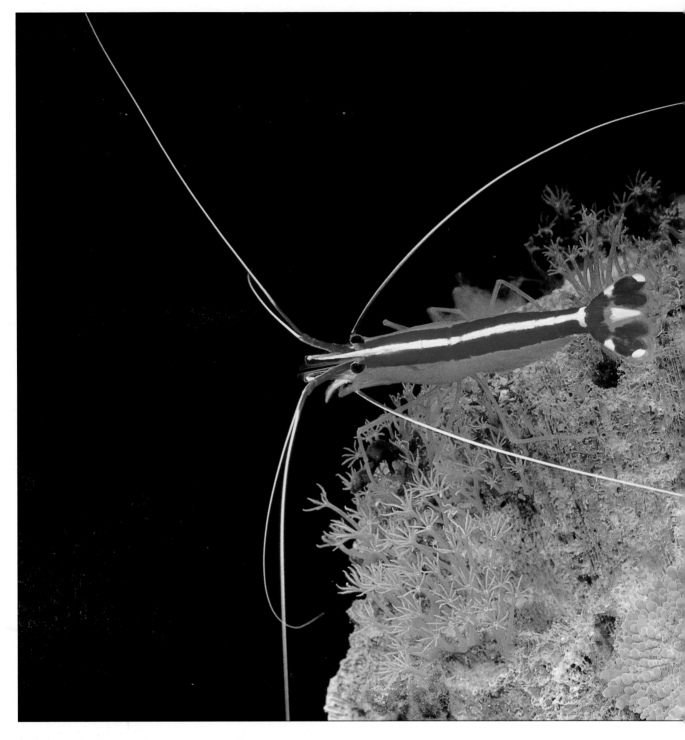

SHRIMPS

Lysmata amboinensis
Cleaner Shrimp

The common name for these very attractive and sociable shrimps derives from their natural cleaning behaviour. In the wild, on Indo-Pacific reefs, they will pick parasites, damaged skin, etc., from many species of fishes, particularly moray eels and large groupers. The fish clearly appreciate these attentions and only rarely eat what would otherwise seem an attractive morsel.

Lysmata lack the dramatic claws of *Stenopus,* but many are attractively coloured. The common Indo-Pacific *L. amboinensis* has a scarlet back with a white stripe running from between the eyes to the base of the tail. The tail is marked with three white patches. The very similar but, in Europe, much rarer Caribbean *L. grabhami*

Above: Lysmata amboinensis
It is easy to see why the Cleaner Shrimp is a favourite with aquarists.

has a white stripe running to the tip of the tail, which is edged, rather than blotched, in white.

Both species reach about 8cm (3.2in) in body length, and make very desirable aquarium inhabitants, being long-lived and easy to care for. They are particularly attractive when kept together in groups of

Lysmata debelius
Blood Shrimp

This intensely red and white spotted species caused a considerable stir within the hobby when it was first introduced in the early 1980s. Early specimens were all imported from Sri Lanka, but they have also been found in Indonesia and now, although this is still one of the more expensive shrimps, it is easily obtainable.

L. debelius is somewhat shyer than *L. amboinensis,* but after a few days adjusting to the aquarium conditions – and with the reassurance that there are plenty of convenient boltholes available – it will soon settle down and do very well.

Several other *Lysmata* species occasionally appear for sale. Most are either banded or striped in combinations of white, pink, beige and red. *L. rathbunae* comes from the Caribbean, while *Lysmata californica* is found on the west coast of the USA. *L. vittata* is an almost transparent Pacific species, while the blotched pink and white *L. kukenthali* comes from the Pacific Ocean. A Mediterranean species, *L. seticaudata,* and *L. wurdemanni* from the western Atlantic share similar red and white longitudinal markings.

Below: Lysmata debelius
The aptly named Blood Shrimp is easy to maintain and long lived.

four or five, when they may perform group cleaning activities on fishes.

Both species regularly breed in the aquarium, the females developing large quantities of green eggs under the abdomen. Unfortunately, the newly hatched shrimps usually provide a welcome addition to the menu of the tank's other inhabitants, or are quickly swept into the filter system. Raising these shrimps to adulthood in captivity is rare and difficult.

Periclimenes brevicarpalis
Anemone Shrimp

P. brevicarpalis is the commonest species in a family of very interesting and easily maintained shrimps. They have acquired their common name from their habit of living among the stinging tentacles of sea anemones for protection against predators, rather like

clownfishes and Anemone Crabs.
These species are rarely more than 2.5cm (1in) long and are often almost totally transparent, which makes them very difficult to see if you are unaware of their presence. *P. brevicarpalis* is transparent with white blotches. It occurs throughout the Indo-Pacific region, where it often shares anemones with clownfishes.

The very delicately formed *P. holthuisi* from the Indo-Pacific, and the Caribbean *P. pedersoni*, are glasslike with fine purple markings. *P. imperator* is found in the Red Sea and Indian Ocean and is very variable in colour. Bright red specimens have been found living among the gill tufts of large, similarly coloured sea slugs, such as *Hexabranchus imperialis*, the

Spanish Dancer (see page 347). All Anemone Shrimps will accept almost any small food items and will thrive in captivity. Do take care not to put these creatures at risk from larger predatory species, however; most crabs and shrimps can stalk unharmed through anemone tentacles and an Anemone Shrimp would make a welcome addition to their diet.

Left: Periclimenes brevicarpalis
This glasslike Anemone Shrimp and its relative P. pedersoni *are delicate species that do not appreciate boisterous neighbours, though they are fairly easy to maintain in captivity.*

Above: Rhynchocinetes uritai
Candy Shrimps are common in the wild, frequently imported and inexpensive. They make interesting aquarium subjects and, as they are quite hardy, are ideal for newcomers to the hobby.

Rhynchocinetes uritai
Candy Shrimp; Dancing Shrimp

This group of shrimps is distinguished by having a movable rostrum (head spine) and very protuberant eyes. There are many attractive species, but *R. uritai* is the only species regularly available. It is found throughout the Indian Ocean and Indo-Pacific regions and exported in large numbers, making them among the cheapest shrimps to buy. They do best in small groups, when they will lose much of the shyness they display if kept singly. Males have large but ineffective claws, so in view of their lack of defensive armament, do not house them with larger, potentially more aggressive crustaceans. On occasion, they will pester corals and anemones, but are generally harmless. Candy Shrimps grow to about 3cm (1.2in) and are ideal specimens for beginners to keep.

Saron rectirostris
Monkey Shrimp

This interesting species and its relatives are comparatively recent introductions to the hobby, with specimens being shipped from Indonesia and Hawaii. They can be distinguished by the strong spines or hooks along the thorax and between the eyes.

S. rectirostris has purple legs and a pale body spotted with brown. S. inermis has greyish spots on the pale front portion, while the rear half of the body is more brown in colour. S. marmoratus is brownish green, with the body heavily coated with hairlike extensions. In all species, the males have a greatly elongated first pair of walking legs. In S. marmoratus these are attractively banded in reddish brown and buff.

They are good scavengers, taking any meaty food, but are essentially nocturnal and thus rarely seen during daylight hours.

Above: Saron marmoratus
All the Saron *species tend to be rather shy and difficult to see; here,* S. marmoratus *is camouflaged among the fronds of a cactus algae,* Halimeda *sp.*

Below: Stenopus hispidus
The brightly coloured and impressive Boxing Shrimp is the most commonly imported, and one of the most popular, shrimp species available to aquarists.

Stenopus hispidus
Boxing Shrimp; Coral-banded Shrimp

The attractive Boxing Shrimp is one of the most commonly imported and justifiably popular species available to hobbyists. *Stenopus* are bold animals and, once settled into a new environment, will rarely hide for long. A large specimen with its 6cm (2.4in)-body looks very impressive as its long claws and delicate white antennae wave in the current. Unless you can be sure of buying a compatible pair, house only one specimen in the tank. However, *S. hispidus* generally mixes quite safely and happily with other species of shrimp.

There are several species of *Stenopus* and as the hobby expands, so more appear on the market, albeit in very limited numbers. *Stenopus spinosus* is found in the Mediterranean and is essentially gold-yellow overall. The Caribbean is home to *S. scutellatus*, which has red-and-white claws and tail, but a yellow thorax. It can be distinguished from the smaller *S. zanzibaricus* by the two scarlet dots near the mouth. *S. tenuirostris* from the Indo-Pacific is another small, and lightly built, species with a purplish blue thorax.

The giant of the family is the rare and very expensive *Stenopus pyrsonotus* from Hawaii, which grows at least 50 percent longer than *S. hispidus*. *S. pyrsonotus* is white throughout, except for a striking, broad, red band down the length of the body. The only similar species is the very small (2-3cm/0.8-1.2in) *S. earlei*, which has a pale body with a red stripe along each side, forming a 'V' shape at the tail. The base of the claws is reddish.

Stenopus are particularly vulnerable when they shed their exoskeletons and thus need plenty of hiding places to enable them to evade predators while the new shell hardens. Legs, claws or antennae are often lost or damaged, but these quickly regrow. All *Stenopus* are omnivorous, taking almost any commercial foods.

Synalpheus sp.
Pistol Shrimp; Snapping Shrimp

Small specimens of Pistol Shrimps are frequently introduced to the marine aquarium by accident, along with pieces of living rock. They often occur in the water canals of sponges. Most are pale brown or green, but there are a few very attractive orange and red species. However, they are all confirmed recluses and of limited interest to most hobbyists.

Their common name comes from the pistol crack sound they are able to produce, a sound so similar to cracking glass that many a hobbyist has had a nasty shock! All Pistol Shrimps have one greatly enlarged claw, usually the right. By snapping this shut they produce a shock wave

Above: Synalpheus sp.
The greatly enlarged snapping claw of this Pistol Shrimp is clearly visible here as it rests on a Tridacna *clam. The reason for the attraction of this species – and for its common name – lies in the loud pistol crack sound that it is able to produce by snapping this huge claw (its victims are stunned by the shock waves when they come within range). Unfortunately, it is normally an extremely secretive species.*

through the surrounding water, stunning the small shrimps that make up a major part of their natural diet. With a maximum body length of about 5cm (2in), they pose little, if any, threat to the other inhabitants of the tank and should not be confused with Mantis Shrimps (see page 383).

Phylum: MOLLUSCA

Bivalves, Cephalopods, Sea Slugs and Univalves

BIVALVES

Lima scabra
Flame Scallop

This very attractive Caribbean species grows to about 6cm (2.4in) in diameter. The shell is unremarkable, but the body flesh is an intense scarlet. In the most popular of the two forms of *L. scabra,* the fringe of tentacles around the lip of the shell is also red, while in the other, the tentacles are off-white.

Position these scallops in a hollow or crevice towards the front of the tank where, hopefully, they will settle and attach themselves with wiry threads. Without an anchoring point, they may gravitate towards the back of the tank and be lost.

Flame Scallops are filter-feeders and have fairly heavy appetites. The commonest cause of death, other than predation, is long-term starvation due to insufficient supplementary feeding or an over-efficient filter system.

Flame Scallops are a very popular food item for many animals, but are able to escape predators by clapping their shells together and using the resultant force to jet through the water. The Indo-Pacific lookalike *Promantellum vigens* has another useful defence system. The scallop's usual tactic is to flee the scene of battle, but if this fails, *P. vigens* defends itself with its tentacles. These are very sticky and easily detached from the body, and thus prove an irritating deterrent to many potentially predatory fish.

Above: Lima scabra
The vivid colouring of the Flame Scallop explains its popularity. This Caribbean species has lookalike relatives throughout the tropics. Feed regularly with a good-quality liquid preparation.

Occasionally, Flame Scallops reproduce in the aquarium and small clusters of spats – miniature, almost transparent 0.5cm (0.2in) diameter scallops – are found in caves and under rocks.

An interesting Philippine species has recently appeared on the market. It is similar in size and coloration to the red form of *Lima scabra* and has rippling luminous lines just inside the shell that flick on and off. We do not yet know what benefit these lines confer on the so-called Flashing Scallop.

Spondylus americanus
Thorny Oyster

As its scientific name would suggest, this is a Caribbean species. Both valves of the shell have long, thornlike extensions and, when cleaned, the shell is very attractive and a favourite with collectors. When the animal is alive, the thorns are often heavily covered with growths of sponge, hydroids and algae, which camouflage the animal.

Spondylus aurantius is a very similar Indo-Pacific species with shorter thorns. This species is often coated with a vivid red sponge and, at 20cm (8in) in diameter, it is twice the size of *S. americanus.*

Both species are found in caves and under overhangs and thus do not appreciate bright lighting. Unfortunately, they appear short-lived in the aquarium and, because of the pressure from shell collectors, who pay high prices for good specimens, they are usually too expensive to appeal to hobbyists.

Very few other bivalves are deliberately added to the aquarium, but they are quite often introduced accidentally. Small specimens are often found growing on sea whips and sea fans, and many burrowing species are introduced with 'living-rock'. Some of the best Caribbean rock is heavily populated with burrowers and mussel species. Take great care to remove any air pockets from this type of rock, as tunnelling bivalves may otherwise die and cause very serious ammonia and nitrite pollution problems.

Tridacna spp.
Giant Clams

The various *Tridacna* species are probably the most popular bivalves, and justifiably so. The shell, particularly that of *T. crocea*, often shows deep flutes along the ridges, while the fleshy mantle of most species can be the most intense, almost fluorescent, blue and green. The largest species is *T. gigas,* the Giant Clam of Hollywood fame, which can reach a weight of over 100kg (220lbs), and is particularly

Above: Spondylus americanus
This species, one of the more expensive marine invertebrates, demands excellent water quality and an adequate supply of particulate food in the home aquarium.

Below: Tridacna crocea
Brightly coloured species such as this are justifiably popular. Usually seen at about 10cm(4in), they can reach twice this length, but are slow growing.

common on the Great Barrier Reef of Australia. Its shell is often covered with algae and coralline growths and the mantle is generally greenish brown. It has been heavily over-fished and is largely protected.

Tridacna maxima and *T. crocea* are of more interest to the hobbyist and are regularly imported from Indonesia and Singapore. *T. maxima* can reach 30cm (12in) in length, and some of the most attractive specimens have green and brown striped mantles, while others may show a chocolate and cream blotched

pattern. The most dramatically coloured blue specimens of *T. crocea* grow to around 15cm (6in).

All clams are filter feeders, drawing water through one siphon, filtering out planktonic organisms, and exhaling the cleaned water through the other. They are all heavily dependent on intense lighting. In fact, much of the colouring in the mantle is due to the zooxanthellae algae living in the tissue. These algae utilize sunlight for photosynthesis and produce the majority of the clam's food.

CEPHALOPODS

Nautilus macromphalus
Nautilus

For many years, Nautilus were thought to be extremely rare and to occur only in very deep waters. Recently, however, large numbers have been found and caught by research teams off Indo-Pacific reefs. It appears that they retreat to the depths, often hundreds of feet down during the day, but float up into shallower water at night and use their many tentacles to catch small fish and shrimps.

Very occasionally, the better wholesalers may obtain specimens of Nautilus, but the supply is so limited and the demand so great that their price is likely to remain beyond the reach of most hobbyists for the foreseeable future. Nonetheless, those specimens that have appeared have proved viable in the aquarium and, despite their essentially nocturnal habits, have many admirers.

Right: Nautilus macromphalus
Nautilus shells are readily available, but the live animal is rarely for sale.

Octopus cyaneus
Common Tropical Octopus

Octopi are the most advanced cephalopods and have lost all trace of their ancestral shells. Most species are less than 60cm (24in) in diameter – and none approach the horror-story dimensions prized by the early Hollywood film makers. *Octopus cyaneus* rarely reaches more than 30cm (12in) across and is typical of the small tropical species shipped in considerable numbers from the Far East. It is common on many reefs and easily captured by overturning stones under which it lurks at the water's edge at low tide. This and other species are able to control not only the colour of their skin, but also its texture. When at rest and relaxed, most octopi are fairly smooth skinned and their colour matches their background. When angered or frightened, they rapidly become much darker or lighter and their skin folds into eruptions resembling algal growth.

Like many octopus species, *O. cyaneus* is an ideal aquarium inhabitant for those prepared to make some effort on its behalf. Although not particularly light sensitive, all octopi demand perfect water conditions and will not tolerate any form of pollution. The aquarium must contain a number of suitable caves as hiding places and a tight-fitting lid, as octopi are notorious explorers, squeezing their boneless bodies through the smallest of openings. Many an octopus has met a dry and dusty end on the carpet, thanks to its owner's carelessness. Suitable tankmates include corals, sponges, featherduster worms, and some echinoderms. Crustaceans and fish, unless intended as food, have no place in the octopus tank.

The octopus should be the last introduction to the tank, as it resents further disturbance. Allow it to become used to the dark, leaving the tank unlit for a day after the animal's release. Take the greatest care during this period; octopi are sensitive creatures and if badly upset will eject sepia ink into the water. This natural defence mechanism can cause major problems in the tank, and if the animal 'inks' in its shipping container it is likely to die.

A happy octopus is a greedy feeder, and laboratory tests have shown that it is quick to learn how to obtain food – even unscrewing bottle tops to get at the food within. Do not overfeed these ever-hungry creatures; they can generate more wastes than the average filter system is capable of dealing with in an acceptably short period. A regular feed of one or two shrimps or small frozen fish per day will comfortably satisfy all but the largest species.

It is not a good idea to house two octopi in any but very large aquariums, as they will usually fight. The female octopus is fertilized internally and occasionally a gravid specimen will produce fertile eggs in the aquarium. She may hang these from the roof of a cave or carry them about with her. The female does not normally feed during the incubation period and generally dies shortly after the eggs hatch, producing miniatures of their parents. The small hatchlings can be kept together and commercial producers in America are now supplying their country's growing number of public and educational aquariums. Breeding and rearing is still a great challenge for the specialist hobbyist and results should be more common as techniques are refined.

Above: Octopus cyaneus
The powerful sucker-lined arms are clearly visible on this Common Octopus.

Below: Sepia plangon
Cuttlefish are efficient hunters, not to be trusted with fish or crustaceans.

Sepia plangon
Cuttlefish

The brittle supporting blade inside cuttlefish of the *Sepia* genus is very familiar to birdkeepers. Unfortunately, cuttlefishbone is the closest most of us will ever come to having a cuttlefish, although *S. plangon* is a common species in the tropical Indo-Pacific, while *S. officinalis* occurs throughout the northeastern Atlantic and the Mediterranean Sea. The latter is a very common and popular animal in European public aquariums, where it can enjoy the large tanks that these very active animals require.

Unlike their relative, the octopus, squids and cuttlefish are fast-swimming and very active hunters. Like an octopus, they have eight arms around the mouth, but they also have two longer and rapidly extendable arms to catch prey.

Very little is known about the behaviour of the vast majority of squid and cuttlefish species because they are difficult to catch undamaged and many live in very deep water. Those few species that have been studied display a variety of interesting behaviour patterns. Many are capable of very rapid colour changes and, in the case of *Sepia*, waves of colour wash over the body when they are excited or agitated. Many squid species show luminescent patches and it is believed that some species communicate with each other with a system of colour codes.

Very occasionally, small species, such as *Loliguncula brevis,* are imported from the Gulf of Mexico. This has very large pigment cells (chromatophores) that produce a kaleidoscope of reds and black. Given plenty of swimming room, it might prove viable in the home aquarium. All squid and cuttlefish require optimum water conditions and are particularly sensitive to low oxygen levels. Most will not tolerate salinities lower than those of their native waters. Being extremely predatory, cuttlefish can be trusted only in a species aquarium, or with sessile invertebrates.

SEA SLUGS

The sea slugs are univalve molluscs that have lost the valve (shell) and have become more mobile. Some of the more primitive sea slugs still have a very tiny, residual shell, but they cannot withdraw into it.

Below: Aplysia dactylomela
The Sea Hare is one of the easiest nudibranchs to maintain, given large quantities of vegetable matter.

Bottom: Chromidoris quadricolor
The bright colours of these common nudibranchs probably serve to warn off predators. Note the retractable tentacles, characteristic of dorids.

Aplysia sp.
Sea Hare

The Sea Hares are an entertaining, if not particularly attractive, group of sea slugs. Typically, they resemble a greenish brown lemon, with continually waving flaps along each side. They have earlike projections on the head, and it is these and their habit of grazing on seaweed that gives them their common name.

Most specimens offered for sale are shipped from the Caribbean, but very similar species are found throughout the world. They are by far the easiest type of sea slug to maintain for any length of time, although the first few days in a new aquarium can be a testing period. They require good water conditions and a continual supply of vegetable matter, preferably in the form of algae. If these few necessities are met, they can survive in captivity for two years or more. Since the invertebrate aquarium occasionally suffers from a plague of green algae, Sea Hares can provide one of the most useful and harmless answers to this problem.

The common Caribbean Sea Hare, *A. dactylomela,* can grow to over 30cm (12in), although smaller specimens of 6-8cm (2.4-3.2in) are most commonly seen. It is a strong swimmer, making good use of the parapodial flaps around its body.

Chromodoris quadricolor
Striped Nudibranch

The dorids are the largest group of nudibranchs and include some of the most vividly coloured animals found in the sea. *C. quadricolor* though striking in its black, white and orange livery, is by no means exceptional. Members of this group typically have two retractable tentacles on the head and a ring of gills, again retractable, towards the rear. Specimens rarely grow more than 6cm (2.4in) long.

Although commonly available and among the cheapest of sea slugs, their life expectancy in captivity is short. Again, the problem is to provide the correct diet. All are predatory, feeding on a wide range of sessile invertebrates, from sponges to barnacles, sea squirts to soft corals, and most seem limited to just one prey species. In the Red Sea, *C. quadricolor* feeds on the red sponge, *Latrunculia.* Until more is known of the requirements of *Chromodoris* species and a suitable alternative diet is available, they are best left in the wild.

Other available dorid species include *Gymnodoris ceylonica,* which lays strings of yellow eggs; *Polycera capensis,* a sea squirt feeder; the green, yellow and black *Tambja affinis,* and the sponge-feeding, white *Casella atromarginata.*

Hexabranchus imperialis
Spanish Dancer

The Spanish Dancer is one of the largest and most dramatically coloured nudibranchs available. It is imported in small but regular quantities from all parts of the Indo-Pacific, in sizes ranging from 6-15cm (2.4-6in). When at rest, or browsing over rocks, the mantle of the Spanish Dancer is folded and marbled red, pink and white in colour. It is not until it swims that its full glory is revealed. The mantle unfolds to reveal an expanse of vivid crimson with a white border and then, with an action like a butterfly-stroke swimmer, it 'flies' through the water, earning its name.

The similar, and even more dramatically scarlet *Hexabranchus sanguineus* can be found on reefs in the Red Sea, but it is some years since this species was available commercially. Both species seem almost immune to fish attack and one small shrimp, *Periclimenes imperator,* takes advantage of this to hide among the gill tufts of *H. sanguineus.*

Spanish Dancers will often lay eggs in captivity, producing a 3-4cm (1.2-1.6in) diameter rosette of pink gelatinous ribbon containing thousands of eggs. Unfortunately, neither these, nor the adult animal, generally succeed in the aquarium. The adults are reputed to be omnivorous scavengers, but from their limited survival rate it seems likely that a significant ingredient is missing from their diet.
H. sanguineus is said to feed on sponges and sea squirts and has been observed feeding on the Elephant Ear Coral, *Sarcophyton trocheliophorum.* These species are not really suited to aquarium life.

Spurilla sp.
Spiny Nudibranch

Spurilla is one of a large group of nudibranchs known as aeolids, most commonly found in cooler waters, although there are some tropical species. Most are fairly small, measuring up to 3cm (1.2in) and, although common as accidental introductions, are rarely available commercially. Small grey species similar to *Spurilla* are common on *Goniopora* coral, but generally go unnoticed. As accidentals they usually fare better than the dorids, as they are usually introduced on their food animals.

This group characteristically has two long tentacles on the head and numerous spikelike protuberances on the back. They are often brightly coloured, but without the striping common in dorids. They generally feed on anemones and corals, but some eat molluscs and fish eggs.

One particularly interesting species is *Glaucus atlanticus.* This blue-grey species is oceanic, living at the surface where it hunts various floating coelenterates, including the notorious Man-of-War Jellyfish. Not only is it immune to the jellyfish's stings, but it can store them within its own body to deter predators, which might eat the otherwise defenceless *Glaucus.*

There are many thousands of species of sea slugs and many occasionally appear on the market. Among the most common are the 'warty slugs', which are often covered in pimples, but show no external gills or tentacles. The duller green-brown species are generally longer lived than the more brightly coloured types.

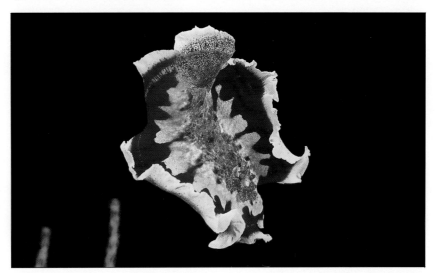

Above: Hexabranchus imperialis
The spectacular Spanish Dancer is, sadly, not long-lived in the aquarium.

Below: Pteraeolidia ianthina
This attractive blue species is one of the many aeolid nudibranchs.

UNIVALVES

Cypraea tigris
Tiger Cowrie

The Tiger Cowrie is one of a large group of sea snails of interest not only to aquarists, but also to shell collectors. Some cowries are extremely rare and command very high prices among collectors, but the Tiger Cowrie is not one of these. All the cowries have characteristic highly polished oval and domed shells, many of which sport very decorative patterns. This colouring is normally well concealed by the fleshy mantle – an extension of the body that folds up and over the shell. The mantle is usually decorated with tufts and tassles and is inconspicuously coloured, providing good camouflage.

Tiger Cowries are commonly imported from Singapore at about 7.5cm (3in) long and are among the easiest invertebrates to keep. They feed avidly on *Caulerpa* species of seaweed and graze the less decorative hairy algae, but must have some animal matter; small pieces of fish and shellfish meat are suitable. A mature environment and good aeration are essential.

Cowries have two major drawbacks; firstly, many are primarily nocturnal and hide under rocks during the day, and secondly, they are somewhat clumsy. They have a powerful foot that is not

Above: Cypraea tigris
The spotted shell of the Tiger Cowrie is largely covered by the fleshy mantle, an extension of its foot.

easily dislodged and often tumble corals and other sessile invertebrates from their allocated position in the aquarium. Additionally, some species and specimens develop a taste for both hard and soft corals.

Among the other regularly available species are *C. arabica* from the Indo-Pacific, a slightly smaller species with a netlike shell pattern; *C. pantherina,* which is somewhat similar to *C. tigris,* and *C. nucleus,* characterized by the furry appearance of its mantle.

Lambis lambis
Spider Shell

A number of Indo-Pacific species are sold under the name Spider, or Millipede, Shell. All are characterized by the five or more long, thin and often sharp extensions to the shell, and all species have a very long and horny foot with which they are able to right themselves if they are inadvertently upturned.

Various other species and families of univalves appear on the market from time to time, but few are of interest to aquarists. Moon shells, olives, murex, tulips and whelks are all predators and have no place in a 'living-reef' set-up. Top shells, limpets and chitons are largely vegetarian and can be treated like cowries and conchs.

Above: Lambis lambis
The attractive underside of this Spider Shell (several species are sold under this common name). The top surface is normally camouflaged with algal growth – effective protection in the wild.

Below: Strombus gigas
When disturbed, S. gigas retracts into its shell, as shown here. The heavy armour is proof against most would-be predators. This large mollusc is likely to appeal most to the specialist.

Strombus gigas
Queen Conch

The Queen Conch is a very important commercial food animal in its native habitats and huge numbers of dead shells are exported for decoration and the curio market. It is one of the largest molluscs available to the hobbyist, reaching 25cm (10in) or more, and is thus likely to be of greater interest to the specialist. At one time, strictly enforced regulations put a limitation on the size at which *S. gigas* could be taken from the sea and no small specimens were available to the hobbyist. In recent years, however, there has been a huge increase in breeding and farming this species, and small specimens, up to 5cm (2in), are now obtainable on the American market and it cannot be long before they are more widely available to hobbyists.

This species is found on sand and sea-grass fields throughout the Caribbean, where it eats algae and detritus. Its attractive brown and pink shell should prove popular with aquarists, but make due allowance for the size of the fully grown animal; it will require an aquarium of some considerable size in which to roam about unrestricted.

Phylum: ECHINODERMATA

Basket and Brittle Stars, Sea Cucumbers, Sea Lilies, Sea Urchins and Starfishes

Above: Astrophyton muricatum
The delicate tracery of the Basket Star's arms is well illustrated here.

Below: Ophiomastix venosa
The small central disc and five slender arms are typical of this group.

BASKET AND BRITTLE STARS

Astrophyton muricatum
Basket Starfish

Most imports of the fascinating Basket Star come from the Caribbean, but similar species are found in other oceans. They are very efficient feeders and require a good supply of food to thrive. They also need plenty of room to spread their 50cm (20in)-long arms, so they are clearly not animals for small tanks.

Given suitable conditions, these are long-lived creatures in the aquarium, but their strictly nocturnal habits limit their popularity to all but the most dedicated hobbyists. Their colours are generally restricted to greys, beige, browns and combinations of these. Provide a suitable liquid feed only when the arms are fully extended; nearly always at night.

Ophiomastix venosa
Brittle Star

Brittle Stars are regularly imported from the tropics and are not expensive, but because they spend most of their time hidden from view, they are not particularly popular. However, they are valuable scavengers, particularly in living reef tanks, which often have many crevices that trap excess food. Furthermore, they will not harm sessile invertebrates, such as corals and featherduster worms.

Brittle Stars have an efficient sense of smell to detect food and a surprising turn of speed. If a promise of food awaits them, they will sprint from one end of the aquarium to the other in seconds.

Ophiomyxa flaccida is a typical smooth-armed Caribbean species that reaches approximately 15cm (6in) in diameter. This species is often seen entangled among the spines of *Diadema* sea urchins, where it is comparatively safe.

SEA CUCUMBERS

Cucumaria miniata
Feather Cucumber

Feather Cucumbers may be among the smallest members of their family, but they are some of the most attractive. There are several similar species, but the most common one, from Singapore, has a bright pink body and five longitudinal rows of bright yellow tube feet. In this species, the mouth tentacles are developed into feathery growths that the animal uses to trap planktonic organisms and other food items drifting in the water. Feed sea cucumbers with live brineshrimp and rotifers.

This and the following species are among the few that may successfully reproduce in a domestic aquarium. Where two or more of the 6cm

Above: Cucumaria miniata
Yellow specimens of this small Sea Cucumber are less often seen than the pink form, which is regularly imported. Both species do much better in an 'established' living reef aquarium. That is to say, one that is at least six months old, if not more.

(2.4in)-long adults are present you may find small clusters of miniatures adhering to the sides of the tank.

Pseudocolochirus axiologus
Sea Apple

This very attractive species of sea cucumber has been given the somewhat confusing common name of 'Sea Apple' to differentiate it from all other cucumbers. It is by far the most strikingly coloured and popular species in the family.

The Indonesian form of *P. axiologus* typically has an ovate, greyish pink body up to 10cm (4in) long, with rows of tube feet defined in pink, orange or yellow. The head of the animal is crowned with a ring of feathery tentacles, which it uses for filter feeding. As the animal feeds, it pushes each tentacle in turn lugubriously into its mouth in the centre of the ring, rather like a child sucking toffee from its fingers. These tentacles vary in colour from pale yellow through to crimson. Do not keep Sea Apples with any fish species that might peck at the feathery tentacles.

P. axiologus is an easy species to maintain in the aquarium, provided it is given plenty of fine food. All too often, however, they slowly starve, becoming progressively smaller until, at about 3cm (1.2in) long, they give up the fight. Feed live brineshrimp and rotifers several times daily.

There is an even more striking giant form of *P. axiologus* from the Great Barrier Reef of Australia. Its body may be 15cm (6in) or more long, and is usually a rich purple colour with tube feet outlined in scarlet. The tentacles are purple and pure white. This desirable form is too expensive for many aquarists.

Stichopus chloronotus
Black Cucumber

Sea cucumbers are a large, varied and largely unprepossessing family within the echinoderms. Looking like a dark, shrivelled and discarded cucumber, *S. chloronotus* is a fairly typical example of the group. It is found on coral rubble throughout the Indo-Pacific region where, like many of its relations, it swallows mouthfuls of gravel and detritus, digesting any organic material and ejecting the residue from its rear.

This species is commonly available from Sri Lanka and, although very hardy, is not sufficiently attractive to appeal to most hobbyists. Nonetheless, it is a useful scavenger, particularly in an aquarium where organic material may begin to accumulate in a thin layer of unfiltered substrate.

Top: Pseudocolochirus axiologus
The Sea Apple has been a mainstay of the aquarium hobby for many years. The red tentacles are efficient traps for particles of food.

Above: Stichopus chloronotus
The Black Cucumber is one of the commonest Indo-Pacific species, though not one of the most popular. It is sometimes host to small fishes, which seek refuge in the cloacal chamber.

SEA LILIES

Himerometra robustipinna
Red Crinoid; Feather Starfish

H. robustipinna, from Singapore and Indonesia, reaches about 18cm (7in) in diameter and is one of the most attractive species.
Lamprometra palmata is one of several common brownish species that achieves a similar size, while others are occasionally available in shades from yellow through orange to black. All these species require very careful acclimatization to the aquarium, as they react badly to rapid changes in salinity and pH. Do not house them with large, boisterous or 'pecky' fishes.

Crinoids feed mainly at night, climbing to a high point on the reef and then extending their arms to trap small particles of food falling from the surface of the sea. Nevertheless, in captivity they are also very decorative during daylight hours when they are unable to retreat to the cavities they would normally seek out in the wild.

Remember that these animals are very brittle. If they are caught by a strong water current or attacked by other animals, one or more arms is easily broken. In the wild, these will regenerate very quickly, but in the aquarium they require perfect water conditions and a great deal of suitable food, in the form of pulverized shrimp or fish, algal fragments and newly hatched brineshrimp, if they are to recover.

In the wild, many small starfishes, shrimps and gobies live a well-camouflaged existence among the arms of crinoids; these 'extra' species are infrequent, but welcome, bonuses in the tank.

Below: Himerometra robustipinna
This vivid red species, one of the most attractive sea lilies, is shown here with its arms extended for feeding.

SEA URCHINS

Diadema savignyi
Long-spined Sea Urchin

D. savignyi is just one member of a large genus found throughout tropical and subtropical seas. It may be uncomfortably familiar to holiday makers who have received painful wounds from treading on these animals. The spines are long, extremely sharp and, in some species, venomous. Despite this, they make good aquarium inhabitants, grazing over algae-covered rocks and surviving for a number of years. Unfortunately, their sharp spines can puncture and damage corals and sea anemones.

    D. savignyi is rather unusual in its genus in that the dark spines become lighter and banded at night. Most *Diadema* species have banded spines as juveniles, but black or dark brown spines when mature.

Echinometra mathaei
Common Urchin

As its common name suggests, this Indo-Pacific species is not only widespread in the wild, but frequently available to hobbyists. Although its spines are shorter and considerably blunter than those of *Diadema savignyi,* take care when handling it. If you are unlucky enough to be stung by *E. mathaei* or *D. savignyi,* bathe the affected area with very hot water to help neutralize the poison. The pain usually subsides in an hour or two.

    E. mathaei is easy to maintain in captivity and will accept a wide variety of food. Unfortunately, it seems to spend a great deal of time hidden behind, or under, rocks and is most active at night. Bear in mind that the tube feet provide the urchin with a very strong grip on the substrate, allowing it to go almost anywhere it pleases, and in doing so it can easily tumble rockwork.

    The shell of the Common Urchin can reach a diameter of about 10cm (4in), but most animals are about half this size and these make the better aquarium specimens.

Above: Diadema savignyi
The long needle-sharp spines typical of this family of urchins are clearly visible on this Indo-Pacific species. They can cause painful injuries; handle this species with care.

Below: Echinometra mathaei
This common species uses its spines and tube feet to amble over the coral rubble of the lagoon as it seeks out algae and detritus. A valuable scavenger in the aquarium and easy to feed.

Eucidaris tribuloides
Mine Urchin

This small species only reaches about 5cm (2in) in diameter and, although not particularly attractive, it is a hardy species. It spends most of the day hidden in crevices in rocks, only emerging to feed at night, and is included here to avoid confusion with the *Heterocentrotus* Pencil Urchins, below.

Left: Eucidaris tribuloides
The Mine Urchin is a common export from Florida. Unfortunately, its nocturnal habits have restricted its popularity within the hobby, though it is a relatively hardy species.

Left: Heterocentrotus mammilatus
The distinctive heavy, thick spines would seem to be rather an encumbrance to the Pencil Urchin as it roams across a Hawaiian reef. This species is expensive but fairly easy to maintain.

Heterocentrotus mammillatus
Pencil Urchin

The Pencil Urchins of the genus *Heterocentrotus* are generally considered the most attractive and desirable sea urchins but, sadly, they are neither as common in the wild nor as regularly imported as other species. This ensures that they command a price roughly twice that of most others. In *Heterocentrotus mammillatus* – and the closely related *H. trigonarius* – the spines are reduced in number, but those that remain are greatly thickened and resemble a 5-7.5cm (2-3in) pencil stub. Indeed, you can use these spines instead of chalk to write on a slate or stone tablet as did the Ancient Egyptians. Unfortunately, Pencil Urchins are often commercially fished to provide component parts for the wind chimes for sale in gift shops.

Pencil Urchins make admirable aquarium inhabitants, but be sure to keep the pH level of the water at the high end of the range. They eat algae, lettuce and small particles of meaty foods.

STARFISHES

Choriaster granulatus
Giant Kenya Starfish

This species is probably the largest starfish available to the hobbyist on a regular basis. It can reach up to 30cm (12in) in diameter and is heavily built, with five thick fleshy arms. Unfortunately, this very attractive red and buff species is generally only imported when it has attained a size too great to appeal to the average hobbyist. This, and the proportionately high air freight costs, ensure that *C. granulatus* remains a species for the specialist.

The animal gets it scientific name from the small gill processes that protrude through the skin to give the arms a granular appearance. Most specimens are imported from Kenya, but it also occurs throughout the Indo-Pacific region, where it feeds on coral polyps and other immobile invertebrates.

Culcita novaeguinea
Bun Starfish

The adult of this large 20cm (8in), Indo-Pacific species has a completely different body form from that of the typical starfish. As a youngster, it shows the typical five-pointed star shape familiar to us all, but as it matures the arms broaden and thicken to produce an almost regular pentagon. The comparatively thick body makes this one of the heaviest starfish in the world.

At their smaller sizes, this and the related *C. schmideliana* make very attractive aquarium inhabitants but, unfortunately, they are not common and only rarely appear in retailers' shops. The dorsal surface is often attractively marked with a pattern of dark raised tubercles.

Bun Stars do best in lightly decorated tanks, as they will often wedge themselves among rocks and risk puncturing their skin, thus opening a route for bacterial infections. They are also somewhat clumsy animals and may dislodge and damage loosely positioned corals as they move around the tank.

Above: Choriaster granulatus
A giant starfish for the specialist.

Below: Culcita novaeguinea
Most Bun Stars have a beige-brown dorsal surface, but colourful undersides.

Fromia elegans
Red Starfish

This bright red species is another common import in shipments from Indonesia and the surrounding region. In view of the sensitivity of the water vascular system, it is important to avoid rapid changes in salinity when acclimatizing starfishes to the aquarium.

F. elegans is one of the smallest starfish species, reaching only 8cm (3.2in) in diameter. Juveniles have black tips to the arms, but these disappear as the animal matures. *F. elegans* requires a similar diet to *F. monilis* but is often eaten by more aggressive tankmates, such as Hermit Crabs and large starfishes.

Right: Fromia elegans
The Red Starfish is commonly imported and easy to keep provided you buy undamaged specimens and avoid more aggressive tankmates.

Below right: Fromia monilis
The Orange Starfish provides a welcome splash of colour among the greens and browns of a well-stocked 'living reef' aquarium. It is inexpensive, easy to maintain and justifiably popular.

Fromia monilis
Orange Starfish

This vivid orange and red species is one of the most popular and commonly imported starfish. Regular supplies from Sri Lanka and Indonesia ensure that they are within the financial reach of all marine hobbyists. They rarely grow more than 6cm (2.4in) in diameter, but can reach up to 10cm (4in).

Bear in mind that starfish's water vascular system is easily damaged, so examine all starfishes before you buy them to ensure that they are in good condition, particularly the tips of the arms, and that the body is not limp and flaccid. A starfish with any such defects is unlikely to survive.

This species will not harm other invertebrates and feeds happily on small pieces of shrimp or shellfish. Highly recommended for the marine invertebrate tank.

Linckia laevigata
Blue Starfish

The Blue Starfish is undoubtedly one of the most dramatically coloured of all marine invertebrates. While blue is a common colour among marine fishes, it is very unusual in invertebrates, and the pure, intense, occasionally almost purple, blue of *Linckia laevigata* makes it look almost artificial.

This species has been regularly imported from Singapore, Sri Lanka and the Philippines almost since the beginning of the marine hobby, and is now often considered a little 'old hat' by experienced hobbyists. Nonetheless, the species has many factors in its favour. It is quite inexpensive, feeds well and, unlike some species, spends a considerable part of the day on view. But there is one particular caution; this species seems particularly prone to parasitization by a small species of bivalve mollusc that burrows into the animal, usually from the underside of one of the arms. If left in place, it will ultimately penetrate the critically sensitive vascular system. These parasites are easily removed with a gentle thumbnail, but it is always wise to examine Blue Starfishes carefully before buying them and to reject any that show evidence of damaged skin.

Below: Linckia laevigata
Three bright Blue Starfishes cross a patch of the Great Barrier Reef in search of edible detritus.

Pentaceraster mammillatus
Common Knobbed Starfish

This species is extremely variable in colour. The background is mostly brown or green, and the knobs – which are substantially smaller than those of *P. lincki* – may be white, yellow, orange, brown or black. Most commonly around 8cm (3.2in), they can reach twice this size.

As its name suggests, this is a common species, regularly included in imports from Singapore, the source of many of the cheapest invertebrates available to many hobbyists. Like its relative *P. lincki*, it is a greedy feeder, capable of everting the stomach in order to digest food. A suitable diet can include shredded prawn and brineshrimp; do not allow uneaten food to pollute the tank.

Above: Pentaceraster mammilatus

This is a common colour form of this most variable species.

Below: Protoreaster lincki

The intense, cherry-red patterning of this species has ensured its popularity.

Protoreaster lincki
Red-knobbed Starfish

The Red-knobbed Starfish is a widely distributed Indo-Pacific species that can reach up to 30cm (12in) in diameter. It is mostly offered for sale at about half this size, and its long, vivid red dorsal spikes ensure that it makes an impact in the aquarium. The most attractive specimens have a red, netlike, pattern on an off-white background, and are quickly snapped up by hobbyists. Unfortunately, most collectors are not aware of a general rule that applies to starfishes, namely that knobbly backed starfish are omnivorous, if slow, predators, while most of the tropical, smooth-armed species are less harmful scavengers. This ignorance often results in Red-knobbed Starfish being introduced to an aquarium well stocked with corals, molluscs and other sessile animals, all of which seem to provide grist to this insatiable animal's mill.

In the right circumstances, Red-knobbed Starfish are rewarding and long-lived aqaurium animals. Feed them by placing them directly onto 1.25cm (0.5in) pieces of fish, squid or shellfish.

Phylum: CHORDATA

Sea Squirts

The phylum Chordata is mainly composed of animals with backbones, but members of two subgroups are generally considered as invertebrates, since they lack a true backbone. The sea squirts *Distomus* spp. (class Ascidiacea) are the only ones of interest to most hobbyists and even these usually arrive by accident. Small specimens are common introductions on pieces of living rock – most often white, beige or reddish species. They thrive with no additional care other than that given to the other inhabitants of the aquarium.

There are more than a thousand species of sea squirts. Some form mats of small specimens, others are large 50cm (20in) individuals. They have a leathery baglike body, with large inlet and outlet siphons. As water is drawn through these tubes, small particles of food are filtered out. Sea squirts are found in every colour and combination of colours, but their general inactivity means that they have never become very popular. Very small, but regular, liquid feeds are generally beneficial.

Above: Cyclosalpa polae
In this large red species the inhalant and exhalant siphons are clearly visible.

Below: Didemnum molle
One of many sea squirts found in large aggregations, typically in deep water.

Right: Pycnoclavella detorta
Several of the smaller sea squirts, such as this semi-transparent species, live in tight colonies in the wild.

Below: Rhopalaea crassa
This attractive blue species is one of the most appealing sea squirts, but is only rarely available.

PART SIX

THE COLDWATER AQUARIUM

Since sea water extends outside the tropical zones to all regions of the world, you should not overlook the possibility of keeping fishes from cooler waters. Because the sun shines less brightly at higher latitudes, less light penetrates the water in temperate regions. In addition, the water is often less clear, due to pollution and the heavy concentration of silt and mud constantly stirred up by coastal traffic. Fishes from these waters are not as brilliantly coloured as their tropical relatives, but they do offer one very real advantage for the hobbyist in temperate parts of the world – they are less expensive to obtain. In fact, if you live fairly near to the seashore, you can collect your own specimens absolutely free! (Coldwater marine species are not generally available from the usual aquatic stores.)

There are also many invertebrates to collect – many sea anemones are very colourful, bearing in mind their murky habitat – and there is also the advantage that should any species outgrow the tank, or outstay its welcome by antisocial adult behaviour, you can return it to the wild to continue its natural lifespan.

Despite the apparent convenience of keeping the local species, you may find problems arising during the summer months; as you enjoy the warm sun, the water temperature in the aquarium may rise uncomfortably high for its occupants and you may need to take steps to cool it down. Blennies, butterfishes, gobies and the marine species of stickleback are all small enough to be suitable for the aquarium in the long term. Juvenile forms of larger species, such as bass, grey mullet and wrasses, outgrow their aquarium before long and must be returned to the wild.

Left: *The Black Scorpionfish* (Scorpaena porcus) *should be handled with extreme care, owing to its venomous spines. However, it does make an unusual and interesting addition to the coldwater marine aquarium.*

Collecting and General Care

Coldwater species require the same aquarium conditions as those described for tropical species, with the obvious omission of heating equipment. Although substrate biological filtration is adequate, you should provide some extra water movement to create surface turbulence and to ensure well-oxygenated water.

As most of the species collected from the wild are likely to be rockpool inhabitants, furnish the aquarium with numerous retreats to recreate their natural habitat.

The biggest problem will be temperature regulation; in summer the average water temperature in the aquarium will be higher than you might expect to find in nature. Provide extra aeration at these times and improvize some kind of cooling system. The serious hobbyist may even consider fitting a cooling plant, or using a refrigerator to cool water in an outside filter system. There are a few 'aquarium' cooling units available, so check this out as well.

Feeding is not usually difficult, as most fishes are more than willing to accept fish and shellfish meats. Only the fishes with the smallest mouths, such as pipefishes and sea sticklebacks, will require copious amounts of tiny live food.

Regular water changes will stabilize the water conditions. If you check the specific gravity, remember that it will give a higher reading at the lower water temperatures, probably about 1.025 at 15°C (59°F).

If you prepare the aquarium before you collect your fish, try to make sure that the water is the same specific gravity as the natural sea water in the rockpools.

You must be well prepared to transport the livestock that you capture. Large plactic buckets with clip-on lids are ideal, although a double thickness of plastic film may be an adequate substitute for a lid. You will find that a battery-operated air pump, supplying air to an airstone in the water, will give the fish a better chance of surviving a long journey home. This is especially important during the summer months, when the journey may take longer.

Collect specimens with care; rocks surrounding the rockpools are usually covered in very slippery seaweeds, so wear suitable footwear. Consult tide timetables in advance to ensure that you get the maximum collecting time. Do not forget the incoming tide. Remember also to leave the rockpool in a fit state for the animals left behind: if you collect invertebrates, such as sea anemones or starfishes, collect site and animal together, replacing any

rocks that you remove with others to restore the number of hiding places in the pool. Transport anemones and other invertebrates separately from the fish; anemones will sting the fish in the close confines of a bucket, and fish may eat small invertebrates, such as shrimps, during transit.

Never over-collect. Not only is this bad practice from the conservation point of view, but it is also false insurance; it is better for the majority of specimens in a small collection to survive than to arrive home with none at all.

Left: *Nearly all coldwater aquariums will need some form of cooling equipment. This one is specially made for the purpose and fully adjustable.*

Below: *Keeping marines from cooler native shores can be very informative, and they can be fascinating to observe. Choose suitable companions with care.*

Coldwater Marine Fishes

Family: BLENNIIDAE
Blennies

These are most common in rockpools, where they are found hiding under overhanging rocks. They are often confused with gobies, but they lack the 'suction cup' formed by the fusion of the pelvic fins. Most blennies have 'tentacles' or a crest, described as 'cirri', positioned over the eye.

Aidablennius sphynx
Sphinx Blenny

☐ **Distribution:** Adriatic and Mediterranean.

☐ **Length:** 80mm (3.2in).

☐ **Diet:** All meaty worm foods.

☐ **Feeding manner:** Bottom feeder.

☐ **Aquarium compatibility:** May be territorial.

Once this fish leaves the security of its favourite bolt-hole, you can see that it is a most attractively coloured fish with dark bands crossing the body. It usually swims with the dorsal fin lowered, but raises it when alarmed. It spawns in caves, males physically cornering any passing female inside.

Lipophrys nigriceps
Black-headed Blenny

☐ **Distribution:** Mediterranean.

☐ **Length:** 40mm (1.6in).

☐ **Feeding manner:** Bottom feeder.

☐ **Aquarium compatibility:** May be territorial.

Whoever gave this fish its common name seems to have disregarded the predominant red of the body, concentrating instead on the darker reticulated pattern of the head region. *B. nigriceps* shares its habitat with the almost identical *Trypterygion minor*, from which it may be distinguished by the absence of a small extra dorsal fin in front of the main dorsal fins.

Lipophrys pavo
Peacock Blenny

☐ **Distribution:** Mediterranean.

☐ **Length:** 100mm (4in).

☐ **Diet:** Worm foods.

☐ **Feeding manner:** Bottom feeder.

☐ **Aquarium compatibility:** Territorial; needs plenty of retreats in which to hide.

Two obvious characteristics identify this species: the helmet-shaped hump above the eye in mature males and the blue-edged black spot just behind the eye. Blue-edged dark bands cross the green-yellow body.

Below left: Aidablennius sphynx
The Sphinx Blenny can be territorial.

Below: Lipophrys pavo
The Peacock Blenny is quite hardy.

Lipophrys pholis
Shanny

☐ **Distribution:** Mediterranean, Eastern Atlantic form West Africa to Scotland.

☐ **Length:** 160mm (6.3in).

☐ **Diet:** Live foods, meat and worm foods.

☐ **Feeding manner:** Bottom feeder.

☐ **Aquarium compatibility:** Gregarious, but can be hard to please when it comes to a choice of hiding places; whelk shells are often acceptable.

Like all blennies, these fish prefer a tank with plenty of hideaways. However, they also like to bask in the light, sometimes emerging from the water to do so. There are no cirri on the head.

Above: Lipophrys pholis
The Shanny is a widely distributed species and is quite easy to collect.

Parablennius gattorugine
Tompot Blenny

☐ **Distribution:** Mediterranean, Eastern Atlantic from West Africa to Scotland.

☐ **Length:** 200mm (8in).

☐ **Diet:** Live foods, meat and worm foods.

☐ **Feeding manner:** Bottom feeder.

☐ **Aquarium compatibility:** Can be territorial and may worry smaller fishes – and themselves be worried by larger ones.

These fishes prefer a tank decorated with medium-sized stones, under which they can hide. They can become tame, quite happy to make friends with you.

Parablennius rouxi
Striped Blenny

☐ **Distribution:** Mediterranean.

☐ **Length:** 70mm (2.75in).

☐ **Diet:** All meaty, worm foods.

☐ **Feeding manner:** Bottom feeder.

☐ **Aquarium compatibility:** Most

Above left: Parablennius gattorugine
The Tompot Blenny has some six attractive dark bars crossing its body vertically. Once established in the aquarium, it can become quite tame.

Above: Parablennius rouxi
The Striped Blenny is distributed mainly throughout the Mediterranean and is easily identified by a dark stripe running horizontally from head to tail.

blennies will live quite happily with sessile invertebrates, but crustaceans will be unsafe.

A distinctive fish with a horizontal dark stripe from head to tail. The fins are colourless.

Spinachia spinachia
Fifteen-spined Stickleback

☐ **Distribution:** Northeastern Atlantic.

☐ **Length:** 200mm (8in).

☐ **Diet:** Very small animal life.

☐ **Feeding manner:** Midwater feeder.

☐ **Aquarium compatibility:** Fin nipper; keep separately.

This species must have frequent meals of tiny live foods; brineshrimp are probably the most useful food for this purpose. The Stickleback lives for only about two years in the wild, and its life expectancy will be even shorter unless the feeding problem is solved.

Above right: Spinachia spinachia
An unusual and fascinating subject.

Lepadogaster candollei
Connemara Clingfish

☐ **Distribution:** Eastern Atlantic, Mediterranean, Black Sea.

☐ **Length:** 75mm (3in).

☐ **Diet:** Worm foods.

☐ **Feeding manner:** Bottom feeder.

☐ **Aquarium compatibility:** Not known.

The common name refers to the ability of the fish to cling to rocks and other surfaces by means of a suction disc formed by the pelvic fins. Colours may vary but generally include reds, browns and greens; males have red dots on the head and on the lower part of the long-based dorsal fin.

Gobius cruentatus
Red mouthed Goby

☐ **Distribution:** Eastern Atlantic, North Africa to southern Ireland.

Family: GASTEROSTEIDAE
Sticklebacks

Although freshwater sticklebacks (*Gasterosteus* spp.) are able to tolerate some degree of salinity, there is one species within the family – *Spinachia spinachia* – that spends its entire life in marine conditions. Like its freshwater relatives, it also builds a nest in which to spawn, fabricating the structure from plant fragments stuck together with a secreted fluid.

Family: GOBIESOCIDAE
Clingfishes

Clingfishes are small, shore or shallow water dwelling species. The Gobiesocidae family has a wide range of body shapes, but in all species the front part of the head is flattened. All members of the family lack scales and have only one dorsal and one anal fin. Their eyes are large and often dorsal, the better for them to see their prey. The papillae on their sucking disc are often used in identifying the species. Clingfishes are sedentary fishes, using their ventral fins, which are specially modified into suckers, to enable them to remain attached to rocks, algae or any other substrate (hence their common name). Despite their apparent immobility, however, they are active carnivores.

Family: GOBIIDAE
Gobies

Gobies have no lateral line system along the flanks of the body, instead, sensory pores connected to the nervous system appear on the head and over the body. Gobies can live quite a long time, records show they have survived for up to ten years. A very large family, gobies inhabit many types of water – tropical and temperate, freshwater, brackish and full salt water.

☐ **Length:** 180mm (7in).

☐ **Diet:** Small crustaceans, worm foods, shellfish meats, small fishes.

☐ **Feeding manner:** Bottom feeder.

☐ **Aquarium compatibility:** Territorial at times.

Gobies are found on both sandy and rocky shores. Sand-dwelling species are naturally camouflaged, whereas rock-dwellers can be much more colourful.

Gobius niger
Black Goby

- ☐ **Distribution:** Mediterranean, Black Sea and eastern Atlantic.

- ☐ **Length:** 150mm (6in).

- ☐ **Diet:** Worm foods, small crustaceans.

- ☐ **Feeding manner:** Bottom feeder.

- ☐ **Aquarium compatibility:** Territorial at times.

A generally dark blotched fish, but how it 'colours up' in captivity depends a great deal on the colour of its surroundings. It is very rarely black! Another scientific name for this fish is *Gobius jozo*.

Pomatoschistus minutus
Sand Goby

- ☐ **Distribution:** Eastern Atlantic, Mediterranean and Black Sea.

- ☐ **Length:** 95mm (3.7in).

- ☐ **Diet:** Worm foods.

Above: Pomatoschistus minutus
The Sand Goby is a shy species. Choose its aquarium companions with care (it is best kept with its own kind).

- ☐ **Feeding manner:** Bottom feeder.

- ☐ **Aquarium compatibility:** Probably shy and likely to be predated upon by other fish. This species is best kept in a tank with its own kind.

Its natural camouflage colouring makes this fish difficult to see when you are collecting it. Being a sand colour, it will 'feel at home' with a similarly coloured covering on the aquarium floor.

Coris julis
Rainbow Wrasse

- ☐ **Distribution:** Mediterranean eastern Atlantic.

- ☐ **Length:** 250mm (10in).

- ☐ **Diet:** Small marine animals, live foods.

- ☐ **Feeding manner:** Bottom feeder, although it will take surface plankton.

- ☐ **Aquarium compatibility:** Peaceful.

The long, slender, green-brown body has a horizontal white-red

---

Family: LABRIDAE
Wrasses

Like their tropical relatives, wrasses from temperate waters can also be brightly coloured. In fact, their colour can lead to identification and sexing problems; colour varies not only between the sexes (that of the male also changing at breeding time) but also depending on the mood of the fish and on the colour of the substrate! Sex reversals are also not uncommon. Juveniles act as cleaner fishes to other fishes, and many species hide away in crevices or bury themselves in the sand at night.

stripe. The eyes are red. These fishes are hermaphrodites, the females turning into fully functional males. Aquarium specimens are active during the day, but bury themselves in the substrate at night. This behaviour has not been observed in this species in the wild. Like their tropical relatives, juveniles act as cleaner fishes.

Below: Coris julis
Coloration of the Rainbow Wrasse varies depending on location and sex. Deeper water fish are red-brown; females have a pale spot on the gill cover base.

Anthias anthias

☐ **Distribution:** Mediterranean, eastern Atlantic as far north as Biscay.

☐ **Length:** 240mm (9.5in).

☐ **Diet:** A varied selection of animal and meaty foods.

☐ **Feeding manner:** Bold.

☐ **Aquarium compatibility:** Peaceful.

The body is golden brown with blue speckling and the facial markings are blue. The long pelvic fins are yellow and blue. In the wild, coloration may appear different because part of the colour spectrum of light is lost in deep waters due to absorption.

Right: Anthias anthias
It is surprising that collectors have not yet given this beautifully coloured fish a popular name.

Pholis gunnellus
Butterfish; Gunnell

☐ **Distribution:** Eastern and western Atlantic.

☐ **Length:** 250mm (10in).

☐ **Diet:** Crustaceans, worms, molluscs, shellfish meats.

☐ **Feeding manner:** Bottom feeder.

☐ **Aquarium compatibility:** Do not keep with small invertebrates.

The eel-like body has a long-based dorsal fin that is twice as long as the anal fin. It may have transverse dark bands on the body and white-edged markings along the base of the dorsal fin. This species is found under stones.

Serranus hepatus
Brown Comber

☐ **Distribution:** Mediterranean, eastern Atlantic (Senegal to Portugal).

☐ **Length:** 130mm (5in).

☐ **Diet:** Animal and meaty foods.

☐ **Feeding manner:** Bold.

☐ **Aquarium compatibility:** Although no reliable information is available, you should not keep this species with smaller fishes.

The reddish brown body has four or five vertical dark bars across it. The undersides are pale. There is a black blotch on the dorsal fin at the junction of the hard and soft rays.

Scorpaena porcus
Black Scorpionfish

☐ **Distribution:** Mediterranean and eastern Atlantic (Biscay and further south).

☐ **Length:** 250mm (10in).

☐ **Diet:** Small fishes.

Family: PHOLIDIDAE
Gunnells

Often seen in the same areas as blennies, gunnells are slender cylindrical fishes with a dorsal fin running the entire length of the back. The anal fin is also long based, occupying almost the rear half of the body. Both the dorsal and anal fins are limited to just one ray. Species found on both sides of the North Atlantic Ocean and also on the northern Pacific coast of America.

Above: Pholis gunnellus
This fish usually hides under rocks.

Below: Scorpaena porcus
Handle the Scorpionfish with care!

☐ **Feeding manner:** Lies in wait for passing prey.

☐ **Aquarium compatibility:** Nocturnal and distinctly unsociable. This is definitely a fish that should be kept in a separate tank.

The reddish brown mottled coloration makes this fish hard to see as it lies on the seabed. Not only is it a danger to other fishes, but also to swimmers who may inadvertently step on it. Use very hot water to bathe any wound, which may turn septic.

Family: SCORPAENIDAE
Scorpionfishes

Although they lack the ornate finnage of the tropical scorpionfishes, species from temperate waters are just as dangerous; the spines on the head are very venomous. When disturbed during the day, these sedentary nocturnal fishes swim only a short distance before settling again to await any passing prey.

Coldwater Marine Invertebrates

Above: Cleona celata
Sponges require excellent conditions.

Below: Actinia equina
A beautiful Beadlet Anemone.

Left: Urticina falina
It is a common fallacy that coldwater invertebrates are dull; this Dahlia Anemone would grace any tank.

Sponges

Sponges are usually difficult to keep in the aquarium as they are very sensitive to adverse water conditions. They must have well-oxygenated, crystal-clear water and are not all compatible with sea anemones. They attach themselves to shells, even those that contain crabs. If this happens, they will devour the shell and in turn become the home of the crab. *Suberites domuncula* is a common Mediterranean and Atlantic species.

Sea anemones

Beadlet anemones (*Actinia equina*) can be found in a variety of colours. The columns can be red, green or brown and the tentacles are usually the same colour, but not always. They move around the aquarium, providing splashes of colour in an ever-changing pattern.

Actinia equina has two sub-species *A. equina var. mesembryanthemum*, the Beadlet Anemone from the North and South Atlantic and the Mediterranean, is a very common sight in coldwater rockpools. The body and tentacles are bright red, but the body contracts to a dull red sphere just as you reach for it. *A. equina var. fragacea*, the strawberry variant, is usually red with green spots – just like a strawberry. Its tentacles are usually red, but can be a paler pink. It is larger than the Beadlet and is found in the slightly deeper waters of the northeastern Atlantic and the Mediterranean.

The long tentacles of *Anemonia vindis (sulcata)*, the Snakelocks Anemone, are not fully retractable. Because it prefers strong light, it is found very close to the water surface in the northeastern Atlantic and Mediterranean. In the same waters you will find *Bunodactis verrucosa*, the Wartlet or Gem Anemone. It has tentacles with ringed markings and vertical rows of wartlike growths on its body, hence the common name.

Cerianthus membranaceus, the Cylinder Rose, is almost a cross between a sea anemone and a tubeworm, with a longer cylindrical body and less stocky in shape. The tube is often partially buried in the sand. It is a delicate animal that needs careful handling, although it may be able to regenerate a damaged tube fairly easily. Its tentacles vary in colour from species to species and are toxic to most fishes; for this reason, too, you should place other sea anemones beyond its reach. Unlike some sea anemones, *Cerianthus* does not move about the aquarium.

Although a fairly large anemone, *Condylactis aurantiaca* from the Mediterranean, has relatively short brown, white-ringed tentacles tipped with violet. Some *Epizoanthus* species are also native to the Mediterranean. They only grow to around 10mm (0.4in), but colonies can be found on rocks just below the waterline, where the constant water movement ensures a regular delivery of food.

Urticina (Tealia) felina var. *coriacea,* the Dahlia Anemone from the North Atlantic and northeastern Pacific, has a body covered with warts, sand and fragments of shell. Tentacles surround the patterned mouth disc. A similar species, *U. crassicornis,* occurs on the east coast of North America. *U. lofotensis* has white and pink tentacles on a red body and, with the larger *U. columbiana,* occurs in the northeastern Pacific. There is also a deepwater species *U. eques.*

Crabs

Although crabs seem to be endearing little creatures, the majority of 'free-swimming' species grow too large and become a disruptive influence in the aquarium. A better choice would be the smaller Hermit Crabs (*Pagarus* spp.), which interestingly shed their adopted shell for larger premises as they increase in size (see page 333 for the tropical species).

Below: Homanus gammanus
The juvenile Common Lobster makes an ideal subject for the coldwater single-species aquarium, but will need quite a large tank as it grows.

Top: Eupagurus bernherdus
A Hermit Crab is ideal for an aquarium.

Above: Leander serratus
Common Prawns are quick-moving.

Prawns and shrimps

It is easy to capture species of *Palaemon, Crangon* and *Hippolyte* – small shrimps and prawns – from rockpools in the northeastern and northwestern Atlantic and the Mediterranean. *Lysmata* is an interesting Mediterranean species, *L. seticaudata* being very similarly marked to the Indo-Pacific species *Rhynchonectes uritae.* Prawns and shrimps are excellent scavengers and often act as cleaners to other fishes. Egg-carrying females may

Above: Chlamys operculans
Don't keep Queen Scallops with starfish!

Right: Copyphella pedata
Nudibranchs need special dietary care.

Above: Echinus esculientes
Edible Sea Urchins are easy to collect.

Below: Luidia ciliaris
A predatory starfish.

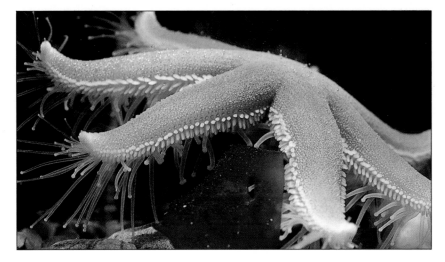

Nudibranchs

Relatively colourful species occur in the Mediterranean and northeastern Atlantic. *Chromodoris* and *Hypselodoris* are typical genera of these molluscs.

Sea urchins

Like their tropical relatives, sea urchins from temperate waters can also make interesting aquarium species. The Black Urchin, *Arbacia lixula*, from the Mediterranean is a purple-black in colour and looks like a short-spined version of the tropical *Diadema antillarum*.

Starfishes

The following species are among the wide range of starfishes found in temperate waters.

Asteria rubens is commonly found in the northeastern Atlantic, where it feeds on mussels and scallops, prising them apart with its feet and introducing its stomach into the shell. The skin of this species is covered with many tubercles.

Astropecten aranciacus, the Red Comb Star from the Mediterranean and northeastern Atlantic, is a large predatory starfish (up to 500mm/20in) with comblike teeth along the edges of its arms.

Echinaster sepositus, a Mediterranean species, grows to 300mm (12in). Fertilized eggs develop directly into small starfishes.

Ophidiaster ophidianus, another red starfish from the Mediterranean, grows to 200mm (8in). The long arms issuing from an almost non-existent central 'body' of this starfish are cylindrical in section rather than flat, with sharply tapering ends.

Sea squirts

Sea squirts are vase-shaped bivalves that draw in water through one valve, trapping suspended minute food on a mucus-covered pharyngeal basket, and then exhale the water through the second siphon. *Halocynthia papillosa,* about 100mm (4in) tall, is red-orange in colour with many bristles around the siphons. It is common in the Mediterranean.

provide extra numbers for the coldwater marine aquarium.

Shellfish

When you are collecting from rockpools, do not forget that there are some surprisingly active shellfish that will add extra interest to the aquarium. Species of limpet (*Patella*) and winkle (*Littorina*) are quite suitable. Do not ignore empty shells; a collection of shells of various sizes make ideal homes for a growing hermit crab.

APPENDIX ONE: MARINE ALGAE

To freshwater hobbyists coming to marines for the first time, the mention of 'algae' is likely to conjure up a picture of greenish brown slime growing over plants and masses of hairlike tendrils choking the aquarium. By contrast, marine aquarists consider the different forms of algae as allies.

The species of alga that live within the tissues of corals and anemones – the zooxanthellae – provide food and help with the elimination of the animals' waste products. Small encrusting, and generally insignificant, algae coat the rockwork, producing a more natural-looking scene and providing a continually available food source for many browsing invertebrates and fishes. The larger species include many decorative forms, and these are the marine equivalent of plants in a freshwater tank.

The encrusting species and those that form furry 'lawns' on rocks generally arrive as accidental introductions with living rock, or as small particles included in the water when you buy invertebrate specimens. The larger, decorative species of algae are usually bought separately, but some are so prolific that you may be able to obtain 'cuttings' from neighbouring aquarists. They may be green, brown or red in colour.

Like most terrestrial plants, all algae species use chlorophyll to synthesize food, so require moderate to strong lighting. Commercial algae fertilizers are available, but these are only rarely necessary. Indeed, if other conditions are less than ideal, they can do more harm than good, by promoting undesirable species and increasing nitrate levels within the tank. Generally speaking, sufficient trace elements are dissolved within the water, and enough phosphates and nitrates are available from the animals' wastes to ensure good algal growth. In fact, by absorbing

Above: *Careful harvesting of algae growing in the aquarium is essential to prevent an impressive invertebrate display from being overrun and to ensure that species are not deprived of beneficial light.*

nitrates and phosphates as plant fertilizers, all forms of algae play an important role in maintaining good water quality in the aquarium.

In the early days of the newly established invertebrate aquarium, you may encounter a problem with excessive growth of unwanted or undesirable forms of algae. This is most common where high-intensity lighting is provided and mats of green, hairlike algae form as a result. Generally speaking, the sequence of events is as follows. The aquarist provides sufficient light for zooxanthellae to function but, when the tank is new, large expanses of rock are exposed to this light. As there are few corals and anemones to utilize the light, algae are able to prosper. However, as soon as more corals and similar invertebrates are added to the tank,

they begin to shade the algae and use the light themselves, eventually causing the algae to decline. Sea urchins and cowries will graze on green algae, but these animals are somewhat indiscriminating about the routes they follow and may damage sessile animals.

Although most marine algae can be safely encouraged to grow, there is a purple-brown alga that rapidly forms a spreading film over everything in the tank (see pages 91-93). Be sure to siphon any spots of this type of alga out of the tank as soon as you notice them. This problem is normally associated with overstocking, incorrect lighting, overfeeding, insufficient water movement within the tank, insufficient frequency (or quantity) of water changes, and excessive levels of nitrates. A series of partial water changes and an improvement in general conditions will usually solve the problem. On no account should you use the algae-killing preparations designed for ponds and freshwater aquariums in a marine system.

Acetabularia spp.
Mermaid's cup

Above: Acetabularia spp.
Very attractive, but rarely available.

Below: Avrainvillea sp.
A tropical Western Atlantic species.

This small and very delicate plant is one of the most attractive of the marine algae, but it is only rarely available to the hobbyist. The thin stipes, only 5cm (2in) long, are topped by pale blue-green caps like inverted toadstool heads. Unfortunately, the plant is easily damaged, both in transit and by other aquarium inhabitants, and is easily swamped by hairlike algae. It requires good light and less water movement than suits most invertebrates and algae.

Avrainvillea nigricans

This species is related to *Halimeda* spp. and *Pencillus capitatus*, but here the stipe ('stalk' or 'stem') is topped by a single, large, flattened blade, roughly circular and up to 10cm (4in) in diameter. Although hardy, the blades are often coated with brown encrusting algae, which detract from the plant's appeal. *Udotea* species are very similar in appearance and, from the aquarist's point of view, can be considered roughly the same.

Caulerpa spp.

The most commonly cultivated algae are the many *Caulerpa* species. Various types are found throughout the Caribbean, Mediterranean and Indo-Pacific regions and all are prolific. *Caulerpa* species may vary in colour from an intense lime green to a bluish brown, and they may grow 50cm (20in) tall or form low mats. Despite this variation, the basic structure is very similar. Growth develops from a main runner (stolan), with leaf stalks being produced from the top of the runner and a rootlike growth from the bottom. This 'holdfast' serves to anchor the plant body in position. It does not absorb water and food in the same manner as a terrestrial plant's roots; in marine algae, nutritional substances are absorbed through the leaves (blades).

The fronds of *Caulerpa* species are very thin-walled and filled with fluid. Because of this, it is important to acclimatize these plant-like blades very slowly to a new environment, especially in terms of specific gravity. If the transition is too sharp, changes in osmotic pressure can rupture the cell walls, causing an attractive green plant to change rapidly into a decomposing translucent slime.

Since some 'body fluid' will leak when pieces are removed for transplanting to other aquariums, it is safer to buy larger, rather than smaller, segments to reduce the proportion lost. This is particularly important if the tank houses fishes or invertebrates that may peck at the plant. A large sample has a much better chance than a small piece of surviving – and outgrowing – this minor pruning process.

In some circumstances, *Caulerpa* species grow so rampantly that they threaten to swamp, or at least severely shade, the various corals and polyps in the tank. Regular thinning of the growth is much more desirable than infrequent but heavy pruning, which may, on occasion, cause the collapse of the whole growth through excess fluid loss from the cut surfaces.

Most *Caulerpa* species are easy to grow, although the fleshier species require greater care than the more leathery types. The species can be distinguished by the form of the 'leaves'. The easiest to identify is probably *C. prolifera*. It has a thin, wiry runner, or stolon, typically up to 30cm (12in) tall and straplike blades up to 3cm (1.2in) wide. Occasionally, the blades develop as chains of heart-shaped sections, usually as a result of persistent damage to the growing tips. It is a good idea to obtain specimens still attached to rock; this will give them a good start in your tank. Marine algae is very brittle and cannot always support its own weight out of water, so handle it carefully.

Caulerpa mexicana and *C. sertularioides* have very attractive featherlike blades and both grow very rapidly. *C. racemosa* has short bunches (stipes) of spherical or ovate, berrylike growths on the vertical stalks and has earned the species the common name Grape Caulerpa. Although very attractive, it is rather slow growing and somewhat more demanding than most species.

Below: Caulerpa prolifera
As the name suggests, this species can be rampant, even in fairly poor conditions. Harvest it regularly to promote fresh, young growth.

Bottom: Caulerpa racemosa
The berry-like leaves of this Caulerpa *species make it easy to identify.*

Above: Caulerpa sertularioides

Another easily recognized species, with alternate blades arranged herringbone fashion from the main stem.

Below: Codiacea spp.

This slow-growing alga is fairly easy to maintain in the reef tank, where it will not be swamped by nuisance algae.

Above: Caulerpa taxifolia

An easy and very fast-growing species, identified by a profusion of divided blades attached to long runners.

Codiacea spp.

Several attractive members of this algae group are imported, primarily from the Caribbean. These species are known as calcareous algae because their 'leaves' are reinforced with calcium, absorbed from the sea water. This makes them much more rigid than *Caulerpa* species and less prone to predation. Given good lighting, a high pH level and regular use of a pH buffer solution, most species are easy to maintain, though often slow to reproduce.

Halimeda spp.

Typical specimens of these algae consist of a stipe, anchored into a soft substrate, from which grow numerous roughly circular or heart-shaped flat plates. New plates grow from the tips of the old ones. *H. discoidea* is one of the commonest species, its numerous 1.5cm (0.6in)-diameter plates giving the impression of a prickly pear-cactus – hence the common name Cactus Alga. *H. goreaui* and *H. opuntia* form very dense mats of tiny plates and look particularly attractive in a small aquarium. *H. copiosa* produces long chains of small plates and is very delicate and elegant.

Right: Halimeda discoidea
Unfortunately, this commonly available alga is easily overcome by filamentous problem algae in the aquarium.

Pencillus capitatus
Shaving brush

This is probably the most commonly imported of the Caribbean *Codiacea* and grows on a wide variety of substrates, its fleshy stipe often buried 7cm (2.75in) deep in the sand or mud. Its common name is very apt, as it perfectly resembles a green shaving brush. Although easily damaged in transit, intact specimens usually flourish, with small new growths appearing like rose suckers.

Above: Pencillus captitatus
The appropriately named Shaving Brush!

Rhipocephalus phoenix is similar in appearance, but has a crown made up of concentric rings of thin, flattened plates.

Above: Rhodophyceae
Red algae make a welcome change to the more common green species. On the whole, red algae grow very slowly.

Below: Valonia ventricosa
One of the more strange species of marine algae. It can quickly multiply to occupy favoured spots in the tank.

Rhodophyceae
Red algae

Several decorative species of maroon-red algae are occasionally available. Typically, these are anchored to a base rock by a thick stipe that rapidly branches to form a bushlike structure. Some species are quite stiff and erect, while others collapse if removed from water. The success rate with these types of algae is very variable. The best specimens are those that remain attached to a small rock and have few or no pale or faded tips to the branches.

Valonia ventricosa
Sailor's Eyeballs

This species produces a cluster of roughly spherical balls up to 5cm (2in) in diameter. Each ball is a single cell and it is this plant's claim to fame as the largest single-celled growth in the world that earns it a place here. *V. ventricosa* usually occurs as an accidental introduction into the aquarium and, given time, can make an attractive feature. The cells are easily punctured, however, so it is important that you handle them with great care.

APPENDIX TWO: SPECIES TO AVOID

TROPICAL MARINE FISHES

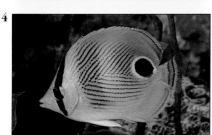

1 Aeoliscus strigatus
(Razorfish; Shrimpfish)
Generally possesses an unnaturally short life expectancy in captivity.

2 Apolemichthys arcuatus
(Bandit Angelfish; Banded Angelfish)
An extremely difficult fish to keep for long periods of time.

3 Centropyge multifasciatus
(Multibarred Angelfish)
An extremely difficult fish to keep for long periods of time.

4 Chaetodon capistratus
(Four-eyed Butterflyfish)
An extremely difficult fish to keep for long periods of time.

5 Chaetodon larvatus
(Red-headed Butterflyfish)
An extremely difficult fish to keep for long periods of time.

6 Chaetodon meyeri
(Meyer's Butterflyfish)
An extremely difficult fish to keep for long periods of time.

7 Chaetodon octofasciatus
(Eight-banded Butterflyfish)
An extremely difficult fish to keep for long periods of time.

8 Chaetodon ornatissimus
(Ornate Butterflyfish)
An extremely difficult fish to keep for long periods of time.

9 Chaetodon trifascialis
(Chevron Butterflyfish)
An extremely difficult fish to keep for long periods of time.

10 Chaetodon trifasciatus
(Rainbow or Redfin Butterflyfish)
An extremely difficult fish to keep for long periods of time.

11 Chaetodon xanthocephalus
(Yellowhead or Goldrim Butterflyfish)
An extremely difficult fish to keep for long periods of time.

12 Dunkerocampus dactyliophorus
(Banded Pipefish)
Generally does not survive for long in captivity.

13 Mirolabrichthys evansi
(Evan's Butterfly perch)
An extremely difficult fish to keep for long periods of time.

14 Mirolabrichthys tuke
(Purple Queen; Butterfly Perch)
An extremely difficult fish to keep for long periods of time.

15 Oxymonocanthus longirostris
(Long-nosed Filefish; Orange-green Filefish; Beaked Leatherjacket)
Rarely survives long in captivity.

16 Rhinomuraenia amboinensis
(Blue Ribbon Eel)
Does not usually survive for long periods in captivity.

17 Synanecja horrida
(Stonefish)
Potentially lethal if mishandled because of its very venomous spines.

18 Zanclus canescens
(Moorish Idol)
Does not generally survive for long periods in captivity.

TROPICAL MARINE INVERTEBRATES

1 Acanthaster planci
(Crown-of-thorns Starfish)
An unsuitable subject for the aquarium.

2 Acropora palmata
(Elkhorn Coral)
Collection of this large reef-building coral is ecologically unsound.

3 Aiptasia sp.
(Rock Anemone)
A very invasive pest in the aquarium.

4 Conus spp.
(Cone shells)
Capable of inflicting lethal injury.

5 Cyphoma gibbosum
(Flamingo Tongue)
Short-lived in captivity.

6 Hapalochlaena maculosa
(Blue Ring Octopus)
A dangerous species, capable of inflicting a lethal bite.

7 Hermodice carunculata
(Bristleworms; Fireworm)
These accidental introductions into the aquarium are carnivorous scavengers, and, in addition, their bristles will cause a painful rash if touched.

8 Hymenocera sp.
(Harlequin Shrimp; Orchid Shrimp)
Unsuitable for the aquarium as starfish (and possibly the tube feet of sea urchins) are their sole source of food!

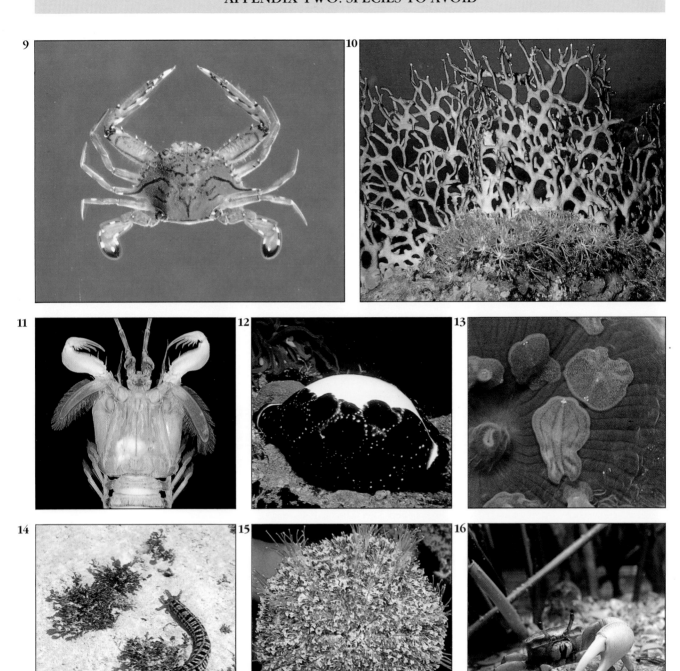

9 Macropipus sp.
(Swimming Crab)
A predatory pest.

10 Millepora spp.
(Stinging Coral)
Can inflict very painful wounds.

11 Odontodactylus spp.
(Mantis Shrimp)
An extremely efficient predator, taking shrimps, crabs and fishes and damaging starfishes and featherduster worms. Can also inflict a serious wound.

12 Ovulum ovum
(Egg Cowrie)
Most specimens demand supplies of soft leather coral if they are to prosper.

13 Pseudoceros sp.
(Brown Flatworm)
A serious pest, capable of multiplying into an overwhelming aquarium plague.

14 Synapta maculata
(Worm Cucumber)
An unsuitable aquarium subject.

15 Toxopneustes pileolus
(Poison Urchin)
The spines are armed with a poison that can produce a very painful reaction.

16 Uca sp.
(Fiddler Crab)
Not a suitable aquarium subject.

GLOSSARY

Absorption The process of taking in and holding physically as a dry sponge takes in water. Liquid vitamins added to flake act in this manner.

Activated carbon Material used in mechanical/chemical filtration systems (external 'power filter' canister types) to remove, by adsorption, dissolved matter.

Adsorption The process by which organic molecules are chemically bonded onto a surface of a medium, such as activated carbon.

Algae Primitive plants, which may be either unicellular or large (e.g. kelp). They have plant characteristics, are almost exclusively aquatic and do not flower.

Ammonia (NH₃) First byproduct of decaying organic material; also excreted by the fishes' gills. Highly toxic to fishes and invertebrates.

Anal fin Single fin mounted vertically below the fish.

Artemia salina Scientific name of brine shrimp.

Barbel Whisker-like growth around the mouth; used for detecting food by taste.

Biological filtration Means of water filtration using bacteria, *Nitrosomonas* and *Nitrobacter*, to reduce otherwise toxic ammonium-based compounds to safer substances such as nitrates.

Bivalve A mollusc or shell-dwelling animal with two respiratory valves.

Brackish water Water containing approximately 10 percent sea water; found in estuaries where freshwater rivers enter the sea.

Brine shrimp Saltwater crustacean, *Artemia salina*, whose dry-stored eggs can be hatched to provide live food for fish or invertebrates.

Buffering action Ability of a liquid to maintain its pH value. Calcareous substrates may assist in this respect.

Cable tidy Commercial 'junction box' for neat and safe connection of electrical supply circuits.

Calcareous Formed of, or containing, calcium carbonate, a substance which may help to maintain a high pH of the aquarium water.

Caudal fin Single fin mounted vertically at the rear of the fish, the tail.

Caudal peduncle Part of fish's body joining the caudal fin to the main body.

Cirri Crestlike growths found above the eyes in some species, such as blennies.

Commensalism Living practical partnership, where one party derives more benefit than the other.

Copper Metal used in copper sulphate form as the basis for many marine aquarium remedies. Poisonous to fishes in excess, and even more so, at trace levels, to invertebrates.

Counter-current More efficient design of protein skimmer where the water flows against the main current of air, thereby giving a longer exposure time for collection of waste or sterilization if ozone is used.

Cover glass Panel of glass to form an anti-condensation, anti-evaporation protection placed on top of the aquarium immediately below the hood.

Cryptocaryon Parasitic infection, often referred to as the marine equivalent of the freshwater White Spot Disease, *Ichthyophthirius*.

Daphnia Freshwater crustacean, the water flea, occasionally used as food in the marine aquarium.

Demersal Term usually applied to eggs or to spawning action of fishes. Demersal eggs are heavier than water and are laid in prepared spawning sites on the sea bed. The fertilized eggs are then guarded by one or both adult fishes until hatching occurs.

Denitrification The removal of nitrate by anaerobic bacteria into nitrous oxide and then into free nitrogen gas. Used as a reliable method of keeping nitrates at a low level in the aquarium.

Diffuser An alternative name for wooden airstones.

Dorsal fin Single fin mounted vertically on top of the fish; some species have two dorsal fins, one behind the other. Many marine species have venomous rays in the dorsal fin, so handle them with care.

Dropsy Disease, where body fluids build up and produce a swollen body.

Filter feeder Animal (fish or invertebrate) that sifts water for microscopic food, e.g. pipefishes, tubeworms.

Filter medium Material used in filtration systems to remove suspended or dissolved organic substances from the water either mechanically, biologically or chemically.

Fin rot Bacterial ailment; the tissue between the rays of the fin rots away.

Foam fractionation Method of separating out proteinous substances from water by foaming action. Also known as protein skimming.

Fry Very young fish (see *Larvae*).

Fungus Parasitic infection, causing cotton-wool-like growths on the body.

Gallon (Imp) Measure of liquid volume (= 1.2 US gallons = 4.55 litres.)

Gallon (U.S.) Measure of liquid volume (= 0.83 Imp gallons = 3.8 litres.)

Gill flukes Trematode parasites, such as *Dactylogyrus*, that in severe infestation cause rapid breathing and gaping gills.

Gills Membranes through which fish absorb dissolved oxygen from the water during respiration.

Gravel tidy Plastic mesh fitted between layers of gravel to protect biological filtration systems from being exposed (and thus rendered ineffective) by digging fishes.

Hood Aquarium cover containing light fittings.

Hydrometer Device for measuring the specific gravity (S.G.) of the salt water, especially useful when making up synthetic mixes. May be either a free-floating or swing-needle type.

Impeller Electrically driven propeller that produces water flow through filters.

Irradiation Method of exposing food to gamma rays to sterilize it.

Larvae Often the first stage of very young marine fish; under-developed fish fry; also first reproductive stage of many invertebrates.

Lateral line Line of perforated scales along the flanks which lead to a pressure-sensitive nervous system. Enables fish to detect vibrations in surrounding water caused by other fishes, or reflected vibrations of their own movement from obstacles.

Length (standard) Length of fish (SL) measured from snout to end of main body; excludes caudal fin.

Litre measure of liquid volume (1 litre = 0.22 Imp gallons = 0.26 US gallons.)

Lymphocystis Viral ailment that causes cauliflower-like growths on skin and fins.

Mercury vapour Type of high-intensity lamp.

Mimicry The close resemblance of one creature to another. Specifically, the resemblance of predatory fishes to 'safe' fishes allowing them to gain unfair advantage over other animals.

Mouthbrooder Fishes than incubate fertilized eggs in the mouth.

Mysis Commercially available marine shrimp used as live and frozen food.

Nauplii Term used generally for the newly hatched form of brine shrimp.

Nitrate (NO$_3$) Less toxic ammonium compound produced by *Nitrobacter* bacteria from nitrite. Nitrate levels can be kept to a minimum by regular partial water changes; anaerobic filters convert nitrate back to free nitrogen.

Nitrification The process by which toxic nitrogenous compounds are converted by aerobic bacteria into less harmful substances, e.g. ammonia to nitrite to nitrate.

Nitrite (NO$_2$) Toxic ammonium compound produced by *Nitrosomonas* bacteria from ammonia. Toxic to fishes, and even more so to invertebrates.

Nitrobacter A species of aerobic bacterium essential in the biological filter to convert nitrite into far less harmful nitrate.

Nitrosomonas A species of aerobic bacterium utilized in the biological filter to convert ammonia into less toxic substances, e.g. nitrite.

Oodinium Single-celled parasite causing coral fish disease. Highly infectious, but curable with proprietary remedies.

Osmosis Passage of liquid through a semi-permeable membrane to dilute a more concentrated solution. Accounts for water losses through the skin of marine fishes, i.e. to the relatively stronger sea water, which they have to constantly drink to replenish these losses.

Ozone (O$_3$) Three-atom, unstable form of oxygen used as a disinfectant. Only to be used in conjunction with a protein skimmer, which prevents ozone coming into direct contact with fishes or invertebrates.

Ozonizer Device that produces ozone by high-voltage electrical discharge. Air from an air pump is passed through the ozonizer on its way to the protein skimmer.

Pectoral fins Paired fins, one on each side of the body immediately behind the gill cover.

Pelagic Strictly meaning 'of the open sea', this term is also applied to eggs and spawning methods. Pelagic eggs are lighter than water and are scattered after an ascending spawning action between a pair of fishes in open water. The fertilized eggs are then carried away by water currents.

Pelvic fins Paired fins on the ventral (lower) surface, usually immediately below the gill covers. Not all marine fishes have pelvic fins.

pH Measure of water acidity or alkalinity; the scale ranges from 1 (extremely acid) through 7 (neutral) to 14 (extremely alkaline). Sea water is normally around pH 8.3 and aquarium water should be kept in the range of pH 7.9 to 8.3. A falling pH indicates a partial water change is necessary or that there is a failure in the filtration sytstem.

Phytoplankton Extremely small plants (e.g. unicellular algae) that drift around in the water.

Power filters External canister-type filtration devices, usually fitted with an electric impeller to drive aquarium water through the enclosed filter media. Often used to prefilter water in 'reverse-flow' biological filtration systems.

Power head Electric impeller system fitted to biological filter return tubes to increase water flow.

Protein skimmer Device that removes proteinous substances from the water by fractionation: may be air-operated or electrically powered. Also used in conjunction with ozonized air for further water sterilization purposes.

Quarantine Mandatory period of separation for new fishes, to screen them from any latent diseases. Quarantine tanks must be maintained to the same high standard as the main aquarium to reduce stress when fishes are moved from one to the other. Can double as a treatment tank.

Rays Bony supports in fins.

Reverse-flow Alternative design of biological filtration system in which water flows up through the base covering instead of the more usual downward direction. Best powered by external power filters.

Salinity Measure of saltiness of the water. Quoted in terms of gm/litre. Natural sea water has a salinity of about 33.7 gm/litre.

Silicone sealant Adhesive used to bond glass or stop leaks. Use it to create rocks and coral formations, caves, etc. Use in well-ventilated conditions; it gives off heavy vapour smelling of vinegar. Allow at least 24 hours for it to cure. Be sure to use proper aquarium sealant, not the type sold for domestic use.

Siphon A length of tube with which to remove water from the aquarium; may also refer to inhalant organ of molluscs.

Spawning Act of reproduction involving the fertilization of the eggs. Many marine species have been observed spawning in captivity, but very few young fishes have been raised. Best chances so far are with clownfishes and Neon Gobies.

Specific gravity Ratio of density of measured liquid to that of pure water. Natural sea water has an S.G. of around 1.025, but marine aquarium fishes are normally kept in slightly lower density water (1.020-1.023) to avoid osmotic stress.

Starter Circuit necessary to initialize ('start') the discharge in fluorescent lighting.

Substrate Term for aquarium base covering.

Swimbladder Hydrostatic organ enabling fish to maintain chosen depth and position in water.

Symbiosis Relationship between two parties, each deriving mutual and indispensable benefit. Advanced form of commensalism.

Total system Term given to aquariums with built-in sophisticated filtration and other management systems providing full water treatment.

Trickle filter Slow filter, often involving inert granules, sand or algal system. Anaerobic types convert nitrates back to free nitrogen.

Tungsten Incandescent filament wire type of lighting. Not recommended for aquarium use: inefficient, 'unbalanced' spectral output and produces too much heat.

Turnover Water flow rate through a filter. For marine aquariums a high turnover is recommended.

Ultraviolet (UV) Type of light used as disinfectant, produced by a special tube usually enclosed in a surrounding water jacket through which aquarium water is passed. DO NOT LOOK AT AN OPERATING UV LAMP WITHOUT PROTECTIVE GOGGLES.

Undergravel filter Alternative name for biological filter acting as the substrate in an aquarium.

Ventral Undersurface of a fish. May be especially flattened in bottom-dwelling species.

Ventral fins Alternative name for pelvic fins.

Water change Regular replacement of a proportion (usually 20-25%) of aquarium water with new synthetic sea water. Helps to maintain low nitrate levels, correct pH levels and replaces trace elements. Aerate any stored synthetic sea water before use.

Wattage Unit of electrical consumption used to classify power of aquarium heater or brightness of lamps.

Zooplankton Extremely small animals that drift around in the water.

GENERAL INDEX

Right: Enoplometopus occidentalis
Red Dwarf Lobster.

SPECIES INDEX

PICTURE CREDITS

Artists

Copyright of the artwork illustrations on the pages following the artists' names is the property of Salamander Books Ltd. The artwork illustrations have been credited by page number.

Paul B. Davies: 103
Rod Ferring: 26, 27, 76, 78, 83, 87, 94, 95, 96, 97, 98, 99, 100, 144, 148
Stephen Gardner: 79, 104, 129, 155
David Holmes (Garden Studio): 30, 31, 33, 37, 39, 41, 45, 47, 48, 49, 51, 52, 53
Bill Le Fever: 24
Hans Wiborg-Jenssen: 71

Photographers

The publishers wish to thank the following photographers and picture libraries who have supplied photographs for this book. The photographs have been credited by page number and position on the page: (B)Bottom, (T)Top, (C)Centre, (BL)Bottom left etc.

David Allison: 182-183(B), 207, 222, 227, 254(T), 259(B), 266-267, 281, 285, 332, 355(T), 360(T), 362-363, 376(T), 379(B)
Dr Chris Andrews: 144, 145, 149(TL, TR, BR), 157
Aquarium Life Support Systems: 101(B)
Aquarium Systems Ltd.: 98
Terry Begg © Practical Fishkeeping: 75
Peter Biller: 234(T)
Biofotos: 259(T) (Ian Took)
Dr James Chubb: 148(L)
Bruce Coleman: 230(B) (Alain Compost), 239(T)
Eric Crichton © Salamander Books Ltd.: 92, 128(T)
Nick Dakin: 42(T), 72, 88, 137, 140, 141, 142, 152(B), 225, 268(B), 282(B), 324(T), 325(B), 339
Andy Dalton: 260(T)
Eheim GmbH: 94
Max Gibbs: Endpapers, Half-title page, Copyright page, 22, 24, 28, 43, 46, 47, 49, 52, 66, 67(T), 77(T), 80(T), 114-115, 117, 123, 124, 125, 126, 134-135, 151, 160-161, 163, 164, 165, 166, 168, 169, 170-171, 171, 172, 173, 175(T), 177(T), 178-179, 180, 181, 183(T), 184, 185, 186, 187, 188, 189, 190(B), 191, 192, 193, 194(T), 195, 196(L), 196-197, 198, 199, 200, 202, 203, 204, 205, 208, 210(B), 211, 212, 213(B), 214(T), 214-215, 217, 218-219, 220, 221, 223, 224, 226, 228, 229, 230(T), 231, 232, 233, 234(B), 235, 236, 237, 238-239, 239(B), 240, 241, 242, 243, 244, 245, 246, 247, 248, 249(B), 250-251, 252(B), 253, 254(B), 255(B), 256, 257, 258, 260(B), 261, 263, 265, 267, 268(T), 269, 270(B), 271, 272-273, 274(B), 275, 277(T), 280, 283, 284, 286, 287, 288, 289, 290, 291, 292, 293(T), 300-301, 302-303, 313, 316(B), 327, 328(T), 331, 335, 336-337,

338-339, 345(T), 346(B), 347(T), 348, 350(B), 351, 352(T), 353, 357(T), 380(TR, CL, BL, BC), 381(TL, TC, CR), 383(CR)
Max Gibbs © Salamander Books Ltd.: 89, 90(B), 105, 108, 109, 110, 111, 112, 116, 130, 138, 139, 154
Robert Harding Picture Library: 60
Martyn Haywood: 321(B), 359(T), 382(TR)
Les Holliday: 12, 13, 14, 15, 17, 19(B), 20, 21(B), 23(B), 35(B), 51(L), 53, 54, 55, 56, 57, 58, 61 (courtesy Operation Raleigh), 62, 71, 79, 91, 100, 101(T), 129, 158(B), 297(TR), 304(T), 314(B), 318(B), 357(B), 374
Andy Horton: 69, 364(B)
Ideas into Print: 74
IKAN: 216 (Kleiter), 277 (Debelius)
Alex Kerstitch: 23(T), 35(T), 36, 42(B), 44, 63, 80(B), 122, 248(T), 279, 312(B), 325(T), 326(B), 328(B), 330(B), 343(T), 354(T), 382(TL, CL, CR, BL, BR), 383(TL, C)
Lahaina Systems Ltd.: 68, 364(T)
Frank Lane Picture Agency: 59 (T. Silvestris)
Jan-Eric Larsson: 201(B), 330(TL), 334, 391
Dick Mills: 250(TL)
N. T. Laboratories Ltd.: 128(B)
Natural Science Photos: I. Bennett: 45, 309(B), 310, 330(TL), 352(B), 356(B), 358, 383(CL, BL); I. Bennett & D. G. Myers: 38(B); Mark Caney: 133; D. Hill: 16, 18, 34, 50, 314(T), 382(TC), 383(TR); Nat Fain: 21(T), 315, 318(T), 330(TR), 342, 383(BR); Paul Kay: 27, 40, 131, 132, 298-299, 366(T, BL), 367, 368, 370, 371, 372, 373; Alan Smith: 10-11, 37, 38(T), 51(R), 182(T), 264(T), 311, 333(B), 340(B), 350(T), 382(C)
Arend van den Nieuwenhuizen: 176, 177(B), 194(B), 206, 209, 213(T), 249(T), 213(T), 249(T), 264(B), 270(T), 274(T), 276, 293(B), 294, 295, 296, 297(BL), 299, 307, 344, 346(T), 354(B), 355(B), 380(TC, C, CR, BR), 381(TR, CL, C, BL), 383(BC)
Oxford Scientific Films: 190(T), 214(B), 252(T) (Steffen Hauser), 282(T) (Max Gibbs), 329, 378(B)
Planet Earth Pictures: John Lithgow: 349(B); J. MackKinnon: 349(T); Christian Pétron: 366(BR), 369
Royal Botanic Gardens, Kew: 375(B), 377(TR) (© Andrew McRobb)
Mike Sandford: 113, 175(B), 201(T), 210(T)
Gunther Spies: 32, 167, 255(T), 304, 305(B), 316(T), 321(T), 324(B), 356(T), 359(B)
Peter Stiles: 9, 67(B), 70, 78(B), 156, 378(T)
Linda Stokoe: 377(B)
Stonecastle Graphics © Salamander Books Ltd.: 84, 85, 90(T), 127, 135
R. & V. Taylor: 381(BR)
William A. Tomey: Title page, 64-65, 73, 77(T), 93, 107, 120, 136, 159, 278(T), 278-279(B), 312(T), 317, 320, 322-323, 333(T), 340(T), 341, 345(B), 365(T), 375(T), 376(B), 377(TL), 379(T)
Brent Whitaker: 146, 148(R), 149(BL), 150, 152(T), 153, 158(T)